T0143495

Artificial Intelligence for Space: AI4SPACE

The new age space value chain is a complex interconnected system with diverse actors, which involves cross-sector and cross-border collaborations. This book helps to enrich the knowledge of Artificial Intelligence (AI) across the value chain in the space-related domains. Advancements of AI and Machine Learning have impactfully supported the space sector transformation as it is shown in the book.

> *"This book embarks on a journey through the fascinating realm of AI in space, exploring its profound implications, emerging trends, and transformative potential."*
>
> *Prof. mult. Dr.med. Dr.rer.nat. Oliver Ullrich*
> *Director Innovation Cluster Space and Aviaton (UZH Space Hub)*
> *University of Zurich, Switzerland*

Aimed at space engineers, risk analysts, policy makers, technical experts and non-specialists, this book demonstrates insights into the implementation of AI in the space sector, alongside its limitations and use-case examples. It covers diverse AI-related topics applicable to space technologies or space big data such as AI-based technologies for improving Earth Observation big data, AI for space robotics exploration, AI for astrophysics, AI for emerging in-orbit servicing market, and AI for space tourism safety improvement.

Key Features:

- Provides an interdisciplinary approach, with chapter contributions from expert teams working in the governmental or private space sectors, with valuable contributions from computer scientists and legal experts;
- Presents insights into AI implementation and how to unlock AI technologies in the field;
- Up to date with the latest developments and cutting-edge applications

Matteo Madi, Ph.D., is an entrepreneur, innovator, business developer and space-tech specialist with many years of experiences in the Swiss and International public and private sectors. He is the founder of Sirin Orbital Systems AG, a Swiss innovative company based in Zurich, focused on the development and commercialization of advanced enabling technologies for the needs of emerging space market and future sustainable space exploration. It also creates innovative solutions for the use of space technologies and satellite-based services for terrestrial applications.

Olga Sokolova, Ph.D., is a risk analyst proficient in critical infrastructure risk assessment to natural and technical hazards. She has been engaged in development and analysis of structural risk-management tools towards sustainable future and has a record in raising social awareness of spaceborne risks and opportunities brought to the society by the "New Space" industry developments. Along with Dr. M. Madi, Dr. O. Sokolova is the Co-Editor of the book entitled, *"Space Debris Peril: Pathways to Opportunities"*, published by CRC Press: Taylor and Francis in November 2020 (ISBN 9780367469450).

Artificial Intelligence for Space: AI4SPACE

Trends, Applications, and Perspectives

Edited by
Matteo Madi and Olga Sokolova

CRC Press
Taylor & Francis Group
Boca Raton London New York

CRC Press is an imprint of the
Taylor & Francis Group, an **informa** business

First edition published 2024
by CRC Press
2385 NW Executive Center Drive, Suite 320, Boca Raton FL 33431

and by CRC Press
4 Park Square, Milton Park, Abingdon, Oxon, OX14 4RN

CRC Press is an imprint of Taylor & Francis Group, LLC

ISBN: 978-1-032-43089-8 (hbk)
ISBN: 978-1-032-43244-1 (pbk)
ISBN: 978-1-003-36638-6 (ebk)

DOI: 10.1201/9781003366386

Typeset in Stix General Regular
by KnowledgeWorks Global Ltd.

Publisher's note: This book has been prepared from camera-ready copy provided by the authors.

to SNG. a true intellect.

Contents

Foreword   ix

Contributors   xiii

About the Contributors   xvii

SECTION I   **Introduction**

CHAPTER 1 ▪ Introduction   3

     OLGA SOKOLOVA

SECTION II   **Trends & Valuse**

CHAPTER 2 ▪ Selected Trends in Artificial Intelligence for Space Applications   21

     DARIO IZZO, GABRIELE MEONI, PABLO GÓMEZ, DOMINIK DOLD, AND ALEXANDER ZOECHBAUER

CHAPTER 3 ▪ Space Systems, Quantum Computers, Big Data and Sustainability: New Tools for the United Nations Sustainable Development Goals   53

     JOSEPH N. PELTON AND SCOTT MADRY

SECTION III   **Applications & Use-Cases**

CHAPTER 4 ▪ Neuromorphic Computing and Sensing in Space   107

     DARIO IZZO, ALEXANDER HADJIIVANOV, DOMINIK DOLD, GABRIELE MEONI, AND EMMANUEL BLAZQUEZ

CHAPTER 5 ■ Artificial Intelligence for Spacecraft Location Estimation based on Craters 160

KEIKI TAKADAMA, FUMITO UWANO, YUKA WARAGAI, IKO NAKARI, HIROYUKI KAMATA, TAKAYUKI ISHIDA, SEISUKE FUKUDA, SHUJIRO SAWAI, AND SHINICHIRO SAKAI

CHAPTER 6 ■ Artificial Intelligence for Space Weather Forecasting 190

ENRICO CAMPOREALE

CHAPTER 7 ■ Using Unsupervised Machine Learning to make new discoveries in space data 214

GIOVANNI LAPENTA, FRANCESCO CALIFANO, ROMAIN DUPUIS, MARIA ELENA INNOCENTI, AND GIORGIO PEDRAZZI

SECTION IV Legal Perspectives

CHAPTER 8 ■ Harnessing Artificial Intelligence Technologies for Sustainable Space Missions: Legal Perspectives 273

STEVEN FREELAND AND ANNE-SOPHIE MARTIN

SECTION V Market Perspectives

CHAPTER 9 ■ Future-ready space missions enabled by end-to-end AI adoption 303

LORENZO FERUGLIO, ALESSANDRO BENETTON, MATTIA VARILE, DAVIDE VITTORI, ILARIA BLOISE, RICCARDO MADERNA, CHRISTIAN CARDENIO, PAOLO MADONIA, FRANCESCO ROSSI, FEDERICA PAGANELLI AZZA, PIETRO DE MARCHI, LUCA ROMANELLI, MATTEO STOISA, LUCA MANCA, GIANLUCA CAMPAGNA, AND ARMANDO LA ROCCA

SECTION VI Look Back from the Future

CHAPTER 10 ■ Commercial Human Space Exploration Assisted by Artificial Intelligence 363

ANOUSHEH ANSARI AND JIM MAINARD

Index 398

Foreword

Artificial Intelligence and Space: Shaping the Future of Humanity

"Exploration is really the essence of the human spirit." It was 13 days after the landing of Apollo 8, the first manned flight to the moon, that Frank Borman, commander of the Apollo 8 mission, said this sentence in his speech to the U.S. Congress on January 9, 1969. He placed the human desire to explore and to discover deep into the human being. Humanity now faces the challenge of integrating the rapid development of Artificial Intelligence (AI), which is a highly complex and evolving relationship: AI, as a tool and technology, has the potential to greatly enhance human capabilities and improve various aspects of our lives. It also raises concerns and fears due to its rapid advancements and the potential for AI systems for decision-making, which raises questions about accountability and the potential for biased or unethical outcomes.

Artificial Intelligence can be an integral part of human existence in a beneficial and symbiotic way: The integration of AI technologies has emerged as a transformative force across sciences and industries and holds the key to unlocking new frontiers, enhancing spaceflight capabilities, and maximizing scientific discoveries. With its ability to process vast amounts of data, make autonomous decisions, and adapt to dynamic environments, AI is poised to revolutionize space exploration, to enable new possibilities for applications and to offer invaluable insights that can profoundly impact our life on Earth.

This book dives deeply into the trends and values of AI in space, exploring its impact on space systems, quantum computers, big data, sustainability, legal perspectives, and into a range of use-cases that span from neuromorphic computing and sensing, spacecraft location estimation, and space weather forecasting. As humanity continues its quest to unravel the mysteries of our solar system and the Universe, the convergence of AI and space exploration has opened new frontiers of discovery and innovation. The synthesis of these two rapidly evolving domains promises to revolutionize our understanding fundamentally.

Artificial Intelligence can empower a new era of human space exploration by developing and operating enhanced autonomous systems for remote and challenging environments, such as Mars or deep space missions. AI-powered systems can make independent decisions, interpret data, and adapt to unforeseen circumstances, enabling spacecraft and rovers to navigate and carry out complex tasks with minimal human intervention, improving mission safety and efficiency. The integration of AI and human spaceflight is not limited to exploration alone, but also fuels the growth of the space economy. AI-driven automation and resource management systems optimize space manufacturing, asteroid mining, management of space habitats and therefore the potential establishment of extra-terrestrial colonies.

The integration of AI into the space economy also holds benefits for life on Earth: Technologies developed for space missions—AI-driven resource management, advanced imaging and sensing systems, and efficient energy utilization—have tremendous potential for addressing the United Nations' Sustainable Development Goals (SDGs) on Earth. AI-enabled data analysis can monitor climate change, biodiversity, agriculture, and manage natural resources, providing actionable insights to safeguard our planet's ecological balance. Thus, AI not only helps us to explore our solar system, but also enriches our lives on Earth, fostering sustainability, collaboration, and technological advancements.

There exist critical limitations for the implementation of AI. Understanding these limitations such as lack of contextual understanding, lack of emotional intelligence, dependency on data quality, vulnerability to attacks and most importantly, lack of creativity and intuition and of ethics and morality. AI cannot understand the inherent dignity of every human person, and any application of AI must respect and promote this dignity, the human rights, and the well-being of individuals. Thus, as human, our relation to AI should be characterized as commitment to human dignity, and the pursuit of the common good in a rapidly evolving technological landscape. By understanding the boundaries of AI, we can better harness its capabilities while also avoiding unrealistic expectations or potential negative consequences. Despite fundamental and irresolvable ontological differences, AI and humans should exist and operate related to each other to realize the full potential of human-created AI: AI can significantly enhance human potential and empower human astronauts on their journeys into space. Intelligent systems can analyse mission-critical data, streamline operations, optimizing resource utilization, aid in decision-making, support health and habitat conditions. If we are able to see AI as our trusted companion in the harsh and unforgiving environments beyond our planet, human can focus on new discoveries, advance scientific knowledge and pushing the boundaries of human endurance.

The challenges of space, coupled to AI's transformative potential could become a catalyst for peaceful cooperation, inspiring harmony and solidarity among nations and cultures. As humanity must work together to build a prosperous space-faring civilization, AI fosters a collective effort in space exploration and a vision of unity that transcends earthly boundaries and propels us towards a future where humanity stands united in its quest for knowledge, understanding and exploration, to cultivate a sustainable and prosperous future for generations to come.

This book embarks on a journey through the fascinating realm of AI in space, exploring its profound implications, emerging trends, and transformative potential.

Zurich, Switzerland, June 2023

Oliver Ullrich, Ph.D., Prof. mult. Dr.med. Dr.rer.nat.
Director Innovation Cluster Space and Aviation (UZH Space Hub)
University of Zurich (UZH)

Contributors

Anousheh Ansari
XPRIZE Foundation
Culver City CA, USA

Alessandro Benetton
AIKO S.r.l.
Torino, Italy

Emmanuel Blazquez
European Space Research & Technology
Centre (ESTEC), European Space
Agency (ESA)
Noordwijk, The Netherlands

Ilaria Bloise
AIKO S.r.l.
Torino, Italy

Francesco Califano
University of Pisa
Pisa, Italy

Gianluca Campagna
AIKO S.r.l.
Torino, Italy

Enrico Camporeale
University of Colorado & NOAA Space
Weather Prediction Center
Boulder CO, USA

Christian Cardenio
AIKO S.r.l.
Torino, Italy

Pietro De Marchi
AIKO S.r.l.
Torino, Italy

Dominik Dold
European Space Research & Technology
Centre (ESTEC), European Space
Agency (ESA)
Noordwijk, The Netherlands

Romain Dupuis
Naval Group Belgium
Brussels, Belgium

Lorenzo Feruglio
AIKO S.r.l.
Torino, Italy

Steven Freeland
Western Sydney University School of
Law
Sydney, Australia

Seisuke Fukuda
Institute of Space and Astronautical
Science (ISAS), Japan Aerospace
Exploration Agency (JAXA)
Kanagawa, Japan

Pablo Gómez
European Space Research & Technology
Centre (ESTEC), European Space
Agency (ESA)
Noordwijk, The Netherlands

Alexander Hadjiivanov
European Space Research & Technology
    Centre (ESTEC), European Space
    Agency (ESA)
Noordwijk, The Netherlands

Maria Elena Innocenti
Ruhr-Universität Bochum
Bochum, Germany

Takayuki Ishida
Research and Development Division,
    Japan Aerospace Exploration Agency
    (JAXA)
Kanagawa, Japan

Dario Izzo
European Space Research & Technology
    Centre (ESTEC), European Space
    Agency (ESA)
Noordwijk, The Netherlands

Hiroyuki Kamata
Meiji University
Kanagawa, Japan

Armando La Rocca
AIKO S.r.l.
Torino, Italy

Giovanni Lapenta
Katholieke Universiteit Leuven (KU
    Leuven)
Leuven, Belgium

Riccardo Maderna
AIKO S.r.l.
Torino, Italy

Matteo Madi
Sirin Orbital Systems AG
Zurich, Switzerland

Paolo Madonia
AIKO S.r.l.
Torino, Italy

Scott Madry
University of North Carolina at Chapel
    Hill
Chapel Hill NC, USA

Jim Mainard
XPRIZE Foundation
Culver City CA, USA

Luca Manca
AIKO S.r.l.
Torino, Italy

Anne-Sophie Martin
Sapienza University of Rome
Rome, Italy

Gabriele Meoni
European Space Research Institute
    (ESRIN), European Space Agency
    (ESA)
Frascati, Italy

Iko Nakari
The University of
    Electro-Communications, Tokyo
Tokyo, Japan

Federica Paganelli Azza
AIKO S.r.l.
Torino, Italy

Giorgio Pedrazzi
Cineca
Bologna, Italy

Joseph N. Pelton
Chair of the Alliance for Collaboration
    in the Exploration of Space (ACES
    WORLDWIDE)
Arlington VA, USA

Luca Romanelli
AIKO S.r.l.
Torino, Italy

Francesco Rossi
AIKO S.r.l.
Torino, Italy

Shinichiro Sakai
Institute of Space and Astronautical
    Science (ISAS), Japan Aerospace
    Exploration Agency (JAXA)
Kanagawa, Japan

Shujiro Sawai
Institute of Space and Astronautical
    Science (ISAS), Japan Aerospace
    Exploration Agency (JAXA)
Kanagawa, Japan

Olga Sokolova
Sirin Orbital Systems AG
Zurich, Switzerland

Matteo Stoisa
AIKO S.r.l.
Torino, Italy

Keiki Takadama
The University of
    Electro-Communications
Tokyo, Japan

Fumito Uwano
Okayama University
Okayama, Japan

Mattia Varile
AIKO S.r.l.
Torino, Italy

Davide Vittori
AIKO S.r.l.
Torino, Italy

Yuka Waragai
The University of
    Electro-Communications
Tokyo, Japan

Alexander Zoechbauer
European Space Research & Technology
    Centre (ESTEC), European Space
    Agency (ESA)
Noordwijk, The Netherlands

About the Contributors

Anousheh Ansari is CEO of the XPRIZE Foundation, the world's leader in designing and operating incentive competitions to solve humanity's grand challenges. Ansari, along with her family, sponsored the organization's first competition, the Ansari XPRIZE, a $10 million competition that ignited a new era for commercial spaceflight. Since then, she has served on XPRIZE's Board of Directors. Prior to being named CEO of XPRIZE, Ansari was a serial Tech entrepreneur cofounding and exiting several Tech companies in the Internet of Things and Telecom software industry. She captured headlines around the world when she embarked upon an 11-day space expedition, accomplishing her childhood dream of becoming the first female private space explorer, first astronaut of Iranian descent, first Muslim woman in space, and fourth private explorer to visit space. Ansari serves on the World Economic Forum's (WEF) Global Future Council and has received numerous honors, including the WEF Young Global Leader, Ellis Island Medal of Honor, and STEM Leadership Hall of Fame, among others. She is a UNESCO Goodwill Ambassador. Ansari also co-founded The Billion Dollar Fund for Women, announced in October 2018 at the Tri Hita Karana (THK) Forum on Sustainable Development in Bali, with a goal of investing $1 billion in investment for women-founded companies by 2020. She published her memoir, My Dream of Stars, to share her life story as inspiration for young women around the world. Ansari holds a bachelor's degree in electronics and computer engineering from George Mason University, a master's degree in electrical engineering from George Washington University; and honorary doctorates from George Mason University, Utah Valley University, and International Space University.

Alessandro Benetton Chief Technological Officer of AIKO and Forbes 30 under 30 in Italy for Science, Alessandro Benetton is responsible for the technological strategy of AIKO, a startup he has been a full-time member of since 2020 and a collaborator since its inception. He received a national award for research on space sustainability and was a former researcher at the National Institute of Nuclear Physics in Italy. Alessandro Benetton has a strong technical background in space operations automation and mission analysis activities in general. He is responsible for defining and pursuing the company's technological roadmap, including strategic activities related to prioritizing product development, ensuring quality, and engaging with the industrial and institutional space ecosystem. With his expertise, he has significantly contributed to designing and developing software applications for autonomous onboard operations of spacecraft within internal research and development projects, product development, and industrial and institutional initiatives.

Emmanuel Blazquez is a research fellow at ESA's Advanced Concepts Team, focusing on mission analysis with an emphasis on on-board real-time optimization assisted by Artificial Intelligence. He was awarded a Ph.D. by the University of Toulouse Paul-Sabatier in 2021 for his work on rendezvous optimization and GNC design on cislunar near-rectilinear Halo orbits, which was the result of a collaboration between the European Space Agency, ISAE-SUPAERO and Airbus Defence and Space. Emmanuel's research interests include autonomous Guidance and Control architectures, multibody Astrodynamics, global trajectory optimization and neuromorphic perception for space systems.

Ilaria Bloise Head of AIKO's Machine Learning Development Department. She holds a Master's in Space Engineering from Politecnico di Milano, Italy, and several specializations in Deep Learning from online courses. She has several years of experience in the development of Deep Learning applications for space applications, starting from research developed at the System and Industrial Engineering department of the University of Arizona (USA) on "A deep learning approach to autonomous lunar landing "for navigation and control of spacecraft in Moon proximity operations. In AIKO, she is responsible for several institutional and commercial projects, and she is managing the technical team working on developing breakthrough Machine Learning technologies applied to the space domain.

Francesco Califano obtained his PhD at the University of Florence in 1993. He is full Professor at the Department of Physics "Enrico Fermi" of the University of Pisa where he teaches Plasma Physics and is responsible for a Dual Master Diploma in Physics with Sorbonne University. During his PhD and in the following period he spent more than three years at the Observatoire de Paris in Meudon and continued to collaborate with several French colleagues as visiting scientist in Paris (Meudon, Ecole Polytechnique, Sorbonne University), Nancy (University of Lorraine), Nice (Observatoire de la Côte d'Azur). His main research activity spans from basic plasma physics, laser-plasma interaction, fusion but is today mainly focused on theoretical space plasma physics. The main topics are plasma turbulence, magnetic reconnection, Vlasov equation, solar wind interaction with the Magnetosphere (Earth and Mercury). He has also worked extensively in computational plasma physics, developing several numerical codes, both fluid and kinetic, in particular the Eulerian Vlasov code in the 1990s with a French colleague. F. Califano has been involved in several EU projects as Work Package leader. From 2017 onwards he is scientific co-I of the "Plasma Wave Investigation" consortium, MMO, spacecraft (JAXA), Bepi-Colombo, ESA-JAXA mission. He is Associated Editor for the Journal of Plasma Physics and project creator of the regular meeting Vlasovia held every three years in Italy and France. He is author or co-author of almost two hundred publications on International scientific peer-reviewed journals.

Gianluca Campagna holds a Master's in Aerospace Engineering from Politecnico di Milano (Italy), with a thesis on Synthetic Aperture Radar data processing. After a brief work experience in the Netherlands, he joined AIKO at the end of 2022 as a deep learning engineer. In pursuit of his dream of supporting the development of the Italian space sector, he is working to increase the autonomy of space missions.

Enrico Camporeale graduated in space plasma physics from the Queen Mary University of London. He has worked at the Los Alamos National Laboratory and the Dutch National Center for Mathematics and Computer Science (CWI). He is currently a research associate with the University of Colorado Boulder and affiliated with the NOAA Space Weather Prediction Center, in Boulder, Colorado. His research activities focus on the use of machine learning and artificial intelligence to improve the forecasting capabilities of space weather models and on data-driven discovery of space physics. Enrico Camporeale is currently an associate editor for the Journal of Space Weather and Space Climate and the president-elect of the Nonlinear Geophysics section of AGU.

Christian Cardenio Head of AIKO's Software Development Department. He graduated in 2016 in Aerospace Engineering at Politecnico di Torino with a thesis on an interplanetary CubeSat case study to increase mission autonomy employing Neural Networks. In 2013 he started a collaboration with the CubeSat Team Polito, focused on designing an event/message-based C&DH system for the 3-St@r 3U CubeSat. After his graduation, he strengthens his skills in Systems Engineering as a research fellow in Thales Alenia Space, dealing with the preliminary design of the STRONG Space Tug Electrical Propulsion System. In 2017 he joined Rolls-Royce in the UK as a Controls Systems Engineer, improving his knowledge of design and verification processes and working on the Trent700 Engine Electronic Controller major update, implementation of automated flight software analysis and mapping, requirements management, and problem reports investigation and solving. With a consolidated background in software and system design, systems engineering, and software development, in 2019, he was hired by AIKO as a Software Engineer. Since 2022, as Head of Software Development, he has been responsible for industry-quality Software Development practices, Software Product Assurance, DevOps practices, such as automated testing and Continuous Integration, and Software Integration for AIKO's technology.

Pietro De Marchi Space Systems Engineer at AIKO. He holds a double degree in Space Engineering (Italian and French titles), with a final thesis developed at NASA-JPL about conjunction analysis optimization methods for LEO orbits (2021). He has a great passion for space and previous experiences in space propulsion system design (academic team COSMOS at Ecole Centrale de Lyon), mission analysis (CSUM Montpellier), and constellation architecture design (Alten Italy).

Dominik Dold is a scientist working on artificial intelligence (AI) and neuromorphic computing. He graduated with a PhD from Heidelberg University and after a Research Residence at the Siemens AI Lab in Munich, he joined the Advanced Concepts Team at ESA as a Research Fellow. In his work, he mainly focuses on biologically inspired AI, graph algorithms, models of self-organization and learning, explainable AI, and applications thereof – especially for space and fundamental science. He believes that the recent progress in AI offers exciting opportunities for the space sector and will be essential in supporting us to further explore and understand our Solar System (and everything that lies beyond!).

Romain Dupuis obtained his PhD at the University of Toulouse, France, in 2019. He contributed to the H2020 AIDA project and the associated Python package aidapy for two years. He is currently Data Scientist at Naval Group Belgium. Romain Dupuis' primary areas of interest are machine learning, deep learning, their application to physics and engineering, devOps, and data engineering.

Lorenzo Feruglio Founder and Chief Executive Officer (CEO) of AIKO. He holds a Ph.D. in Aerospace Engineering at Politecnico di Torino, Italy, with a thesis on Artificial Intelligence to enhance Small Satellites Mission Autonomy (2017). He has been visiting researcher both at NASA JPL and MIT, with contributions on small satellite mission design and telecommunication architecture modeling. He has over eight years of developer experience in small satellite flight software and Artificial Intelligence application and past working experience in SES, Luxembourg, a leading space telecommunication company. In 2017 he founded AIKO, and under his guidance, the company grew to over 30 employees and established itself as a leading company for upstream space applications, with offices in Italy and France.

Steven Freeland is Emeritus Professor of International Law at Western Sydney University and Professorial Fellow at Bond University, Australia. He is also Visiting Professor at the University of Vienna, Permanent Visiting Professor at the iCourts Centre of Excellence for International Courts, University of Copenhagen, Adjunct Professor at the University of Hong Kong, Senior Fellow at the London Institute of Space Policy and Law, Visiting Professor at Université Toulouse1 Capitole, and Associate Member at the Centre for Research in Air and Space Law, McGill University. He has represented the Australian Government at United Nations Conferences on space, and has advised various governments on issues related to the national and international regulation of space activities and the development of a national space-industry strategy. He has been appointed by the United Nations Committee on the Peaceful Uses of Outer Space to co-lead multilateral discussions regarding the exploration, exploitation and utilization of space resources, and by the Australian Government as a Member of the Australian Space Agency Advisory Group. Among other appointments, he is a Director of the International Institute of Space Law, and a member of the Space Law Committees of both the International Law Association and the International Bar Association.

Seisuke Fukuda received the B.E. degree in 1995, the M.E. degree in 1997, and the Dr.Eng. degree in electronic engineering in 2000 from the University of Tokyo, Japan, respectively. Since 2000, he has been with the Institute of Space and Astronautical Science (ISAS), Japan, where he is currently a Professor in ISAS, Japan Aerospace Exploration Agency (JAXA). His research interests include satellite and spacecraft systems, signal and image processing, and microwave remote sensing. He has been involved in several scientific spacecraft projects including INDEX (REIMEI), SPRINT-A (HISAKI), ERG (ARASE), and SLIM. He is a member of The Institute of Electrical and

Electronics Engineers (IEEE), Japan Society for Aeronautical and Space Sciences (JSASS), and Institute of Electronics, Information and Communication Engineers (IEICE), Japan.

Pablo Gómez is a research fellow in ESA's Advanced Concepts Team and currently seconded to the national institute for artificial intelligence AI Sweden. He received his PhD from the Friedrich-Alexander-Universität Erlangen-Nürnberg in 2019 (supervisor Prof. Döllinger) and his M.Sc. in computer science from the Technical University Munich in 2015. Research topics of interest to him range from machine learning and inverse problems to numerical methods and high-performance computing.

Alexander Hadjiivanov is currently a research fellow with the Advanced Concepts Team. His PhD from the University of New South Wales tackled adaptation and structural plasticity in neural networks. Straddling the space between academia and industry, he has worked on topics such as morphology induction, interactive AI, computer vision, neuroevolution and spiking neural networks. He is currently working on models of adaptation and homeostasis for artificial neurons, specifically in the context of neuromorphic perception and continual learning. His long-term research goal is to develop a generic 'plug-and-play' artificial neuron model that remains responsive, plastic and robust in the presence of perturbations and concept drift, paving the way for stable online learning with spiking neural networks in dynamic environments.

Maria Elena Innocenti obtained her PhD at the University of Leuven, Belgium, in 2013. She has then been awarded two FWO (Flemish Research Foundation) Postdoctoral Fellowships at the University of Leuven and a NASA Postdoctoral Program Fellowship at Jet Propulsion Laboratory, Pasadena, US. She is currently junior-professor in computational plasma physics at the Ruhr University Bochum, Germany. Her research interests include the development of advanced methods for fully kinetic Particle-In-Cell, PIC, simulations (semi-implicit, adaptive, expanding box simulations), investigation of collisionless processes in space plasmas (kinetic instabilities, magnetic reconnection, turbulence), space weather, High Performance Computing, HPC, and Physics Informed Machine Learning. M.E.I. has been work package leader of the H2020 AIDA project and team leader of an ISSI project. She is currently project PI for the DFG (German Research Foundation)-funded Collaborative Research Center SFB1491, which includes 3 universities and 17 PIs in Germany. She is also currently PI of DFG research projects and of HPC projects, including a large-scale project granted by the Gauss Center for Supercomputing.

Takayuki Ishida received the master's degree in engineering from Keio University, Japan, in 2015. He joined Japan Aerospace Exploration Agency (JAXA) and has been part of the Smart Lander for Investigating Moon (SLIM) since 2015 in Institute of Space and Astronautical Science (ISAS), developing navigation cameras and image navigation algorithms. His research interest includes terrain relative navigation for planetary precision landing.

Dario Izzo graduated as a Doctor of Aeronautical Engineering from the University Sapienza of Rome (Italy). He then took a second master in Satellite Platforms at the University of Cranfield in the United Kingdom and completed his Ph.D. in Mathematical Modelling at the University Sapienza of Rome where he lectured classical mechanics and space flight mechanics. Dario Izzo later joined the European Space Agency and became the lead of its Advanced Concepts Team. He devised and managed the Global Trajectory Optimization Competitions events, the ESA Summer of Code in Space and the Kelvins innovation and competition platform. He published more than 170 papers in international journals and conferences making key contributions to the understanding of flight mechanics and spacecraft control and pioneering techniques based on evolutionary and machine learning approaches. Dario Izzo received the Humies Gold Medal and led the team winning the 8th edition of the Global Trajectory Optimization Competition.

Hiroyuki Kamata received the B.E. degree in 1982, M.E. degree in 1984, and the Dr. Eng. degree in electrical engineering in 1987 from Meiji University, Japan, respectively. He became a lecturer in 1990 and an associate professor in 1995 at school of science and technology, Meiji University. Since 2000, he has been a professor. He served as a visiting professor at Japan Aerospace Exploration Agency (JAXA) / Institute of Space and Astronautical Science (ISAS), from 2010 to 2012 and as vice president for information education at Meiji University from 2018 to 2021. His research interests include the digital signal processing using computers, FPGA and DSP. In particular, he is interested in practical signal processing suitable for low computing resource environments. He is a member of Information Processing Society of Japan (IPSJ), Institute of Electronics, Information and Communication Engineer (IEICE) IEE, and The Institute of Electrical and Electronics Engineers (IEEE).

Armando La Rocca has a solid educational background in computer engineering and data science. He obtained his Bachelor's degree in "Ingegneria Informatica e dell'Informazione" from the "Università degli Studi di Siena". He pursued a Master's in "Data Science and Engineering" from "Politecnico di Torino". He is currently employed at AIKO as a Deep Learning Engineer, where he is involved in developing AI applications for Earth Observation and focuses on autonomous satellite navigation. His work involves advanced technologies to improve satellite data analysis and enhance satellites' navigation capabilities like Deep Neural networks and Generative models.

Giovanni Lapenta obtained his PhD at the Politecnico di Torino (Italy) in 1993 and has worked in the United States of America at the Massachusetts Institute of Technology, at the Los Alamos National Laboratory, at the University of California, Los Angeles and at the Space Science Institute, Boulder. Lapenta is currently Full Professor at the University of Leuven in Belgium. Giovanni Lapenta research interests include the study of astrophysical, solar and space plasma physics using simulation and theoretical methods. Lapenta is the grantee of the European Research Council Advanced Grant TerraVirtulE, the coordinator of the project ASAP on the deployment of artificial intelligence in situ in space missions of exploration and the coordinator of the Belgian Defence project AIDefSpace on the forecast of space weather threats using machine learning. Lapenta has been the coordinator of the AIDA project on artificial intelligence methods for space data analysis. Lapenta was the leader of several international efforts funded by the European Commission (Soteria, eHeroes, Swiff), by the US Department of Energy, by NASA and by NSF with the goals of gathering and organising the most modern sources of data relative to space and coupling fluid and kinetic models using implicit time differencing and adaptive grids. Lapenta developed massively parallel high performance computer codes for space simulation that were awarded one of the most prestigious prizes in software development, the RD100 prize, and that are being used on some of the largest supercomputers in the world. Giovanni led the development and currently maintains the massively parallel computer code iPic3D in use by many teams around the world. Lapenta is the current editor of the scientific journal Nonlinear Processes in Geophysics, of Frontiers in Fusion Research and of the journal Plasma. He has authored and co-authored around 300 peer-reviewed scientific papers.

Riccardo Maderna Head of AIKO's Autonomous Systems Department. He received his Ph.D. in Information Technology from Politecnico di Milano in 2020. His research focused on human-robot collaboration for flexible manufacturing spanning dynamic task scheduling, human activity monitoring and prediction, and digital twin modeling. After that, he joined AIKO S.r.l., an Italian deep-tech company that develops artificial intelligence solutions for space missions. Since 2022, he has been the Head of Autonomous Systems, overseeing the design and development of autonomous systems and algorithms.

Matteo Madi is an entrepreneur, business developer and space-tech specialist with over thirteen years of work experiences in the Swiss and International public and private sectors. He participated in development processes for various space missions as lead R&D engineer or project manager. Matteo Madi received his M.Sc. (2011) in Electrical Engineering and Space Technologies and his Ph.D. (2016) in compact optical space instrumentations from the Swiss Federal Institute of Technology (ETH domain). Dr. Madi worked at the European Space Research and Technology Centre (ESTEC) in 2016 in the frame of ESA's Networking/Partnering Initiative (NPI) program focusing his research on the key technologies enabling compact imaging spectrometers. He is designated as inventor by the European Patent Office for three granted patents, one of which is a joint patent supported by the Technology Transfer & Business Incubation Office (TTBO) at ESTEC, ESA. As business developer, he has been active in providing consultation, strategic advice, and market development support to venture capitals, start-ups, Swiss and international institutions, and contributed to their efforts to align policies with the latest market/technology trends. In response to the new demands of the emerging space market, in 2019, Matteo Madi founded Sirin Orbital Systems AG based in Zürich, Switzerland, which develops viable innovative products, services and solutions for upstream and downstream applications based on cutting edge terrestrial technologies and space assets. Sirin Orbital Systems AG focuses on the development of advanced technologies notably for On-Orbit Servicing (OOS) of satellites, In-Situ Resource Utilization (ISRU), future sustainable space exploration and notably enabling Wireless Power Transfer (WPT) technologies in support of Space-Based Solar Power (SBSP) systems. Sirin Orbital Systems AG also creates innovative solutions for the use of space technologies and satellite-based services for sustainable ground-based applications. Dr. Madi also leads the SIRIN AI R&D Centre, expertise in extended reality (XR) and natural language processing (NLP) domains. Matteo Madi is the Editor of the book entitled, "*Space Debris Peril: Pathways to Opportunities*", published by CRC Press: Taylor and Francis in November 2020 (ISBN 9780367469450).

Paolo Madonia holds a Masters's degree in Astrophysics and one in Aerospace Engineering from Sapienza - University of Rome. In 2021, following a few years of experience in the Observational Cosmology groups at the California Institute of Technology and Sapienza University, he joined AIKO as a Mission Autonomy Engineer, later transitioning into the Product Management Department to oversee the development of AIKO's software products for autonomous space missions.

Scott Madry, Ph.D., is a professor emeritus the International Space University in Strasbourg, France. He was a long-time member of the faculty. Including being at the founding ISU conference in 1987, and he has served as faculty in over 35 ISU programs around the world, and as the program director of the ISU Southern Hemisphere Summer Space Program for four years in Adelaide, Australia. He is currently a Research Associate Professor of Archaeology at the University of North Carolina at Chapel Hill, specializing in the applications of geomatics technologies for regional environmental and cultural research. He is also the founder and President of Informatics International, Inc., a global geospatial services company. In 2020 he was elected a corresponding member of the International Academy of Astronautics. He received his Ph.D. from the University of North Carolina in 1986 and then worked for 3 years at the Institute for Technology Development, Space Remote Sensing Center at the NASA Stennis Space Center, where he was very involved in the early GRASS GIS community, delivering the software and holding one of the annual user conferences. He then took the position of Senior Associate Director of the Center for Remote Sensing and Spatial Analysis at Rutgers University, where he taught for 9 years in the Anthropology, Geography, and Natural Resources departments. He is widely published, is the author of nine books[1] and over 75 articles and papers, and was an editor for the 1,300 page, two-volume *Handbook of Satellite Applications* for Springer Press which has had over 400,00 downloads. He was also the associate editor for the two-volume *Handbook of Small Satellites* for Springer Press.

Dr. Madry is a three-time Fulbright Scholar, having taught in France and twice in South Africa, in Cape Town and Johannesburg. Continuing his work there, he was a visiting lecturer at the University of Cape Town's SpaceLab program for four years. He has conducted field research in North America, Europe, and Africa, and has given over 200 short courses and seminars in over 30 countries on six continents around the world. He has consulted for numerous governments, corporations and non-profits, and has received over

---

[1]See: http://scottmadry.web.unc.edu

US$7.5 million in grants and contracts. Dr. Madry has acted as a consultant on satellite remote sensing for a major Hollywood film, and has been a consultant for major museum exhibitions on the subject. Some of his research was featured in the *McGraw Hill Yearbook of Science and Technology* in 1991. In 1997 he was awarded the Russian Tsiolkovsky Gold Medal for his international research and teaching activities. He has conducted field research in the Burgundy region of France for some 45 years. He is very active in the American Red Cross Disaster Services and has won several awards for his activities on behalf of the Red Cross. In 2012 he was awarded, along with the other members of the GISCorps, the President's Volunteer Service Award by President Barak Obama for his work in applying geomatics technologies to disaster management. and he received a second, gold award for his work with the Red Cross in 2017.

Jim Mainard is CTO, EVP Deep Technology at XPRIZE Foundation where he oversee prize development in emerging / deep technologies including AI/ML, remote sensing haptic robotics, and quantum computing. Prior, Jim was the President of the global technology acceleration at Shiseido to disrupt the beauty and skincare industry by adapting emerging technologies. Prior, Mainard held several executive roles at DreamWorks and DreamWorks Animation as EVP Digital Strategy fostering, developing, and investing in digital technologies to further business interests, SVP Production Development where he launched new visualization technologies for film making (motion capture, virtual sets, and stereoscopy), and SVP R&D in which he drove the company's substantial investment in visual effects tools and pipelines. He was a Cofounder and former Chairman of the Advanced Imaging Society serving the Entertainment industry, and CTO of MGO, which was acquired by, and became Fandango's online movie business Fandango Now. Credited on more than twenty major films, a Broadway musical, and several made for television shows, Mainard has deep roots developing and adapting technology in service to creative expression. Mainard made the transition to the entertainment industry after spending more than a decade in Aerospace where he was proud to serve as Senior Systems Engineer at TRW for the Hubble Space Telescope Ground Station. Among many roles during his tenure at TRW, Jim redesigned tactical mapping software relied upon by the U.S. Army and Navy and was principle in the creation and establishment of the Tactical Aircraft Mission Planning System (TAMPS) used in all fixed and rotary wing aircrafts.

Luca Manca holds a Master's in Space Engineering from Politecnico di Milano (Italy). He co-authored his master's thesis with the European Space

Agency, where he applied Deep Learning algorithms to model the spacecraft's thermal subsystem. In 2021, he joined AIKO as a Deep Learning Engineer to work on algorithms to enhance satellite autonomy. His main areas of interest are telemetry analysis in Space Operations and onboard data processing.

Anne-Sophie Martin is Research Fellow, Sapienza University of Rome. She received her LL.M. in Space Law and Telecommunications Law from the University of Paris-Saclay (France) and her Ph.D. from Sapienza University of Rome (Italy). Between 2016 and 2019, she was an observer within The Hague Space Resources Governance Working Group. On August 2017, she attended the Centre for Studies and Research of The Hague Academy of International Law. Since 2021, she is an observer in the Global Expert Group on Sustainable Lunar Activities (GEGSLA). Since 2022, she is a member of the Policy Hub of the IAU's Centre for the Protection of the Dark and Quiet Skies from Satellite Constellation Interference. She participates in the UN-OOSA Space4Women initiative as a mentor. Member of the International Institute of Space Law, Space Generation Advisory Council, European Centre of Space Law, French Society of Air and Space Law, and Institute of Space and Telecommunications Laws. She is also Member of the Legal Council of the organisation 'For All Moonkind' and a Fellow in the For All Moonkind's Institute of Space Law and Ethics.

Gabriele Meoni, PhD, is an internal research fellow in the Φ-lab division of the European Space Agency. He is a former member of the ESA Advanced Concepts Team, which he joined in the 2020 after receiving his PhD from University of Pisa (supervisor Prof. Luca Fanucci) in information engineering. From October 2022 to March 2023, he was seconded to AI Sweden to conduct research on distributed edge learning onboard Earth Observation satellites. His research topics of interest include satellite onboard processing, embedded computing systems, edge computing, and neuromorphic computing.

Iko Nakari received the B.E. degree in 2019 and the M.E. degree in 2021 from the University of Electro-Communications, Japan, respectively. He has been a DC2 research fellow for young scientists in the Japan Society for the Promotion of Science since 2022 and is currently the Ph.D. candidate at the University of Electro-Communications. His main interests include autonomous rovers and AI technology in space. He developed CanSat for Japanese CanSat competition and the world CanSat competitions in A Rocket Launch for International Student Satellites (ARLISS) in 2018 and 2019, and won several awards in both competitions. He organized two CanSat competitions in

Japan to provide a chance for a team for ARLISS. He is a member of The Institute of Electrical and Electronics Engineers (IEEE), The Association for the Advancement of Artificial Intelligence (AAAI), and space- and AI-related research societies in Japan.

Federica Paganelli Azza graduated from Politecnico di Milano with a Master's in Automation and Control Engineering. She worked on a thesis in collaboration with Ferrari, designing, implementing, and testing an autonomous steering system for high-performance vehicles. She joined AIKO in 2021 as an Autonomous Systems Engineer, where she is currently working to improve satellite autonomy and Artificial Intelligence capabilities. Her main areas of interest focus on spacecraft guidance, control, and autonomous maneuvering.

Giorgio Pedrazzi holds a Doctorate in Statistical Methodology for Scientific Research from the Faculty of Statistical Sciences at the University of Bologna, which he earned in 1995. Currently, he works as a Data Scientist at Cineca within the HPC (High Performance Computing) department. Alongside his role at Cineca, Giorgio coordinates the Big Data Laboratory module in the master's program for Data Science and Business Analytics at BBS (Bologna Business School). Giorgio Pedrazzi's primary areas of interest revolve around data analysis, machine learning, and deep learning. Within the HPC department, he actively collaborates on various European projects, including AIDA on artificial intelligence methods for space data analysis focusing on the analysis of heliophysics data. He contributes to EUHUBS4DATA, a project aimed at creating a federation of Data Innovation HUBs to provide data-related services to small and medium-sized enterprises, startups, and web entrepreneurs. He is currently involved in the Graph-Massivizer project to develop a high performance, scalable and sustainable platform for information processing and reasoning based on the massive graph representation of extreme data.

Joseph N. Pelton, Ph.D., is the Chairman of the Board of the Alliance for Collaboration in the Exploration of Space (ACES Worldwide[2]). He received his degrees from the University of Tulsa, New York University and from Georgetown University, where he received his doctorate. Dr. Pelton is the Dean emeritus and former Chairman of the Board of Trustees of the International Space University. He is the Founder of the Arthur C. Clarke Foundation and the founding President of the Society of Satellite Professionals International—now known as the Space and Satellite Professionals International

---

[2]See: https://acesworldwide.org

(SSPI). Dr. Pelton currently serves on the Executive Board of the International Association for the Advancement of Space Safety. He is the Director Emeritus of the Space and Advanced Communications Research Institute (SACRI) at George Washington University, where he also served as Director of the Accelerated Masters' Program in Telecommunications and Computers from 1998 to 2004. Previously, he headed the Interdisciplinary Telecommunications Program (ITP) at the University of Colorado-Boulder. Dr. Pelton has also served as President of the International Space Safety Foundation and President of the Global Legal Information Network (GLIN). Earlier in his career, he held a number of executive and management positions at COMSAT and INTELSAT, the global satellite organization where he was Director of Strategic Policy. A prolific author and futurist, Dr. Pelton has now published over 60 books and over 400 articles, encyclopedia entries, op-ed pieces and other research publications during his career. He has been speaker on national media in the U.S. (PBS News Hour, Public Radio's All Things Considered, ABC, and CBS) and internationally on BBC, CBC, and FR-3. He has spoken and testified before Congress, the United Nations, and delivered talks in over 40 countries around the world. His honors include the Sir Arthur Clarke, International Achievement Award of the British Interplanetary Society; the Arthur C. Clarke Foundation Lifetime Achievement Award; the ICA Educator's award; the ISCe Excellence in Education Award; and being elected to the International Academy of Astronautics. Most recently, in 2017, he won the Da Vinci Award of the International Association for the Advancement of Space Safety and the Guardian Award of the Lifeboat Foundation.

Dr. Pelton is a member of the SSPI Hall of Fame, Fellow of the IAASS and Associate Fellow of the AIAA. Pelton's *Global Talk* won the Eugene Emme Literature Award of the International Astronautics Association and was nominated for a Pulitzer Prize. His most recent books are: *Space Systems and Sustainability*, *Preparing for the Next Cyber Revolution*, *Space 2.0: Revolutionary Advances in the Space Industry*, *The New Gold Rush: The Riches of Space Beckon*, *The Handbook of Small Satellites*, *Global Space Governance: An International Study*, and the second editions of *The Handbook of Satellite Applications and The Farthest Shore: A 21st Century Guide to Space*.

Luca Romanelli has gained a solid foundation in reinforcement learning through his work at AIKO since 2019, allowing him to successfully apply cutting-edge algorithms in various projects. His computer engineering degree also equips him with a solid understanding of daily software tasks and efficient problem-solving techniques. He is always looking for ways to broaden his knowledge and skills through continuous learning and staying current on the latest reinforcement learning advancements and applications.

Francesco Rossi Since 2021, he has been a Senior Deep Learning Engineer at AIKO, leveraging his decade of professional experience in AI and Computer Vision. He holds a Ph.D. in Computer and Control Engineering from Politecnico di Torino, Italy. Previously, he has applied his skills as a Research Engineer, developing Machine Learning-based innovative assistive technologies for the biomedical sector, and as a technical leader in a sports industry startup. His focus is on harnessing the power of deep learning for visual navigation, particularly within space-oriented applications.

Shinichiro Sakai received the B.E. degree in 1995, the M.E. degrees in 1997, and the Ph.D. degrees in electrical engineering in 2000 from the University of Tokyo, Japan, respectively. He joined The Institute of Space and Astronautical Science (ISAS) in 2001, became an associate professor in 2005, and has been a professor in ISAS, Japan Aerospace Exploration Agency (JAXA) since 2019. His research fields are the spacecraft guidance, navigation and control issues and electro-magnetic formation flying. He also participated to the attitude control system design and development for several scientific satellites, such as REIMEI (launched in 2005), ASTRO-G, HISAKI (2013) and ARASE(2016). From 2016, he is also a project manager of JAXA's the Smart Lander for Investigating Moon (SLIM) for pin-point landing demonstration. He is a member of Japan Society for Aeronautical and Space Sciences (JSASS), and Institute of Electrical Engineers of Japan (IEEJ).

Shujiro Sawai received the M.E. degree in 1991 and the Doctor of Engineering degree in 1994 from the University of Tokyo, Japan, respectively. He joined Institute of Space and Astronautical Science (ISAS) as a research associate from 1994 to 2003. As ISAS was merged to Japan Aerospace Exploration Agency (JAXA), he was engaged as an associate professor of JAXA from 2003 to 2018. He has been a professor from 2018. From 2021 he has served concurrently as director of Department of Space Flight Systems of ISAS/JAXA. His research interests include control theory for spacecraft, chemical propulsion as an actuator of controller, and spacecraft system design. He has been involved in several scientific spacecraft projects including MUSES-B (HALKA), MUSES-C (HAYABUSA), ASTRO-E2 (SUZAKU), ASTRO-F (AKARI), SOLAR-B (HINODE), PLANET-B (NOZOMI), PLANET-C (AKATSUKI) and SPRINT-A (HISAKI), as well as development of M-V launch vehicle. He is a member of the Japan Society for Aeronautical and Space Science (JSASS), and an academician of International Astronautical Federation (IAF).

Olga Sokolova is a Risk Analyst with expertise in the domain of ground- and space-based critical infrastructure risk assessment exposed to natural and man-made hazards. She obtained her PhD in 2017 for her research in the field of "solar storm impact on power grids in accordance with interconnected critical infrastructures taking into account economic and juridical aspects of power system design". The project was initially funded by Swiss Reinsurance Company Ltd. (Swiss Re), hosted in the Technology Assessment group at Paul Scherrer Institut (PSI). Based on the results of Dr. Sokolova's research, in 2014, Swiss Re published a lead-authored technical brochure for increasing the social awareness to space weather. Olga Sokolova has worked so far in various academic and industrial settings including Swiss Federal Institute of Technology in Lausanne (EPFL), Paul Scherrer Institute (PSI), and since 2020 at Sirin Orbital Systems AG as CTO / Risk Analyst. Olga Sokolova is skilled in developing and promoting risk management products among governmental and private sectors and even to non-practioners. Olga has a record in raising social awareness of spaceborne risks and opportunities brought to society by the "New Space" industry development. She regularly shares her thoughts on space industry risks at corresponding events. Ms. Sokolova contributes in developing ad-hoc risk models validations, challenging the appropriateness of assumptions and modelling processes, and risk governance. She is the Co-Editor and Author of two books respectively published in 2020 and 2021 by CRC Press: Taylor & Francis entitled, "Space Debris Peril: Pathways to Opportunities" and "Geomagnetic Disturbances Impact on Power Systems: Risk Analysis and Mitigation Strategies".

Matteo Stoisa born and raised in Turin. His greatest passions are electronic music, innovative technologies, and many sports activities. After earning his master's in computer science engineering, he focused on applied Artificial Intelligence. He works at AIKO, collaborating with a highly interdisciplinary team to create autonomous algorithms for space missions.

Keiki Takadama received the M.E. degree in 1995 from Kyoto University, Japan, and the Doctor of Engineering degree in 1998 from The University of Tokyo, Japan, respectively. He joined Advanced Telecommunications Research Institute (ATR) International from 1998 to 2002 as a visiting researcher and worked at Tokyo Institute of Technology as a Lecturer from 2002 to 2006. He moved to The University of Electro-Communications as an Associate

Professor in 2006. He has been a Professor since 2011. His research interests include multiagent systems, distributed artificial intelligence, autonomous systems, evolutionary computation, machine learning, and space application. He is a member of The Institute of Electrical and Electronics Engineers (IEEE), Association for Computing Machinery (ACM), The Association for the Advancement of Artificial Intelligence (AAAI) and a major space-, AI-, informatics- related academic societies in Japan. He served the chair of Genetic and Evolutionary Computation Conference (GECCO) in 2018 and organized the several international conference including the world congress on social simulation (WCSS), the international symposium of AAAI, and the international workshop of GECCO and International Joint Conference on Autonomous Agents and Multi-Agent Systems (AAMAS).

Fumito Uwano received the B.E. degree in 2015, the M.E. degree in 2017, and the Doctor of Engineering degree in 2020 from the University of Electro-Communications, Japan, respectively. He became a DC1 research fellow for young scientists in the Japan Society for the Promotion of Science from 2017 to 2020. He has been an assistant professor at the Graduate School of Natural Science and Technology at Okayama University since 2020 and also belonged to the Faculty of Environmental, Life, Natural Science and Technology since 2023. He was a visiting fellow at Queensland University of Technology, Australia, for collaborative research in a half year in 2022. His main interests include distributed AI in robotics, and evolutionary machine learning, especially learning classifier systems. His work focuses on analyzing knowledge structures in AI and developing theories for them. He is a member of (The Institute of Electrical and Electronics Engineers (IEEE), Association for Computing Machinery (ACM), and major space-, AI- and informatics-related research societies in Japan.

Mattia Varile Chief Innovation Officer of AIKO. He is responsible for exploring and experimenting with cross-sectorial disruptive technologies that can be transferred to the space sector. Mattia holds a degree in Aerospace Engineering from Politecnico di Torino and has been part of CubeSat Team Polito, working as a systems engineer. Mattia has been involved in AIKO since 2018 and, during this time, has specialized in Deep Learning, Reinforcement Learning, and Artificial Intelligence. Mattia was previously involved in other research projects and startup experiences.

Davide Vittori Chief Operating Officer of AIKO. He is responsible for leading the business and finance efforts and supports the CEO in defining and executing the company vision. Davide comes from the world of finance at Goldman Sachs, where he was part of the Investment Banking Division. Previous experience also includes consulting at McKinsey & Co. Davide holds a degree from the World Bachelor in Business, a university program that unfolds across three continents - Asia, Europe, and North America. As a proper Italian, Davide can cook dishes worthy of Michelin-star restaurants. Davide was also an accomplished national-level swimmer and still enjoys jumping in the pool once in a while.

Yuka Waragai received the B.E. degree in 2020 and the M.E. degree in 2023 from the University of Electro-Communications, Japan, respectively. Her research interests include the space application, artificial intelligence, spacecraft location estimation. She developed the small rover and launched the rocket which loaded the rover in the the world CanSat competition in A Rocket Launch for International Student Satellites (ARLISS) in 2018.

Alexander Zoechbauer graduated as a Master of Science in Information Technology from the Swiss Federal Institute of Technology (ETH). He then joined the Advanced Concepts Team as a young graduate trainee, where he was working on image demosaicing algorithms for the HERA mission and differentiable 3D shapes from LiDAR measurements. He is currently a fellow with the European Organization for Nuclear Research (CERN) working on Digital Twins for physics and earth observation powered by high-performance computing.

I

Introduction

CHAPTER 1

Introduction

Olga Sokolova

Sirin Orbital Systems AG, Switzerland

T HE global space industry is at a pivotal point as reported by KPMG [1]. Already many multinational businesses from various sectors adjusted their activities by integrating values brought by the space sector. It is forecasted that by 2030 every business is going to be a space-business. Moreover, long established terrestrial industries will build a presence in space. Case in point is Microsoft that launched Azure Space in 2020 – a platform connecting the possibilities of space with the power of cloud – for providing the benefit of space for every client.

Beyond the traditional space faring nations, the development of advanced technologies invited new countries to enter the game. Currently, 70 countries carry active space programs, and 20 of them have a budget more than USD 100 million. Commercial space activity is at the focal point of the modern space race. There have been many technological breakthroughs shaping the modern space sector. Advances in manufacturing technologies, satellite miniaturization, artificial intelligence (AI), education in payload and launch cost – all drove the space commercialization. For instance, the cost for heavy launches in the Low Earth Orbit (LEO) has decreased from USD 65,000 to USD 1,500 per kilogram (reference rates for 2021) [2]. As more companies embark on unique opportunities and advantages of space, more innovative solutions are being elaborated.

Artificial intelligence is one of the key technologies assisting a revolutionary change in the market landscape. It became an essential horsepower of progress. According to the European Space Agency (ESA) [3], AI comprises all techniques that enable computers to mimic intelligence, for example, computers that analyse data or the systems embedded in an autonomous vehicle.

DOI: 10.1201/9781003366386-1

Usually, artificially intelligent systems are taught by humans – a process that involves writing an awful lot of complex computer codes. AI supported technologies and services help to reduce complexity of interdisciplinary projects and automating routine tasks. However, AI-assisted products present both opportunities and challenges. It pushes industry to rethink their low-risk models and perceive new risk acceptance models.

Overall, the value creation is moving from hardware towards data acquisition and analysis. The potential for innovative space technologies and solutions is immense. Until now the most value is delivered through satellite services. Different end-users across the continents may obtain space data for various industries and purposes. Therefore, it is viable to address the users' perspective for understanding the benefit of space as space service/data *per* se is worthless. The lack of interaction between end-users and space organizations may result in missed progress opportunities. Although the value is generated by enhancing or enabling activities on Earth using space assets (Space-for-Earth economy), the future market development is as well associated with services that will occur completely in space (Space-for-Space economy). One can say that the services or technologies – which are emerging or promising today – may become critical tomorrow.

According to the OECD[1], the space economy is the full range of activities and the use of resources that create value and benefits to human beings in the course of exploring, researching, understanding, managing, and utilising space [4]. Space is no longer an inclusive domain of national governments. The modern space has branches into distinct sectors. Modern agriculture, pharmaceuticals, consumer goods and tourism are the bold examples of space penetration. SpaceTech Industry [5] concludes that there are now more than ten thousand SpaceTech related companies, five thousand leading investors, one hundred and fifty R&D hubs, and one hundred and thirty governmental organizations . Almost one third of the companies are in the Navigation&Mapping sector, whilst circa 1.5% are directly involved in the AI solutions development. However, the real number is larger, since artificial intelligence and machine learning (ML) techniques impact the whole spectrum of economic sectors. Advanced technologies development has enabled a wider range of companies to enter the space sector and mature new business models, such as constellations. It is likely that satellite constellations will drive the space market in the coming years. For instance, it will stimulate even broader need for satellite integration, components, and launch vehicles.

---

[1]OECD – Organisation for Economic Co-operation and Development

The modern space value chain is a complex interconnected system with many actors that involves cross-sector and cross-border collaborations.The space value chain is traditionally segmented in three components: upstream (activities that contribute to an operational space system), midstream (activities related to satellite operation and the lease or sale of satellite capacity and data) and downstream (activities related to space-infrastructure exploitation and the provision of space-based products and services to end-users). Major players are space companies, government agencies, non-space companies that are impacted by space commercialization and activities, and academic institutions. The chain is driven by R&D supported by the policies and the agency's mandates. We are currently observing a rapid evolution in the value chain driven by technological advances and the need to monetize technologies beyond their traditional scope. Correspondingly, there are multiple examples of horizontal and vertical integrations. ICEYE[2] or Orbital EOS[3], are integrating horizontally by expanding their activities across new sectors within their business line. For example, Orbital EOS provides business support using artificial intelligence for infrastructure monitoring, deforestation, ship detection and others. However, some satellite producers are going through vertical integration by expanding their data processing capacities. For instance, Maxar Technologies acquired the Radiant Group in 2016. In other words, it was consolidation of the additional parts of the supply chain in an attempt to capture a higher share of value. The Big Tech companies' arrival to the space market should not be taken for granted. Digital titans like Amazon with its Amazon AWS and Google with its Google Cloud began by offering data storage and cloud computing services to players at every stage of the value chain, and are now investing in additional functionalities [6].

Overall, one may distinguish four ways for value creation using space: data, capabilities, resources and markets (Figure 1.1). Frontier technologies such as robotics, machine learning and artificial intelligence are among the key players for fostering the usage of space. However, space data integration is currently the dominant focus, either gathering data from space or data transmitting through space. Private companies, research institutions and

---

[2]ICEYE is a Finnish microsatellite manufacturer, which was founded in 2014 as a spin-off of Aalto University's University Radio Technology Department, and is based in Espoo. It owns the world's largest synthetic-aperture radar (SAR) satellite constellation, enabling data-driven decisions for customers in sectors such as insurance, natural catastrophe response and recovery, security, maritime monitoring and finance.

[3]Orbital EOS is a Spanish Maritime Safety & Rescue Agency (SASEMAR), which provides satellite-based solutions to challenges in the maritime domain, like oil spill monitoring and ship detection.

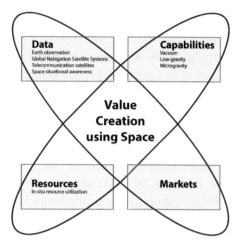

Figure 1.1 Ways for value creation using space.

government agencies are all using space data in support of their activities. The custom data is delivered upon subscriptions for tailored use-cases. GPS technology is a representative case. The impact study of the National Institute of Standards and Technology (NIST) of the United States Department of Commerce found that USD 1.4 trillion were gained by the U.S. private sector industries in economic benefits since GPS was made available to the private sector in the 1980s [7]. About 90% of the benefits have been reached in the past decade. New kind of companies have been build on its back. Ten economic sectors were under scope: location-based services, mining, surveying, telecommunications, telematics, electricity, maritime navigation, and use in oil and gas industries. The case of GPS shows the importance of the relationship among private sector investments, science and time for technology development.

Key sources are the data from Earth Observation (EO), global navigation satellite system (GNSS) and telecommunication satellites together with the data on space situational awareness that refers to the knowledge of space objects characterization. In the Space Traffic Study (2021) report, it is highlighted that the amount of data to/from space will reach more than 500 exabytes of information[4] from 2020 to 2030 [10]. In other words, the data volume

---

[4] 1 exabyte is equivalent to 10^{18} bytes. For comparison: according to Cisco Systems [8], an American multinational technology conglomerate, the global IP traffic achieved an estimated 1.2 zettabytes (an average of 96 exabytes (EB) per month, whereas 1 zettabyte is equivalent to 10^{21} bytes) in 2016. Global IP traffic refers to all digital data that passes over an IP network which includes, but is not limited to, the public Internet. The largest contributing factor to the growth of IP traffic comes from video traffic, and as of December 2022 almost half (48%) of

will increase 14× in the next decade. Advances in AI are one of the enablers for space data processing. Spacecraft navigation, satellite systems control and monitor, image and signal processing, scientific data analysis are the bold examples of AI utilization. For instance, AI is used for large amounts of scientific data processing or satellite health monitoring.

Capabilities for value creation brought by space include unique conditions of the outer space environment such as vacuum, low- and microgravity. For example, the vacuum of space allows to reduce the energy consumption for semiconductor manufacturing by 60% [11]. High degree of purity is also used for producing fluoride glass fiber ZBLAN or synthetic retina manufacturing by LambdaVision. The benefit of space vacuum and solar electricity is planned to be used for data center cooling that Lonestar Data Holdings Inc.[5] hopes to install on the Moon by 2026. Pharma and healthcare are using the space conditions in support of cutting edge research. The drop in cost of launching has brought closer the possibility of in-space manufacturing, however, the industry feels that the current cost of launching equipment is still high. In addition, insufficient legal framework and lack of developed infrastructure in space are slowing down the process. Additive manufacturing, which allows building of complex objects layer by layer from 3D model data, is directly addressing the cost of space missions. For instance, additive manufacturing is used in new liquid rocket engines R&D – the PROMETHEUS project. Up to 70% of engine weight is intended to be constructed using additive manufacturing [12].

The value creation through in-situ space resource utilization is in its infancy compared to the previous two ways. It is expected that the Moon will be the first target. 70 commercial lunar missions are identified as being prepared in the next decade [13]. The market development is quite uncertain. A host of start-ups propose mining of valuable minerals, water and other materials. As an example, the AstroForge raised USD 13 million in seed funding for asteroid materials processing development and is planning to launch the first two missions in 2023. The sheer amount of these materials on the Earth will abide the interest.

Internet traffic is in India and China, while North America and Europe have about a quarter of global internet traffic. It has been forecasted that the amount of data generated each year will grow to 103 zettabytes by 2023 and 175 zettabytes by 2025. [9]

[5]Founded by a proven team of experts from Cloud and Space verticals, Lonestar Data Holdings Inc. is pioneering the future of cislunar services. Lonestar ventures to extend terrestrial data services to the Moon for both the terrestrial archival market and to support space operations with data analytics, archival services and broadband relay to terrestrial public and private clouds.

The space tourism market is predicted to reach USD 400 billion in the next decade although it is currently rather limited. In the distant future its development will stimulate the hospitable and economically thriving terrestrial environments with corresponding market development on its heels. The "people is space" will demand for comport creation as on Earth. This Space-for-Space market evolution will vitalize expansion of global brands from the whole spectrum of economic sectors to new markets. Despite the fact that the perspective is rather faded, the companies are already collaborating with the leading artists for space humanization. For instance, Axiom Space collaborated with the architect Philippe Starck, and on the other hand the Bjarke Ingels Group was enrolled by ICON Technology, inc.[6] to help with the NASA's Artemis mission design [14].

The space economy is at a flex point similar to what internet development brought in the 1990s. For instance, the global space economy was valued at USD 447 billion in 2020 that is 55% higher than a decade ago. Above all, the commercial R&D spending has risen by 22% annually over the past five years. Since 2013 until late 2022, the global space sector has attracted private equity investments of about USD 272 billion into 1,791 unique companies [15]. Moreover, the number of space related start-ups funded annually increased more than twofold from 2010 to 2018 [16]. However, the main portion is represented by the Space-for-Earth economy that includes telecommunication infrastructure, observation capacities and other services or goods produced in space for use of Earth. Though back in the 1970s, NASA evaluated the rise of Space-for-Space economy – goods and services produced in space for the use in space – in their report [17].

The enormous potential of space development is accentuated in a host of reports published by notable financial institutes. Goldman Sachs states that the sector will reach the value of USD 1 trillion by 2040[7]. Comparable number is given by Morgan Stanley (USD 1,1 trillion) and even more optimistic predictions were speculated by Bank of America – Merryl-Lynch (USD 2,7 trillion

---

[6]ICON develops advanced construction technologies that advance humanity by using 3D printing robotics, software and advanced materials.

[7]For comparison: as of June 1, 2023, Apple is the world's most valuable company with a market value of USD 2.78 trillion, followed by Microsoft (USD 2.44 trillion), Saudi Aramco (USD 2 trillion), Google's parent company Alphabet (USD 1.56 trillion), Amazon (USD 1.23 trillion) and then Nvidia (USD 0.93 trillion). Apple became the first U.S. stock to hit USD 1 trillion in 2018, and the first to reach USD 3 trillion in 2022. Tesla and Facebook's parent company Meta briefly dipped above the USD 1 trillion mark, but have both fallen back around USD 600 billion. Meta was in the USD 1 trillion club for about three months in 2021, while Tesla has been in and out a couple times, most recently falling out in April 2022 after a poor quarterly report. [18]

USD in 2050). However, the growing number of IPOs in the SpaceTech industry is evidence of high investors' interest in the sector.

The space sector is not only an emerging sector within itself – it's also proving to be a key enabler of growth and efficiency in other sectors. ESA specifies that the space sector's evolvement positively impacts energy, telecommunication, transport, maritime, aviation, and smart cities development. Further in the book, certain areas are going to be deeper described. Here, we would like to highlight those areas that are techno-economically viable and foster growth over the next few years, but are not covered in the chapters to follow.

- **Agriculture**. The precision agriculture development is driven by the need to make the management more efficient by aiming for cost and environmental impacts reduction and output increase. Space data is intensively used for crop monitoring together with potential threats to harvest. The remote sensors collect a multitude of data about weather patterns, electromagnetic waves parameters and images. NASA is partnering with farmers to deliver new data and new services using artificial intelligence and machine learning technologies, i.e. the NASA's Harvest program. NASA is collaborating with universities, private companies and government agencies for converting space data into actionable information that is easily perceived by a non-scientific community. According to McKinsey's annual digital farmer adoption survey, 29% of row-crop farmers and 45% of speciality-crop farmers already rely on such data or plan to do so [19]. It is evaluated that the space data unavailability during the planting season (April-May) in the US could result in the economic loss 50% higher than in other seasons due to the widespread adoption of the precise agriculture technologies.

  The global Crop Monitoring market is projected to reach 4,4 billion USD by 2025. The market includes remote sensing imagery and systems, AI and robotics, variable rate technology (VRT) and more. According to a European Commission report, the European agriculture sector will be valued at USD 2,295 billion by 2025. The widespread use of AI-based analytical systems, robotics and the Internet of Things (IoT) may remarkably enlarge investment in the market. A range of cutting-edge farming technologies have been provided since 2005. For instance, the Canadian Farmers Edge established Farm Command – a monitoring platform that allows clients monitoring state of the fields, forecast yield, morbidity, fertilization level.

The EOS Data Analytics is one of the largest players that assists users to filter information by zones, separately highlighting water irrigation systems, topographic elevation data, and different vegetation indices. The remarkable example of both horizontal and vertical integration in the market is the development of the Farmstar platform to capture the agricultural market by Airbus since early 2000s. Airbus is expanding its offering across the value chain right the way to the downstream, by creating software which enables end-clients to easily access and analyze the data from their Pleiades and SPOT constellations [20]. At the same time the company has been involved in multiple SAR and optical satellite projects.

- **Communication**. Space industry development is also shaping up the internet connectivity as a social and economic progress catalyst [21]. The lack of network infrastructure is considered to be the main reason why over one billion people are living in areas without internet connection and over three billion are yet to get online [22]. It leads to vital services being undersupplied such as medicine or education resulting in social inequity. Traditional mobile network operators' infrastructure is often uneconomical to install in rural areas. Declining revenue per user made it harder for mobile network operators to deploy new infrastructure in hard-to-reach regions. The satellite internet technology has been available for more than two decades, but relatively high cost and poor performance compared to terrestrial networks restrained its deployment. LEO satellite constellations for delivering high-capacity wireless broadband connection is one of the technologies that received a lot of attention in the past years. This shift from the traditional Geosynchronous Equatorial Orbit (GEO) has capabilities to overcome latency issues all at reasonable cost. Techniques ranging from spectrum sharing to beamhopping are needed for ensuring high-capacity wireless broadband connection [23]. The goal is to achieve lower cost per bit for serving. From an economic perspective, LEO constellations are rather scalable as a new unit can be added without established broadband services disruption. Companies like SpaceX with its Starlink, OneWeb, Telesat and Amazon's Project Kuiper are pressing ahead. Currently, Starlink operates almost half of all active satellites in this business line.

Clients now expect robust high-speed connection in all locations, whilst businesses need more bandwidth to support Internet of Things and other areas. The usage of LEO gives advantage for providing broadband in high-speed trains or aircrafts [24]. However, such a technology with low round trip time may be sensitive for connectivity provision for video-gaming or future real-time Internet of Things. Nevertheless, companies together with governmental agencies are upping their efforts. For instance, the Bipartisan Infrastructure Law in the US calls for USD 65 billion funding. Overall, the capital cost of deploying satellite constellations is substantial, though the technological concerns loom large. None of the current players has experience of operating such a large constellation. It requires capabilities beyond traditional control and telemetry in situational awareness, route management, inter-satellite links deployment and others. A review was carried out on the challenges that satellite constellations have to become more sustainable, with a focus on three categories: constellation management, communication and space traffic [25]. These are the areas where AI-supported solutions will play a premier role. For example, operators should maximize the number of users in areas with high density by providing acceptable data throughput. The analysis of the capacity limits spread across the users in each satellite coverage area is given by Ogutu B. Osoro et al. [26]. Service providers may succeed by providing just connectivity services, but they will likely offer a range of value-added services such as bundled computing, logistics, entertainment.

- **Disaster risk reduction**. Satellites have been supporting climate observation already for decades. Roughly 60% of the World Meteorological Organization's essential climate variables are addressed by satellite data. The implementation of the Global Climate Observing System (GCOS) should enable the provision of information about the total climate system (Table 1.1).

  Satellite monitoring of Earth 's environment works on a similar principle as space-aided agriculture. It assesses the climate observation and detects activities, i.e. illegal logging, fishing and mining. Satellites monitor environmental changes related to ocean water, clouds, ice sheets and help government actions to perform active mitigation. The space-based products are maturing from raw data to products comprehensible by clients without specific scientific background. The incorporation of AI in data from satellites will only enhance the impact of space.

  For instance, the beneficial applications of AI have advanced forest protection. Precision forestry provides detailed business analytics based on space data on a 24/7 basis. Additionally, it also supports forest conversation. Back in 2019, the Norwegian Ministry of Climate and

TABLE 1.1 Data on Essential Climate Variables

Satellite data	Re-Analysis	Ambition
Cryosphere		
Ice sheets & Ice shelves	–	Snow
Glaciers	–	Permafrost
Atmosphere		
Surface radiation budget	Surface pressure	Lightning
Precipitation	Surface temperature	Aerosols & ozone precursors
Upper water vapour	Surface water vapour	–
Earth radiation budget	Surface wind speed/ direction	–
Clouds	Upper air temperature	–
Aerosols	Upper air wind speed/ direction	–
CO_2, CH_4 & other GHGs	–	–
Ozone	–	–
Land		
Soil moisture	Evaporation from land	Ground water
Lakes	–	River discharge
Albedo	–	Land surface temperature
Fire	–	Above-ground biomass
FAPAR	–	–
Land cover	–	–
Ocean		
Surface currents	–	Surface stress
Sea surface temperature	–	Ocean surface heat flux
Sea ice	–	Sea surface salinity
Sea level	–	Sea state
Ocean colour	–	–

Environment together with the Global Forest Observation Initiative decided to give free access to high-resolution data of all world's tropical forests. The price tag for this initiative is USD 50 million; but environmental organizations and private entities donated by believing that the dividends in combating deforestation will be much larger.

The dynamic risk assessment frameworks based on the fusion of satellite data and terrestrial sensors help industries to develop optimal flood risk reduction protocols. Satellite data is treated as a key aid in flood risk assessment that is the costliest natural hazard. Globally, there were more than fifty severe floods in 2021, that caused more than 2,300 victims – the second deadliest peril after earthquakes. The proportion of the world's population exposed to floods has increased by 24% since the turn of the century. Floods often occur as a secondary peril in an already vulnerable area causing a great deal of damage. The Global Flood database[8] hosted by Cloud of Street is a representative example of how satellite observations combined with ground observations using AI allow generating seamless, near real-time flood maps.

This manuscript has a strong interdisciplinary emphasis. Each chapter is meant to be understood relatively independent from the others. The chapters are written by leading experts from many fields dealing with the problems of AI integration in space industry. The contents of the book can be briefly described as follows:

CHAPTER 2 – by D. Izzo, G. Meoni, P. Gómez, D. Dold, and A. Zoechbauer describes the trends in AI adoption in the context of space applications. The authors focus on the emerging and potentially disruptive trends in the field illustrated by the activities carried out in the Advanced Concepts Team (ACT) at the European Space Agency (ESA) and show how the traditional procedures are revised.

CHAPTER 3 – by J. Pelton and S. Madry shows how AI-assisted space systems and quantum computers support achieving the 17 Sustainable Development Goals for 2030 of the United Nations. Space systems are frequently addressed as an enabler towards a sustainable future. The chapter addresses how space systems – linked to Big Data and AI – empower positive global development in new ways that are more rapid, accurate and actionable.

---

[8]See https://www.floodbase.com/.

CHAPTER 4 – by D. Izzo, A. Hadjiivanov, D. Dold, G. Meoni and E. Blazquez concerns on the potential opportunities of neuromorphic technologies for edge computing and learning in space focusing on preliminary results in the areas of event-based sensing, spiking neural networks and neuromorphic hardware for onboard applications.

CHAPTER 5 – by K. Takadama, F. Uwano, Y. Waragai, I. Nakari, H. Kamata, T. Ishida, S. Fukuda, S. Sawai and S. Sakai discusses the Triangle Similarity Matching method – an AI-based technique – for spacecraft self-location estimation. The method contributes to improving accuracy of spacecraft's self-location and preventing wrong estimation of location. The method is employed in the Smart Lander for Investigating Moon (SLIM) mission of Japan Aerospace Exploration Agency (JAXA).

CHAPTER 6 – by G. Lapenta, F. Califano, R. Dupuis, M. E. Innocenti, and G. Pedrazzi investigates the status of space data analysis transformation using unsupervised machine learning techniques. The unsupervised machine learning as the method where learning is not guided by previous knowledge can truly find new discoveries and open the ways to escape from traditional approaches.

CHAPTER 7 – by E. Camporeale stresses the problem of space weather forecasting, that shares much with forecasting weather on Earth, though the field is relatively new. Space weather prediction has far to go before it develops to the level of terrestrial weather forecasting in accuracy and lead-time. Nevertheless it is rapidly advancing by adopting AI technologies. Wrong information can lead to human failure, technical solutions misoperation or significant economic losses.

CHAPTER 8 – by S. Freeland and A. S. Martin reviews the current states of legal and policy issues related to AI technologies. The legal issues brought by AI are detailed and questions about how AI advances interact with existing legal concepts and technical standards are entailed.

CHAPTER 9 – by L. Feruglio, A. Benetton, M. Varile, D. Vittori, I. Bloise, R. Maderna, C. Cardenio, P. Madonia, F. Rossi, F. P. Azza, P. De Marchi, L. Romanelli, M. Stoisa, L. Manca, G. Campagna, and A. La Rocca gives an overview of business cases for space industry ranging from established Earth Observation and telecommunications verticals to the emerging in-orbit servicing market. The key limitations of the status-quo, the possible solutions details and mission/use-cases scenario are outlined.

CHAPTER 10 – by A. Ansari and J. Mainard looks back into the future. The advanced technologies will shape the future of humanity in space for protecting the body and mind from the hazardous environment and isolation in space. However, making space as our home, will need re-analysis of how much we can rely on technological innovations. The chapter questions "what it means to become intelligent!".

This book brings together interdisciplinary perspective in the topic of AI implementation for space sector. It gives an up to date information on the topic across various industries in different geographical regions. This book will not specifically address the basic AI techniques.

BIBLIOGRAPHY

[1] M. Kalms, J. Hacker, J. Mabbott, and S. Lanfranconi. 30 voices on 2030 – the future of space. *92. KPMG*, Australia, 2020. https://assets.kpmg.com/content/dam/kpmg/au/pdf/2020/30-voices-on-2030-future-of-space.pdf.

[2] Thomas G Roberts and Spencer Kaplan. Space launch to low earth orbit: How much does it cost. *Civil and Commercial Space Space Security*, 2020.

[3] European Space Agency (ESA). Artificial intelligence in space (accessed on 15 june 2023). https://www.esa.int/Enabling_Support/Preparing_for_the_Future/Discovery_and_Preparation/Artificial_intelligence_in_space.

[4] OECD. Publishing, Organisation for Economic Co-operation, and Development Staff. *OECD handbook on measuring the space economy*. OECD publishing, 2012.

[5] SpaceTech Analytics. Spacetech industry 2021 / q2 landscape overview (accessed on 15 june 2023). https://analytics.dkv.global/spacetech/SpaceTech-Industry-2021-Report.pdf, 2021.

[6] Aleks Buczkowski. *Understanding The Earth Observation Value Chain*. GeoAwesomeness, 2023.

[7] Alan C O'Connor, Michael P Gallaher, Kyle Clark-Sutton, Daniel Lapidus, Zack T Oliver, Troy J Scott, Dallas W Wood, Manuel A Gonzalez, Elizabeth G Brown, and Joshua Fletcher. Economic benefits of the global positioning system (gps). 2019.

[8] Cisco Systems (2016). Cisco report – the zettabyte era: Trends and analysis. https://files.ifi.uzh.ch/hilty/t/Literature_by_RQs/RQY\%20102/2015_Cisco_Zettabyte_Era.pdf.

[9] 27 November 2018 Forbes. 175 zettabytes by 2025 (accessed on 15 june 2023). https://www.forbes.com/sites/tomcoughlin/2018/11/27/175-zettabytes-by-2025/.

[10] NSR: Northern Sky Research. Space traffic study (accessed on 15 june 2023). https://www.nsr.com/?research=space-traffic-study-2nd-edition/, 2021.

[11] Space Forge. In-space manufacturing (accessed on 15 june 2023). https://www.spaceforge.com/in-space-manufacturing.

[12] Alessandra Iannetti, Nathalie Girard, Nicolas Ravier, Emmanuel Edeline, and David Tchou-Kien. Prometheus, a low cost lox/ch4 engine prototype. In *53rd AIAA/SAE/ASEE Joint Propulsion Conference*, page 4750, 2017.

[13] Moon markets analysis (accessed on 15 june 2023). https://www.nsr.com/tag/moon-markets-analysis/, 2021.

[14] ICON. Icon to develop lunar surface construction system with $57.2 million nasa award (accessed on 15 june 2023). https://www.iconbuild.com/newsroom/icon-to-develop-lunar-surface-construction-system-with-57-2-million-nasa-award.

[15] Space Capital. Space investment quarterly reports (accessed on 15 june 2023). https://www.spacecapital.com/quarterly, 2023.

[16] Carissa B Christensen, Raphael G Perrino, and Hunter J Garbacz. Startup space 2018: Update on investment in commercial space ventures. In *2018 AIAA SPACE and Astronautics Forum and Exposition*, page 5295, 2018.

[17] William M Brown and Herman Kahn. Long-term prospects for developments in space (a scenario approach). Technical report, NATIONAL AERONAUTICS AND SPACE ADMINISTRATION WASHINGTON DC, 1977.

[18] 25 May 2023 Forbes. Nvidia nears $1 trillion market capitalization—closing in on these other companies (accessed on 15 june 2023). https://www.forbes.com/sites/katherinehamilton/2023/05/25/nvidia-nears-1-trillion-market-capitalization-possibly-joining-these-other-companies/.

[19] D Fiocco, V Ganesan, L Harrison, and J Pawlowski. Farmers value digital engagement, but want suppliers to step up their game. *McKinsey & Company. Available at*, 2021.

[20] Airbus. Farmstar: Crop monitoring and precision farming insight (accessed on 15 june 2023). https://www.intelligence-airbusds.com/markets/agriculture/precision-farming/farmstar/.

[21] Ruth Kennedy-Walker, Nishtha Mehta, Seema Thomas, and Martin Gambrill. Connecting the unconnected. 2020.

[22] J Garrity and A Garba. The last-mile internet connectivity solutions guide sustainable connectivity options for unconnected sites. *International Telecommunication Union Development Sector*, 2020.

[23] Yan Yan, Kang An, Bangning Zhang, Wei-Ping Zhu, Guoru Ding, and Daoxing Guo. Outage-constrained robust multigroup multicast beamforming for satellite-based internet of things coexisting with terrestrial networks. *IEEE Internet of Things Journal*, 8(10):8159 8172, 2020.

[24] Xinmu Wang, Hewu Liy, Wenbing Yao, Tianming Lany, and Qian Wu. Content delivery for high-speed railway via integrated terrestrial-satellite networks. In *2020 IEEE Wireless Communications and Networking Conference (WCNC)*, pages 1–6. IEEE, 2020.

[25] Giacomo Curzi, Dario Modenini, and Paolo Tortora. Large constellations of small satellites: A survey of near future challenges and missions. *Aerospace*, 7(9):133, 2020.

[26] Ogutu B Osoro and Edward J Oughton. A techno-economic framework for satellite networks applied to low earth orbit constellations: Assessing starlink, oneweb and kuiper. *IEEE Access*, 9:141611–141625, 2021.

II

Trends & Valuse

CHAPTER 2

Selected Trends in Artificial Intelligence for Space Applications

Dario Izzo

Advanced Concepts Team (ACT), European Space Research & Technology Centre (ESTEC), Keplerlaan 1, 51014 AG Noordwijk, The Netherlands

Gabriele Meoni

Advanced Concepts Team (ACT), European Space Research & Technology Centre (ESTEC), Keplerlaan 1, 51014 AG Noordwijk, The Netherlands
Φ-lab, European Space Research Institute (ESRIN), Via Galileo Galilei, 1, 00044 Frascati RM, Italy

Pablo Gómez

Advanced Concepts Team (ACT), European Space Research & Technology Centre (ESTEC), Keplerlaan 1, 51014 AG Noordwijk, The Netherlands

Dominik Dold

Advanced Concepts Team (ACT), European Space Research & Technology Centre (ESTEC), Keplerlaan 1, 51014 AG Noordwijk, The Netherlands

Alexander Zoechbauer

Advanced Concepts Team (ACT), European Space Research & Technology Centre (ESTEC), Keplerlaan 1, 51014 AG Noordwijk, The Netherlands

CONTENTS

2.1 Introduction 22
2.2 Differentiable Intelligence 23
    2.2.1 G&CNets 24
    2.2.2 Implicit representations of irregular bodies 25
    2.2.3 Inverse Material Design 31
2.3 Onboard Machine Learning 35

DOI: 10.1201/9781003366386-2

2.3.1   Why Have Machine Learning on Board? 36
2.3.2   Hardware for Onboard Machine Learning 37
2.3.3   Onboard Inference 37
2.3.4   Onboard & Distributed Learning 40
Bibliography 44

T HE development and adoption of artificial intelligence (AI) technologies in space applications is growing quickly as the consensus increases on the potential benefits introduced. As more and more aerospace engineers are becoming aware of new trends in AI, traditional algorithms and procedures are revisited to consider potential applications of AI in the context of space missions. Already, the scope of AI-related activities across academia, the aerospace industry and space agencies is too wide to review in-depth within this chapter. Thus we chose to focus on emerging trends in the field illustrated by research activities carried out within the European Space Agency's Advanced Concepts Team. These trends focus on advanced topics that go beyond the simple transposition of established AI techniques to space problems as they require fundamental cross-fertilization between the two areas.

2.1 INTRODUCTION

Two main trends we believe capture the most relevant and exciting activities in the field of AI for space application: differentiable intelligence and on-board machine learning. Differentiable intelligence, in a nutshell, refers to works making extensive use of automatic differentiation frameworks to learn the parameters of machine learning or related models. Automated differentiation has been an under-researched topic for quite some decades and is only now finally getting the attention it deserves from researchers and practitioners. Onboard machine learning considers the problem of moving inference as well as learning of machine learning models onboard.

Within these fields, we discuss a few selected projects originating from the European Space Agency (ESA)'s Advanced Concepts Team (ACT), giving priority to advanced topics going beyond the transposition of established AI techniques and practices to the space domain, thus necessarily leaving out interesting activities with a possibly higher technology readiness level. We start with the topic of differentiable intelligence by introducing Guidance and Control Networks (G&CNets), Eclipse Networks (EclipseNETs), Neural Density Fields (geodesyNets) as well as the use of implicit representations to learn differentiable models for the shapes of asteroids and comets from LiDAR data. We then consider the differentiable intelligence approach in the context of inverse problems and showcase its potential in material science research. In the following section we investigate the issues that are introduced when porting generic machine learning algorithms to function onboard spacecraft and discuss current hardware trends and the consequences of their memory and power requirements. In this context we present the cases of the European Space Agency satellites Φ-sat and OPS-SAT, introducing preliminary results obtained during the data-driven competition "the OPS-SAT case" on the real-time onboard classification of land use from optical satellite imagery. All in all, we wish this chapter to contain a list of recent results and ideas able to stimulate, in the coming years, the interest of practitioners and scientists and to inspire new research directions to look further into the future.

2.2 DIFFERENTIABLE INTELLIGENCE

The term "differentiable intelligence" or "differential intelligence" has been recently used to indicate, in general, learning algorithms that base their internal functioning on the use of differential information. Many of the most celebrated techniques in AI would not be as successful if they were not cleverly exploiting differentials. The backpropagation algorithm, at the center of learning in artificial neural networks (ANNs), leverages the first and (sometimes) second order derivatives of the loss in order to update the network parameters [1]. Gradient boosting techniques [2] make use of the negative gradients of a loss function to iteratively improve over some initial model. More recently, differentiable memory access operations [3, 4] were successfully implemented and shown to give rise to new and exciting neural architectures.

Even in the area of evolutionary computations, mostly concerned with derivative-free methods, having the derivatives of the fitness function is immensely useful, to the extent that many derivative-free algorithms, in one way or another, seek to approximate such information (e.g. the covariance matrix in CMA-ES [5] is an approximation of the inverse Hessian). At the very core of any differentiable intelligence algorithm lie "automated differentiation" techniques, whose efficient implementation determine, ultimately, the success or failure of a given approach. For this reason, in recent years, software frameworks for automated differentiation have been growing in popularity to the point that software frameworks such as JAX, PyTorch, Tensor-Flow and others [6] are now fundamental tools in AI research where lower order differentials are needed and an efficient implementation of the backward mode automated differentiation is thus of paramount importance. Higher order differential information, less popular at the time of this writing, is also accessible via the use of dedicated frameworks such as pyaudi [7] or COSY infinity [8]. In the aerospace engineering field, many of the algorithms used to design and operate complex spacecraft can now leverage these capabilities, as well as make use of the many innovations continuously introduced to learning pipelines. A number of applications deriving from these advances have been recently proposed and are being actively pursued by ESA's ACT: a selection is briefly introduced in the following sections.

2.2.1 G&CNets

The acronym Eclipse Networks (EclipseNETs)G&CNet [9] stands for Guidance and Control Networks and was introduced to refer to a learning architecture recently proposed in the context of real-time optimal spacecraft landing. The basic idea is to go beyond a classical G&C scheme, which separates the spacecraft guidance from its control, and have instead an end-to-end ANN able to send optimal commands directly to the actuators using only the state information computed from the navigation system as input. Indicating the spacecraft state with $\mathbf{x}$ and its control with $\mathbf{u}$, it is known how, under mild hypothesis, such a mapping exists (i.e. the optimal feedback $\mathbf{u}^*(\mathbf{x})$) and is unique [10]. The structure of the optimal feedback $\mathbf{u}^*$ is, in general, that of a piece-wise continuous function, which suggests the possibility to use a neural network as its approximator. We are thus allowed to write $\mathbf{u}^* = \mathcal{N}_\theta(\mathbf{x}) + \epsilon$, where $\mathcal{N}$ is any neural model with parameters θ and ϵ an error that can, in principle, be sent to zero as per the Universal Approximation Theorem [11]. We are now left with the task to find (learn) the values of θ as to make ϵ sufficiently small. The use of imitation learning has been shown, in this context, to be able to deliver satisfactory results in a variety of scenarios including spacecraft landing [9], interplanetary spacecraft trajectory design [12, 13] as well as drone racing [14]. In all cases, we first generate a large number of optimal flight profiles and we record them into a training dataset containing optimal state-action pairs $(\mathbf{x}_i, \mathbf{u}_i^*)$. We then use the dataset to learn the network parameters θ solving the resulting nonlinear regression problem (i.e. in a supervised learning approach). The problem of efficiently assembling the dataset to learn from is solved introducing an ad hoc technique called backward propagation of optimal samples [12, 13]. At the end, the G&CNet $\mathcal{N}_\theta(\mathbf{x})$ is used in simulation, and later on real hardware, to produce optimal flights.

In Fig. 2.1, we reproduce a plot from [14] where a G&CNet is used to optimally control the two-dimensional dynamics $\mathbf{f}(\mathbf{x}, \mathbf{u})$ of a quadcopter.

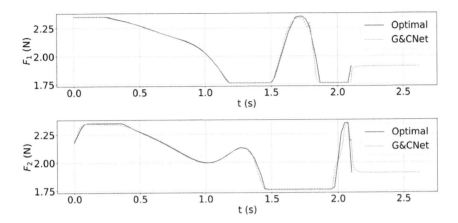

Figure 2.1 History of the forces (called F_1, F_2) controlling a quadcopter during a simulated aggressive flight (i.e. with a strong emphasis on time optimality). The ground truth optimal solution is shown as well as the one produced using a G&CNet. Reproduced from [14].

Starting from some initial condition, the results from the following simulations are compared: $\dot{x} = f(x, u^*)$ (optimal) and $\dot{x} = f(x, \mathcal{N}_\theta(x))$ (G&CNet). The figure compares the control history during both simulations and well illustrates, in this specific case, several generic aspects of the method: the satisfactory approximation accuracy, the stability of the resulting flight as well as the sensible behaviour once the optimal flight is over (in this case after $t^* \approx 2.1$ seconds). The stability of the system dynamics under the influence of a G&CNet is studied in details in [15], while the learning pipeline optimization and tuning is discussed in general in [16].

2.2.2 Implicit representations of irregular bodies

The expressivity of ANNs, mathematically discussed in [11], found recently a remarkable showcase in [17] where an original machine learning pipeline was used to learn a so-called NEural Radiance Field (NERF): a relatively small, albeit deep, feedforward neural network mapping Cartesian coordinates (x, y, z) and ray direction (θ, φ) to the local volume density and directional emitted color of a 3D scene (e.g. an object like a digger). After training, the network was able to render two-dimensional views of the target scene with unprecedented levels of details. We argue that an important take-away message from this work – and the many that followed and introduced the term *implicit representations* to refer to this type of networks – is the ability of a relatively simple network architecture to encode highly complex objects with great precision in its parameters. The availability of a shape model of an irregular astronomical body is important when planning orbital maneuvers and close-proximity operations, but is also of great interest to scientists trying to reconstruct the complex history of the Solar System dynamics and understand its current state. Following this remark, it is natural to ask whether the shapes of irregular bodies in the solar system, such as comets, asteroids and even spacecraft, could be a) represented with sufficient accuracy by an ANN and b) learned from measurements typically available in some space mission context. These scientific questions are at the core of several innovative projects the ACT is currently developing towards higher technology readiness levels.

2.2.2.1 GeodesyNets

Any body orbiting in the Solar System can be described, for the purpose of modelling the resulting gravitational field, by its density $\rho(x, y, z)$. Such a function will be zero outside the body and discontinuous in its interior as to model possible fractures and material heterogeneity, and thus a great candidate for being represented by an ANN: a geodesyNet [18]. We thus set $\rho(x, y, z) = \mathcal{N}_\theta(x, y, z) + \epsilon$ and, again, learn the parameters θ as to make ϵ vanish. Since we do not have access to the actual values of the body's density we cannot set up a standard supervised learning pipeline. Instead, we assume to be able to measure the gravitational acceleration $\mathbf{a}_i$ at a number of points $\mathbf{X}_i = [X_i, Y_i, Z_i]$, for example along a putative spacecraft orbit. At any of those points, the network will predict $\hat{\mathbf{a}}_i(\mathbf{x}_i) = -\iiint_V \frac{\mathcal{N}_\theta(x,y,z)}{r_i^3} \mathbf{r}_i dV$, where $\mathbf{r}_i = [X_i - x, Y_i - y, Z_i - z]$. The difference between the predicted and the measured acceleration can then be used to learn the network parameters θ. As detailed in the work introducing geodesyNets, the final representation of the body density (i.e. the so-called neural density field) has several advantages over more classical approaches such as spherical harmonics, mascon models or polyhedral gravity models [18]. Most notably it has no convergence issues, it maintains great accuracy next to the asteroid surface, it can be used to complement a shape model and propose plausible internal structures and it is differentiable. In Fig. 2.2 the overall scheme and an example of the results obtained after training a GeodesyNet is shown. A model of the comet 67p/Churyumov-Gerasimenko is used to produce a synthetic scenario where to test the training. The comet is assumed to be perfectly homogeneous and all units are non-dimensional (see [18] for details). The results reported here are qualitatively similar for other irregular bodies and the main take away is that geodesyNets are able to represent the body shape, its density and the resulting

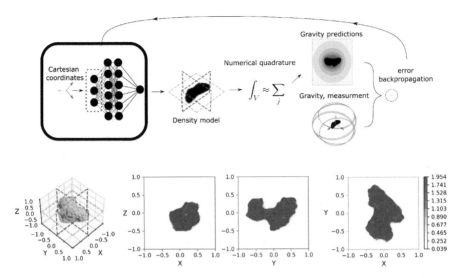

Figure 2.2   Top: generic training setup for geodesyNets. Bottom: visualization of the learned neural density field in the specific case of a homogeneous model of the comet 67p/Churyumov-Gerasimenko. Reproduced from [18].

gravity field with great accuracy. Furthermore, the model parameters can be learned from the body's gravitational signature.

An interesting question, addressed in recent works [18, 19], is the behaviour of geodesyNets when the interior of the target irregular body is not uniform and the gravitational signature is noisy (for example being contaminated by non-gravitational effects). Under the additional assumption that a shape model has already been obtained for the body, it is possible to modify the geodesyNet pipeline as to have the network directly predict the deviation from a homogeneous density distribution, thus targeting a differential density $\partial\rho(x, y, z) = \rho - \rho_U = \mathcal{N}_\theta(x, y, z)$. In this case, internal details of the body's heterogeneous density become clear and sharper [18]. The approach, in this case utilized to see inside the body, cannot, however, evade the constraints imposed by Newton's shell theorem (i.e. the impossibility to invert gravity fields uniquely) and the effects of measurement sensitivity and noise (i.e. the possibility to measure the gravitational signature of heterogeneity).

2.2.2.2   EclipseNets

During an eclipse event (e.g. when the Sun light gets obscured as an orbiting object enters the shadow cone of some other body) the orbital dynamics changes significantly, in particular for small area-to-mass ratio objects such as orbiting dust, i.e., debris pieces or pebbles detached from rotating asteroids. Due to the possible irregular shape of the eclipsing body as well as diffraction and penumbra effects, modelling these events with precision is in most cases computationally expensive.

Ignoring diffraction and penumbra effects, the neural architecture EclipseNet Eclipse Networks (EclipseNETs) introduced in [20] is a differentiable model capable of reproducing the complex shape of shadow cones. An EclipseNet is a relatively simple feedforward neural architecture trained to predict the value of the complex and discontinuous eclipse function $\mathbf{F}(\mathbf{r}, \hat{\mathbf{i}}_S)$. The eclipse function maps the Cartesian position vector $\mathbf{r}$ and the light direction $\hat{\mathbf{i}}_S$ into a positive number if $\mathbf{r}$ is outside the shadow cone (cast along the direction $\hat{\mathbf{i}}_S$) and negative otherwise. The idea here is that the zero-level curves of such a function, once $\hat{\mathbf{i}}_S$ is fixed, determine the shape of the body's shadow cone. The concept is similar to that of the signed distance function (SDF) [21] used widely in image reconstruction, but is defined on a 2D projection and parameterized by the direction of the light source. One possible way to have such a function is to define it as the distance of $\mathbf{r}$ from the shadow cone (degenerated into a cylinder in this case) whenever $\mathbf{r}$ is outside the cone. The eclipse function can otherwise (when inside the shadow) be defined as the length of the ray portions inside the body. Both quantities are easily computed using the Möller-Trumbore algorithm [22] starting from a polyhedral shape model of the body. Hence, a standard regression pipeline can be applied to learn the network parameters θ so that we can write $\mathbf{F}(\mathbf{r}, \hat{\mathbf{i}}_S) = \mathcal{N}_\theta(\mathbf{r}, \hat{\mathbf{i}}_S) + \epsilon$. In Fig.2.3, we report the results obtained from training an EclipseNet in the case of the irregular shape of the comet 67p/Churyumov-Gerasimenko being the source of the eclipse. A feedforward neural network with 6 layers of 50 tanh units is used in this case for a dataset containing 10000 values of an eclipse function computed for 300 distinct views forming a Fibonacci spiral.

The remarkable expressivity of such a relatively simple network can be clearly seen in the figure where the zero level curve of the network outputs is plotted for a fixed direction. The network thus appears to be able to represent well the shadow cone.

To model the influence of eclipses over the orbiting dynamics, one can use a simple model. Assume a reference frame attached to the main body, which is uniformly rotating with an angular velocity ω. The solar radiation pressure is introduced via an acceleration η acting against the Sun direction $\hat{\mathbf{i}}_S$ and regulated by an eclipse factor $v(\mathbf{r}) \in [0, 1]$ determining the eclipse-light-penumbra regime. In particular, $v = 1$ when the body is fully illuminated and $v = 0$ when fully eclipsed (umbra). Formally, the following set of differential

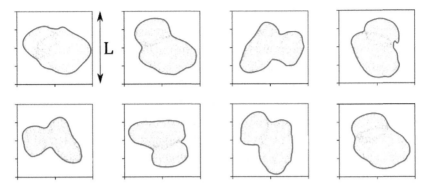

Figure 2.3   Shadow cone predicted by an EclipseNet (red) for the comet 67p/Churyumov-Gerasimenko along eight different random directions. The vertices of the polyhedral model of the comet used to compute the ground truths are shown in blue. During training, non-dimensional units (i.e. L) are used

equations are considered:

$$\ddot{\mathbf{r}} = \mathbf{a}_g - 2\boldsymbol{\omega} \times \mathbf{v} - \boldsymbol{\omega} \times \boldsymbol{\omega} \times \mathbf{r} - \eta v(\mathbf{r})\hat{\mathbf{i}}_S(t).$$ (2.1)

To numerically solve these equations, one needs to know the value of $v(\mathbf{r})$. To do so, one can detect the eclipse function's sign change as an event using the EclipseNet. This avoids more complex raytracing computations (e.g. running the Möller-Trumbore algorithm at each instance) as to determine whether the solar radiation pressure is active. This also avoids the need to store and access the body polyhedral model compressing that information into the network parameters. Furthermore, the differentiability of the EclipseNet model allows, in this use case, the usage of a guaranteed event detection scheme based on Taylor integration 2.3.

2.2.2.3   Differentiable 3D shapes from LiDAR measurements

Terrestrial-based observations are often used to determine the preliminary shape model of irregular bodies and use it during the design of a space mission targeting operations in its proximity. Such a model, used mainly for navigation purposes and sufficient for that purpose, is obviously inaccurate and gets refined at a later stage for scientific purposes – thanks to the data produced by the various scientific payloads typically on board. One is often a LiDAR (LIght Detection And Ranging), a technique that can be used to measure surface features as well as to support scientific mapping and ultimately close operations. Such is the case, e.g., for the upcoming HERA mission [23] (ESA) as well as the past Hayabusa (JAXA) and Osirix-Rex (NASA) missions. In this context it is of interest to study whether AI-based techniques can be used to introduce new algorithms able to build a full 3D shape model directly from LiDAR data. Such algorithms, accelerated by edge computing solutions, can be considered to be run onboard and thus be able to leverage incoming LiDAR data to continuously improve the shape model available to the spacecraft. With this vision in mind, and inspired by previous work in the field of computer vision, in particular [21, 24], in [25] we developed and preliminary tested novel pipelines able to reconstruct a triangle mesh, representing the surface of the irregular astronomical body from a point cloud consisting of LiDAR points $\mathbf{x}_i$. A first idea is to use neural implicit representations to learn the surface information encoded in the Signed Distance Function [21] computed from LiDAR data as:

$$\Phi(\mathbf{x}, \Omega) = \begin{cases} \inf_{\mathbf{y} \in \partial\Omega} ||\mathbf{x} - \mathbf{y}||^2 & \text{if } \mathbf{x} \in \Omega, \\ -\inf_{\mathbf{y} \in \partial\Omega} ||\mathbf{x} - \mathbf{y}||^2 & \text{if } \mathbf{x} \in \Omega_c. \end{cases} \quad (2.2)$$

The SDF describes the minimum distance between any point $\mathbf{x}$ to the surface $\partial\Omega$ of a three-dimensional geometry Ω. Notably the points where $\Phi(\mathbf{x}) = 0$ is the surface. It is then possible to train a feedforward neural network to predict $\Phi(\mathbf{x})$ knowing that $\Phi(\mathbf{x}_i) = 0$ at the LiDAR points $\mathbf{x}_i$ and adding the condition $|\nabla\Phi(\mathbf{x})| = 1$ everywhere. A shape model can subsequently be obtained using a marching cubes algorithm, which constructs a mesh from the zero level set of $\Phi(\mathbf{x})$. A second method studied emerged through recent advances in 3D scene reconstruction made by [24] where a Differential Poisson Solver (DPS) for surface reconstruction is introduced. The solver is used to compute the Indicator Function defined as:

$$\chi(\mathbf{x}, \Omega) = \begin{cases} 0.5 & \text{if } \mathbf{x} \in \Omega/\partial\Omega, \\ 0 & \text{if } \mathbf{x} \in \partial\Omega, \\ -0.5 & \text{if } \mathbf{x} \in \Omega_c. \end{cases} \quad (2.3)$$

The DPS solves the Poisson equation defined as $\nabla^2 \chi(\mathbf{x}) = \nabla \cdot \upsilon(\mathbf{x})$ for $\chi(\mathbf{x})$, where $\upsilon(\mathbf{x})$ is a vector field of points and normal vectors of a predefined point cloud, by using spectral methods (Fast Fourier Transforms). From the zero level set of $\chi(\mathbf{x})$, a mesh can be obtained using the marching cubes algorithm and subsequently comparing the Chamfer distance to the LiDAR data to inform a loss and thus update – thanks to the differentiability of such a pipeline – the shape model iteratively. In this second case, no intermediary ANN is used as the position of the vertices in the shape model are

(a) DPS (b) Implicit Representation (c) Shape model from images

Figure 2.4   Difference between the reconstructed shape models of the asteroid Bennu and the LiDAR ground truth. a), b) Shape model derived from the proposed differentiable methods. c) Shape model reconstructed from camera images (publicly available online from the Osirix-Rex mission). Regions with high/low error are encoded in red/blue.

directly updated. We report in Fig. 2.4 a visualization of the error one can obtain running these algorithm in the case of the LiDAR reading available for the asteroid Bennu from the Osirix-Rex mission. In the figure, the ability of the proposed method to construct an accurate body shape compatible with the LiDAR measurements appears evident in comparison to, for example, the shape model constructed using camera images.

2.2.3   Inverse Material Design

Modeling physical processes via differentiable code, involving also highly parameterized neural models, allows for the development of new powerful inversion algorithms as well as solvers. The approach has been recently applied to complex phenomena such as fluid mechanics [26–28] and more generally those defined by complex PDEs [29]. In the ACT this is being explored also in the context of material design, e.g., [30] and [31], which will be discussed in the following sections.

2.2.3.1   *Neural Inverse Design of Nanostructures*

The open-source module called *Neural Inverse Design of Nanostructures* (NIDN) has been released and allows for the design of nanostructures targeting specific spectral characteristics (Absorptance, Reflectance, Transmittance). NIDN has applications, e.g., in designing solar light reflectors, radiators to emit heat, or solar sails. Naturally, the applications expand

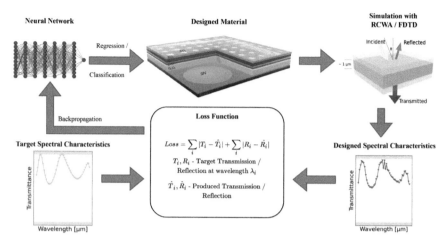

Figure 2.5   Training setup for NIDN.

beyond the space domain. In contrast to previous works [32,33] NIDN utilizes internal, differentiable Maxwell solvers to build a completely differentiable pipeline. Among other advantages, this eliminates the need for any form of training dataset and simplifies the inversion as the training process in NIDN aims to find the best solution for a specific set of spectral characteristics, and not to find a general inverse solution of the Maxwell equations.

On the inside, NIDN relies on two different solvers for the Maxwell equations depending on desired complexity and wavelength range. The first, TRCWA, is based on rigorous coupled-wave analysis (RCWA) [34] and a highly efficient solver for stacked materials. The second one is a finite-difference time-domain (FDTD) [35] solver, which is computationally more expensive but a highly accurate method. A detailed description of the pipeline for designing a material with NIDN is given in Fig. 2.5. Fundamentally, the idea is to iteratively improve the material's permittivity based on the gradient signal backpropagated through the Maxwell equations solver.

As exemplary practical applications, we demonstrated two test cases, one of these is displayed in Fig. 2.6 using stacked uniform layers with RCWA. The first showcases creating a 1550nm filter [36] with NIDN. As can be seen, NIDN is able to produce an almost perfect replica of the spectral characteristics of the desired filter. In the second application demonstrated in the work, results for designing a perfect anti-reflection material are shown. This type of application can, e.g., be relevant to avoid creating satellites which produce specular highlights in astronomical observations [37]. The spectral

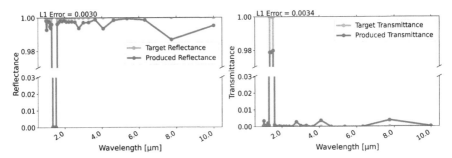

Figure 2.6   Spectral characteristics with an 1550nm exemplary filter designed with NIDN.

characteristics obtained with NIDN once again were able to match the specified target spectrum extraordinarily well.

However, there are still some challenges remaining for this type of approach. In NIDN, currently it is difficult to map the produced permittivities to existing materials as this is a non-differentiable operation. Secondly, parallelization of the utilized simulations is desirable to improve computational efficiency, especially of FDTD. Overall, this line of research seems promising, especially given the low requirements. No training data is needed, limitations on the conceivable structures are only due to chosen Maxwell equations solver and the approach is agnostic to both, the concrete application (e.g. solar sails or filters) and modular with regards to the used Maxwell solver (FDTD, RCWA, etc.) as long as it can be implemented in a differentiable way.

2.2.3.2   Inverse Design of 3D-printable Lattice Materials

The approach can also be applied to other types of materials. In an ongoing project [31], a similar inverse design approach is being developed for 3D-printable lattice materials. In this case, a differentiable finite element solver is implemented to predict the mechanical properties of a lattice, and the lattice structure is adjusted using gradient information to optimize for certain material properties like in-plane stiffness and Poisson's ratio. The described methods are available as an open-source Python module called *pyLattice2D*.

More precisely, we represent lattices as mathematical graphs: edges represent lattice beams and nodes represent the spatial locations where beams cross. Hence, a lattice $\mathcal{L}$ can be described as a collection of nodes $\mathcal{N}$ and edges $\mathcal{E}$, $\mathcal{L} = (\mathcal{N}, \mathcal{E})$. In addition, both nodes and edges can be assigned attributes $\mathcal{N}_A$ and $\mathcal{E}_A$, respectively. For instance, nodes are by default characterized by their spatial location, while edges can hold information such as

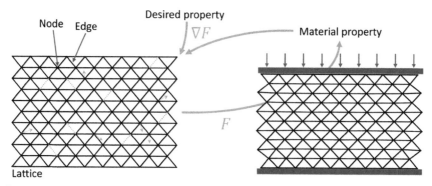

Figure 2.7 We represent a lattice as a graph with nodes and edges. Inverse design is enabled through a differentiable forward function F that predicts material properties (e.g. the response to loading the lattice from above and below, red arrows indicate loading) from the graph structure alone. Gradient information ∇F is used to adjust elements of the graph, i.e., adjust the lattice structure by changing node positions and removing or adding edges (yellow dashed lines) with the goal of designing a lattice with desired material properties.

beam cross-area, beam shape and Young's modulus. On this graph structure, we can define functions $F(\mathcal{L}, \mathcal{N}_A, \mathcal{E}_A)$ that take the graph and its attributes as input and return its material properties, e.g., effective in-plane stiffness, Poisson's ratio or the deformation under certain loading conditions.

To directly work on the graph, F implements functions using message passing which – at least in most modern approaches – generalizes the well-known convolution operator from images to graphs [38]. We found that direct stiffness methods as well as loading experiments can be formulated in a fully differentiable way using this technique. In this case, F calculates lattice properties exactly while its gradient provides sufficient information to iteratively adjust both node and edge attributes until a lattice structure with the desired target property has been found (Fig. 2.7) – completely without having to train on data. To change the beam structure of the lattice, we assign each edge e a masking value m_e, from which the existence of a beam is derived using thresholding, $\vartheta_e = 1$ if $m_e > 0$ and $\vartheta_e = 0$ otherwise, i.e., $\vartheta_e = \theta(m_e)$ with $\theta(\cdot)$ being the Heaviside function. This way, contributions from edge e are either added ($\vartheta_e = 1$) or removed ($\vartheta_e = 0$) from F by multiplying edge messages with this term, allowing us to change the beam connectivity via gradient information for m_e. However, gradient-based optimization of m_e is incredibly slow since the derivative of $\theta(\cdot)$ is a Dirac delta distribution. This is solved by using the surrogate gradient method (see [39] for details), enabling geometric changes (moving nodes, removing or adding edges) during inverse design. Finally, instead of a finite element solver, the function F used for inverse design can also be a neural network – in this case a graph neural network, an architecture that generalizes ANNs to graph-structured input – trained to predict material properties from lattice graphs.

Materials used to build future space infrastructure, e.g., for habitats on other celestial bodies, will require very unique, environment-specific properties with the constraint of being easy to process and shape. The geometric freedom of 3D-printed lattice materials allows for a huge range of customizable properties while satisfying this constraint. Thus, it is not surprising that at ESA, research on 3D printing lattices from a variety of space-relevant polymers and metals has been growing over the past decade [40, 41]. Computational inverse design approaches as presented here will greatly assist the design of novel lattice materials for future space applications – especially by extending the design space to unintuitive structures such as irregular lattices as found in nature.

2.3 ONBOARD MACHINE LEARNING

Many of the emerging trends using machine learning for space applications, in particular all the differentiable intelligence methods introduced in the previous sections, are greatly empowered if inference and training can be performed directly the spacecraft and not on the ground. The use of machine learning (ML) methods onboard spacecraft (pioneered in 2003 by JPL's Earth observation mission EO-1 [42]) is receiving renewed attention as the next generation of spacecraft hardware is targeted at supporting specifically these methods and their applications [43–46]. With launch costs decreasing and the consequent increase in number of payloads sent – especially to low-Earth orbit (LEO) [47, 48] – operating these spacecraft has become more accessible and commercially viable. The first applications of ML methods for inference have already been demonstrated and, increasingly, efforts are being undertaken to move also the training on board. This section will first explore the motivations for performing ML onboard a spacecraft. Next, the transition from "classical"spacecraft hardware to current and next-generation hardware [49] is presented in the context of current spacecraft hardware trends. After that, a few examples are given to showcase actual in-orbit demonstrations of ML applications, and a particular emphasis is given to "The OPS-SAT Case", a competition to develop models suitable to inference onboard the OPS-SAT satellite from the European Space Agency. We conclude by touching upon preliminary results from our work on onboard and decentralized training in space.

2.3.1 Why Have Machine Learning on Board?

From an ML perspective, the space environment is particularly challenging if the methods developed must run onboard the spacecraft [49]. Hardware that is typically used for advanced ML techniques – such as graphics cards – has high power consumption, creates excess heat and is usually not radiation-hardened. All three of these aspects are detrimental to the viability of ML in space, with heat dissipation being a critical problem for spacecraft in general [50]. Power is usually only available in the form of solar power – the most common power source in Earth orbit – and most satellites are designed with strict power constraints (in the case of small satellites often in the range of few Watts [49,51]). Finally, typical spacecraft hardware has to be at least radiation-tolerant and, if orbits beyond LEO are considered, radiation-hardened. With all these issues, it is only natural to question why there is such a strong push to bring heavy ML computations to space in the first place. A satisfactory answer is revealed, once again, by considering the constraints of operating spacecraft, most prominently communication: downlinks to Earth – especially in LEO – depend on short communication windows with ground stations. Thus, depending on the orbit and number of ground stations, a satellite often has only a few minutes to communicate with the ground on a day. These limited windows introduce a high latency between observations made by a satellite and when the data is available and processed on ground. Depending on the application, this high latency is a critical issue. For example, the detection of natural disasters, such as wildfires or volcano eruptions [44–46], in practice requires a rapid response, and minimizing delays is imperative. Secondly, downlinks to Earth are always limited in terms of bandwidth. For instance, the 6U CubeSat HYPSO-1 (HYPer-spectral Smallsat for ocean Observation) is equipped with 1 Mbps downlink in the S-band [52], while the 6U CubeSat OPS-SAT is equipped with an X-band 50 Mbps downlink. It is thus often not practical to transmit all the data captured by the onboard sensors to the ground segment and wait for processing to be performed on the ground. At the beginning of our millenia, the satellite EO-1 pioneered the use of onboard ML [42] showcasing the potential advantages in autonomous detection and response to science events occurring on the Earth. More recently, the work by Bradley and Brandon [53] showcases to what extent leveraging onboard processing could ensure better scalability and reduced need for additional ground stations while keeping the same capability to download data for future constellations.

2.3.2 Hardware for Onboard Machine Learning

The improvement in hardware technology is fundamental to enabling ML on-board spacecraft. Size, weight and power constraints have a significant impact on the choice of hardware and the algorithms that can be implemented on board. Given that some ML methods, such as those based on neural networks, are computationally intensive, the choice of energy-efficient hardware becomes fundamental to enabling the use of ML onboard [51]. Given the strict safety requirements and the limited need for computing power, the design of space components has been much more geared towards resilience against harsh thermal cycles, electrostatic discharges, radiation and other phenomena typical of the space environment, rather than ensuring high performance and energy efficiency [54,55]. Because of that, numerous works have been investigating the use of commercial off-the-shelf (COTS) devices for non-mission-critical space applications, which generally offer more convenient trade-offs in terms of power consumption, performance, cost and mass [43,49,52,56,57].

For neural network inference, such devices include AI processors, such as the Intel® Movidius™ Myriad™ 2 used in the Φ-Sat-1 mission [43], field-programmable gate arrays (FPGAs) and system-on-a-chip FPGAs used onboard the OPS-SAT [57] and HYPSO-1 [51,52] missions, and graphics processing units (GPUs) [53,56]. To assess the usability of COTS hardware in space, some devices have been tested [49,56] under radiation, demonstrating sufficient resistance for short-term non-critical missions. ESA's ACT is investigating novel hardware solutions that might enable future applications. In the frame of *Onboard & Distributed Learning* research, presented in Section 2.3.4, the onboard availability of space qualified GPUs is discussed as an enabler for the distributed training of machine learning models [56]. In the frame of the research on *Neuromorphic sensing & processing*, described in Chapter 4, neuromorphic hardware is suggested as an alternative to conventional von Neumann computing architectures to enable complex computations under extreme low-power constraints.

2.3.3 Onboard Inference

As mentioned in Section 2.3.1, numerous researchers have been investigating the possibility of using AI onboard satellites to extract actionable information with reduced latency. In 2020, the Φ-Sat-1 mission demonstrated the possibility of discarding cloud-covered images onboard the satellite through inference enabled by a convolutional neural network (CNN) running on a COTS processor [43]. Similarly, the 6U CubeSat HYPSO-1 exploits self-organizing maps to perform onboard real-time detection of harmful algal bloom [52].

Figure 2.8 Examples of post-processed patches for each class (Snow, Cloud, Natural, River, Mountain, Water, Agricultural, Ice) in the "The OPS-SAT case" competition.

Recent work [57] describes the use of ML onboard the 6U ESA OPS-SAT CubeSat, which is an in-orbit laboratory that enables the testing of new software and mission control technologies. In particular, [57] demonstrates the ability of modern AI techniques to detect and discard "unwanted" images for the purpose of reducing the amount of data to download.

2.3.3.1 The OPS-SAT case

A recent event dedicated to advancing onboard machine learning was a competition organized via ESA's Kelvins platform[1] called "The OPS-SAT case" [58]. In the "The OPS-SAT case", the competitors were given a (quantized) neural model – an *EfficientNet-lite-0* [59] – that was tested and passed all the requirements for inference onboard the European Space Agency's OPS-SAT satellite. The challenge consisted in tuning the parameters of the model to enable it to predict one of eight classes for cropped patches coming directly from the spacecraft's raw imaging sensor data. The most performing models trained in the context of the competition were selected to run onboard OPS-SAT during a dedicated flight campaign.

Three main aspects were targeted by the competition setup. The primary one was the use of a limited number of labeled patches. This fact is of major importance to future onboard learning applications where labels are expected to be scarce because of the high effort and cost of labeling Earth observation images [60] as well as the need to reduce the operation time required for labeling when a novel sensor is used [58]. The second aspect concerned the use

---

[1]https://kelvins.esa.int/. Accessed 23/11/22

of raw satellite images. The need to post-process images increases the number of operations to be performed onboard and consequently the energy and the total time to process one image. Finally, the use of quantization-aware training to enable the use of quantized models (16-bit floating point). Quantization is, in fact, fundamental to decreasing the model size in order for it to match the satellite uplink requirements [58].

To learn the model weights, a dataset was manually created containing patches of 200×200 pixels produced cropping the raw onboard sensor data. The dataset was then split into *training* and *evaluation*, of which only the first one was released to the competitors. The *training* dataset [61] contained 10 labeled patches per class and 23 unprocessed original images, while the evaluation dataset consisted of 588 labeled patches. Given the different number of available evaluation patches for each of the classes (Agricultural: 36, Cloud: 114, Mountain: 99, Natural: 58, River: 31, Ice: 37, Snow: 106, Water: 107), the evaluation dataset was unbalanced. Because of that, model accuracy was not an appropriate metric for evaluating the competition results since it would not have significantly penalised submissions underscoring on the "River "and 'Ice"classes. Hence, the metric used for the competition was devised as follows:

$$\mathcal{L} = 1 - \kappa \qquad (2.4)$$

where κ is *Cohen's kappa* coefficient, which is used extensively for the evaluation of classification performance on unbalanced datasets in remote sensing scenarios [62, 63]. To provide a baseline for the competition, we trained an *EfficientNet-lite-0* model by using *MSMatch* [60], which is a machine learning approach developed by the ACT for the classification of MultiSpectral and RGB remote sensing data. Fig.2.9 shows the results of the competition baseline model on reconstructing images from the *training* set. The 41 teams that participated in the three-month-long competition produced a total of 891 submissions. The highest score on the evaluation dataset was $\mathcal{L} = 0.367140$. For reference, the competition baseline model obtained a score of $\mathcal{L} = 0.539694$ and would have ranked $12th$. Fig.2.10 shows the confusion matrix produced by the leading submission on the evaluation dataset. The confusion matrix for the submission from the winning team *inovor* shows that the winning model is capable of distinguishing between the "Cloud"and "Ice"classes, which have respectively 85% and 89% true positive rates. The "River" class had the worst true positive rate (50%), with misclassified "River"patches split equally between the "Natural'and "Snow"classes (17% each). For comparison, only 12.8% of the other classes were misclassified as "River". Finally, it should be noted that 21% of "Mountain"patches

Figure 2.9 (Left) A post-processed image from the camera onboard OPS-SAT. (Right) Classification of the correspondent 200×200 patches performed by the competition baseline model.

were classified as "Snow", while 18% of "Snow"patches were classified as "Mountain". This could be due to the presence of common features between the two classes, such as the presence of mountains in the "Snow "patches or the presence of snow (or other elements that cannot be easily distinguished from snow in an RGB image) in the "Mountain" patches. A dedicated publication authored with the winning teams will contain a detailed analysis of the competition preparation, its results, the machine learning approaches developed and the inference results obtained onboard OPS-SAT.

2.3.4 Onboard & Distributed Learning

At the time of writing, to the authors' knowledge, no space mission has yet demonstrated end-to-end training of a large machine learning model onboard a spacecraft. There are many reasons for this, including the lack of suitable hardware on most spacecraft and the large amounts of power and data required for training machine learning methods, especially large neural networks. As a reference, the training of current large-scale language models can require power on the order of gigawatthours [64]. However, small and efficient models, such as the EfficientNet architecture [65], can be trained even on small edge devices, such as those described in the previous section. Although power consumption is still a relevant factor, onboard training of large

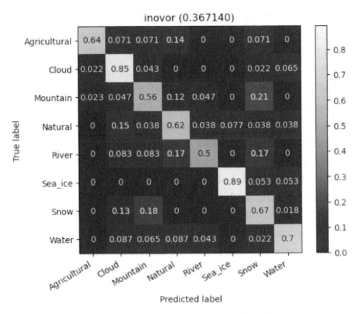

Figure 2.10 Confusion matrix of the best-performing submission on the evaluation dataset for "The OPS-SAT Case".

machine learning models is entering the realm of possibility. In fact, there are several factors that make this relevant and interesting from a practical point of view. Similar to inference, performing training onboard (as opposed to doing so on the ground and then updating the ML model parameters), reduces the communication bandwidth required while potentially contributing to the mission autonomy significantly. Depending on the location of the spacecraft, sending to and receiving data from the ground may even be completely infeasible. From that perspective onboard training may become essential for mission scenarios where autonomy is of paramount importance. For instance, this is the case for deep-space missions where communication becomes increasingly difficult as the spacecraft moves further away from Earth. Furthermore, assuming learning can be performed onboard, the application of decentralized learning approaches [66] are enabled for fleets or constellations of spacecraft. This adds a requirement on the existence of inter-satellite links, one that is in synergy with proposals advocating for advanced optical inter-satellite communications (see for example [67]).

Distributed learning has lately generated an increasing interest due to the combination of the number of commercial activities around satellite constellations [68] and the broader availability of hardware capable of running machine learning algorithms in space [69]. In particular, the potential of (a distributed learning paradigm) is being explored for joint training of satellite constellations [70, 71].

Together with several collaborators, the ACT is supporting this effort with the design of an package called PASEOS (PAseos Simulates the Environment for Operating multiple Spacecraft) [72]. [2] PASEOS aims to model the operational constraints of spacecraft with respect to decentralized learning and similar activities. It operates in a fully decentralised manner, where each device (or 'actor' in PASEOS terminology) runs an instance of the PASEOS software that provides it with a model of the simulated spacecraft's power, thermal, radiation, communication and orbital constraints. The motivation behind the software is that, in comparison to the well-explored decentralized and federated learning approaches on Earth [73], the operational constraints for training ML models onboard spacecraft are different and unique. Thus, in order to realistically explore these applications, it is important to find efficient and feasible solutions to these challenges by first specifying the constraints relevant to space.

ACRONYM

ACT Advanced Concepts Team

AI Artificial Intelligence

ANN Artificial Neural Networks

CNN Convolutional Neural Network

COTS Commercial Off-The-Shelf

EO Earth Observation

ESA European Space Agency

FDTD Finite-difference time-domain

FPGA Field-Programmable Gate Array

GPU Graphics Processing Unit

IADC Inter-Agency Space Debris Coordination Committee

---

[2]`https://github.com/aidotse/PASEOS` Accessed: 2022-10-28

JPL Jet Propulsion Laboratory

LEO Low-Earth Orbit

ML Machine Learning

NIDN Neural Inverse Design of Nanostructures

PDE Partial Differential Equation

RCWA Rigorous Coupled-Wave Analysis

GLOSSARY

Differentiable Methods: Methods that enable the computation of gradients with regard to input variables. This enables neural network training and other methods relying on gradient-based optimization.

Ground station: A station on Earth used to communicate with satellites and spacecraft.

Inference: Running a machine learning model after a training process computing the models output / prediction for some input

Inter-satellite Links: Communication links between satellites, often relying on optical communications.

Inverse Problem / Solution: A problem where we seek the initial state or input to a system, equation or similar that led to an observed output.

Partial Differential Equations (PDEs): Equations involving partial derivatives of of multivariable functions. They are often suitable to describe physical processes.

3D printing: Manufacturing technique for printing structures from one or several base materials.

Lattice material: Material composed of a lattice structure, where mechanical properties depend on the base material and geometry of the lattice.

Stiffness: Mechanical property for characterizing a material. Measures how strongly the material resists deformation.

Poisson's ratio: Mechanical property for characterizing a material. It indicates how the material deforms perpendicular to the direction it is being compressed in.

Graph: Mathematical structure composed of nodes (vertices) and connections between nodes (edges).

Message passing: Method for performing calculations on graph-structured data. For instance, graph neural networks are based on message passing.

Radiation-hardening: Changes made to devices to make them endure and operable in an environment with higher radiation load, such as space, especially beyond low-Earth orbit.

Training: Iterative optimization process of the parameters of a machine learning model, such as a neural network, to minimize the error on a training data set afor some loss function describing the error in the task the model is designed to solve.

FURTHER READING

Izzo, D., Märtens, M., & Pan, B. (2019) "A survey on artificial intelligence trends in spacecraft guidance dynamics and control". *Astrodynamics*, 3, 287-299.

Izzo, D., & Gómez, P. (2022) "Geodesy of irregular small bodies via neural density fields". *Communications Engineering*, 1(1), 48.

Gómez, P., Toftevaag, H. H., Bogen-Storø, T., Aranguren van Egmond, D., & Llorens, J. M. (2022) "Neural Inverse Design of Nanostructures (NIDN)". *Scientific Reports*, 12(1), 22160.

Dold, D., & van Egmond, D. A. (2023) "Differentiable graph-structured models for inverse design of lattice materials". *arXiv preprint*, arXiv:2304.05422.

Derksen, D., Meoni, G., Lecuyer, G., Mergy, A., Märtens, M., & Izzo, D. (2021) "Few-Shot Image Classification Challenge On-Board". In *Workshop-Data Centric AI, NeurIPS*.

BIBLIOGRAPHY

[1] David E Rumelhart, Richard Durbin, Richard Golden, and Yves Chauvin. Backpropagation: The basic theory. *Backpropagation: Theory, architectures and applications*, pages 1–34, 1995.

[2] Alexey Natekin and Alois Knoll. Gradient boosting machines, a tutorial. *Frontiers in neurorobotics*, 7:21, 2013.

[3] Alex Graves, Greg Wayne, Malcolm Reynolds, Tim Harley, Ivo Dani-helka, Agnieszka Grabska-Barwińska, Sergio Gómez Colmenarejo, Edward Grefenstette, Tiago Ramalho, John Agapiou, et al. Hybrid computing using a neural network with dynamic external memory. *Nature*, 538(7626):471–476, 2016.

[4] Alex Graves, Greg Wayne, and Ivo Danihelka. Neural turing machines. *arXiv preprint arXiv:1410.5401*, 2014.

[5] Nikolaus Hansen. The cma evolution strategy: A tutorial. *arXiv preprint arXiv:1604.00772*, 2016.

[6] Bart Van Merriënboer, Olivier Breuleux, Arnaud Bergeron, and Pascal Lamblin. Automatic differentiation in ml: Where we are and where we should be going. *Advances in neural information processing systems*, 31, 2018.

[7] Dario Izzo and Francesco Biscani. Opensource code: pyaudi, 2016.

[8] Kyoko Makino and Martin Berz. Cosy infinity version 9. *Nuclear Instruments and Methods in Physics Research Section A: Accelerators, Spectrometers, Detectors and Associated Equipment*, 558(1):346–350, 2006.

[9] Carlos Sánchez-Sánchez and Dario Izzo. Real-time optimal control via deep neural networks: study on landing problems. *Journal of Guidance, Control, and Dynamics*, 41(5):1122–1135, 2018.

[10] Martino Bardi, Italo Capuzzo Dolcetta, et al. *Optimal control and viscosity solutions of Hamilton Jacobi Bellman equations*, volume 12. Springer, 1997.

[11] Ingo Gühring, Mones Raslan, and Gitta Kutyniok. Expressivity of deep neural networks. *arXiv preprint arXiv:2007.04759*, 2020.

[12] Dario Izzo and Ekin Öztürk. Real-time guidance for low-thrust transfers using deep neural networks. *Journal of Guidance, Control, and Dynamics*, 44(2):315–327, 2021.

[13] Dario Izzo and Sebastien Origer. Neural representation of a time optimal, constant acceleration rendezvous. *Acta Astronautica*, 2022.

[14] Shuo Li, Ekin Öztürk, Christophe De Wagter, Guido CHE De Croon, and Dario Izzo. Aggressive online control of a quadrotor via deep network representations of optimality principles. In *2020 IEEE International Conference on Robotics and Automation (ICRA)*, pages 6282–6287. IEEE, 2020.

[15] Dario Izzo, Dharmesh Tailor, and Thomas Vasileiou. On the stability analysis of deep neural network representations of an optimal state feedback. *IEEE Transactions on Aerospace and Electronic Systems*, 57(1):145–154, 2020.

[16] Dharmesh Tailor and Dario Izzo. Learning the optimal state-feedback via supervised imitation learning. *Astrodynamics*, 3(4):361–374, 2019.

[17] Ben Mildenhall, Pratul P Srinivasan, Matthew Tancik, Jonathan T Barron, Ravi Ramamoorthi, and Ren Ng. Nerf: Representing scenes as neural radiance fields for view synthesis. *Communications of the ACM*, 65(1):99–106, 2021.

[18] Dario Izzo and Pablo Gómez. Geodesy of irregular small bodies via neural density fields: geodesynets. *arXiv preprint arXiv:2105.13031*, 2021.

[19] Moritz von Looz, Pablo Gomez, and Dario Izzo. Study of the asteroid bennu using geodesyanns and osiris-rex data. *arXiv preprint arXiv:2109.14427*, 2021.

[20] Francesco Biscani and Dario Izzo. Reliable event detection for taylor methods in astrodynamics. *Monthly Notices of the Royal Astronomical Society*, 513(4):4833–4844, 2022.

[21] Stanley Osher and Ronald Fedkiw. Signed distance functions. In *Level set methods and dynamic implicit surfaces*, pages 17–22. Springer, 2003.

[22] Tomas Möller and Ben Trumbore. Fast, minimum storage ray-triangle intersection. *Journal of graphics tools*, 2(1):21–28, 1997.

[23] Patrick Michel, Michael Küppers, and Ian Carnelli. The hera mission: European component of the esa-nasa aida mission to a binary asteroid. *42nd COSPAR Scientific Assembly*, 42:B1–1, 2018.

[24] Songyou Peng, Chiyu "Max" Jiang, Yiyi Liao, Michael Niemeyer, Marc Pollefeys, and Andreas Geiger. Shape as points: A differentiable poisson

solver. In *Advances in Neural Information Processing Systems (Neur-IPS)*, 2021.

[25] Alexander Zoechbauer, Dan Kelshaw, Dario Izzo, and Jai Grover. Differentiable representations for asteroid surface reconstruction. *In preparation.*

[26] Shengze Cai, Zhiping Mao, Zhicheng Wang, Minglang Yin, and George Em Karniadakis. Physics-informed neural networks (pinns) for fluid mechanics: A review. *Acta Mechanica Sinica*, pages 1–12, 2022.

[27] Nils Thuerey, Philipp Holl, Maximilian Mueller, Patrick Schnell, Felix Trost, and Kiwon Um. Physics-based deep learning. *arXiv preprint arXiv:2109.05237*, 2021.

[28] Pablo Gómez, Anne Schützenberger, Marion Semmler, and Michael Döllinger. Laryngeal pressure estimation with a recurrent neural network. *IEEE journal of translational engineering in health and medicine*, 7:1–11, 2018.

[29] Liu Yang, Xuhui Meng, and George Em Karniadakis. B-pinns: Bayesian physics-informed neural networks for forward and inverse pde problems with noisy data. *Journal of Computational Physics*, 425:109913, 2021.

[30] Pablo Gómez, Håvard Hem Toftevaag, Torbjørn Bogen-Støro, Derek Aranguren van Egmond, and José M Llorens. Nidn: Neural inverse design of nanostructures. *arXiv preprint arXiv:2208.05480*, 2022.

[31] Dominik Dold and Derek Aranguren van Egmond. Differentiable graph structured models for inverse design of lattice materials. *arXiv preprint arXiv:2304.05422*, 2023.

[32] Wei Ma and Yongmin Liu. A data-efficient self-supervised deep learning model for design and characterization of nanophotonic structures. *Science China Physics, Mechanics & Astronomy*, 63(8), June 2020.

[33] Christian C. Nadell, Bohao Huang, Jordan M. Malof, and Willie J. Padilla. Deep learning for accelerated all-dielectric metasurface design. *Optics Express*, 27(20):27523, September 2019.

[34] MG Moharam and TK Gaylord. Rigorous coupled-wave analysis of planar-grating diffraction. *JOSA*, 71(7):811–818, 1981.

[35] Dennis M Sullivan. *Electromagnetic simulation using the FDTD method*. John Wiley & Sons, 2013.

[36] Sheng-Kai Liao, Hai-Lin Yong, Chang Liu, Guo-Liang Shentu, Dong-Dong Li, Jin Lin, Hui Dai, Shuang-Qiang Zhao, Bo Li, Jian-Yu Guan, et al. Long-distance free-space quantum key distribution in daylight towards inter-satellite communication. *Nature Photonics*, 11(8):509–513, 2017.

[37] Susanna Kohler. Astronomy impacts of satellite megaconstellations. *AAS Nova Highlights*, page 6432, 2020.

[38] Thomas N Kipf and Max Welling. Semi-supervised classification with graph convolutional networks. In *J. International Conference on Learning Representations (ICLR 2017)*, 2016.

[39] Emre O Neftci, Hesham Mostafa, and Friedemann Zenke. Surrogate gradient learning in spiking neural networks: Bringing the power of gradient-based optimization to spiking neural networks. *IEEE Signal Processing Magazine*, 36(6):51–63, 2019.

[40] Advenit Makaya, Laurent Pambaguian, Tommaso Ghidini, Thomas Rohr, Ugo Lafont, and Alexandre Meurisse. Towards out of earth manufacturing: overview of the esa materials and processes activities on manufacturing in space. *CEAS Space Journal*, pages 1–7, 2022.

[41] A Mitchell, U Lafont, M Ho lyńska, and CJAM Semprimoschnig. Additive manufacturing—a review of 4d printing and future applications. *Additive Manufacturing*, 24:606–626, 2018.

[42] Stephen G Ungar, Jay S Pearlman, Jeffrey A Mendenhall, and Dennis Reuter. Overview of the earth observing one (eo-1) mission. *IEEE Transactions on Geoscience and Remote Sensing*, 41(6):1149–1159, 2003.

[43] Gianluca Giuffrida, Luca Fanucci, Gabriele Meoni, Matej Batič, Léonie Buckley, Aubrey Dunne, Chris van Dijk, Marco Esposito, John Hefele, Nathan Vercruyssen, Gianluca Furano, Massimiliano Pastena, and Josef Aschbacher. The phi-sat-1 mission: The first on-board deep neural network demonstrator for satellite earth observation. *IEEE Transactions on Geoscience and Remote Sensing*, 60:1–14, 2022.

[44] Gonzalo Mateo-Garcia, Joshua Veitch-Michaelis, Lewis Smith, Silviu Vlad Oprea, Guy Schumann, Yarin Gal, Atılım Güneş Baydin, and Dietmar Backes. Towards global flood mapping onboard low cost satellites with machine learning. *Scientific reports*, 11(1):1–12, 2021.

[45] Vít Růžička, Anna Vaughan, Daniele De Martini, James Fulton, Valentina Salvatelli, Chris Bridges, Gonzalo Mateo-Garcia, and Valentina Zantedeschi. RaVÆn: Unsupervised change detection of extreme events using ML on-board satellites. *Scientific Reports*, 12(1):16939.

[46] Maria Pia Del Rosso, Alessandro Sebastianelli, Dario Spiller, Pierre Philippe Mathieu, and Silvia Liberata Ullo. On-board volcanic eruption detection through cnns and satellite multispectral imagery. *Remote Sensing*, 13(17):3479, 2021.

[47] Stijn Lemmens and Francesca Letizia. Esa's annual space environment report. Technical report, Technical Report GEN-DB-LOG-00288-OPS-SD, ESA Space Debris Office, 2021.

[48] Harry Jones. The recent large reduction in space launch cost. 48th International Conference on Environmental Systems, 2018.

[49] Gianluca Furano, Gabriele Meoni, Aubrey Dunne, David Moloney, Veronique Ferlet-Cavrois, Antonis Tavoularis, Jonathan Byrne, Léonie Buckley, Mihalis Psarakis, Kay-Obbe Voss, et al. Towards the use of artificial intelligence on the edge in space systems: Challenges and opportunities. *IEEE Aerospace and Electronic Systems Magazine*, 35(12):44–56, 2020.

[50] Derek W Hengeveld, Margaret M Mathison, James E Braun, Eckhard A Groll, and Andrew D Williams. Review of modern spacecraft thermal control technologies. *HVAC&R Research*, 16(2):189–220, 2010.

[51] Radoslav Pitonak, Jan Mucha, Lukas Dobis, Martin Javorka, and Marek Marusin. Cloudsatnet-1: Fpga-based hardware-accelerated quantized cnn for satellite on-board cloud coverage classification. *Remote Sensing*, 14(13):3180, 2022.

[52] Aksel S Danielsen, Tor Arne Johansen, and Joseph L Garrett. Self-organizing maps for clustering hyperspectral images on-board a cubesat. *Remote Sensing*, 13(20):4174, 2021.

[53] Bradley Denby and Brandon Lucia. Orbital edge computing: Nanosatellite constellations as a new class of computer system. In *Proceedings of the Twenty-Fifth International Conference on Architectural Support for Programming Languages and Operating Systems*, pages 939–954, 2020.

[54] George Lentaris, Konstantinos Maragos, Ioannis Stratakos, Lazaros Papadopoulos, Odysseas Papanikolaou, Dimitrios Soudris, Manolis Lourakis, Xenophon Zabulis, David Gonzalez-Arjona, and Gianluca Furano. High-performance embedded computing in space: Evaluation of platforms for vision-based navigation. *Journal of Aerospace Information Systems*, 15(4):178–192, 2018.

[55] Gianluca Furano and Alessandra Menicucci. Roadmap for on-board processing and data handling systems in space. In *Dependable Multicore Architectures at Nanoscale*, pages 253–281. Springer, 2018.

[56] Fredrik C Bruhn, Nandinbaatar Tsog, Fabian Kunkel, Oskar Flordal, and Ian Troxel. Enabling radiation tolerant heterogeneous gpu-based on-board data processing in space. *CEAS Space Journal*, 12(4):551–564, 2020.

[57] Georges Labrèche, David Evans, Dominik Marszk, Tom Mladenov, Vasundhara Shiradhonkar, Tanguy Soto, and Vladimir Zelenevskiy. Opssat spacecraft autonomy with tensorflow lite, unsupervised learning, and online machine learning. In *2022 IEEE Aerospace Conference*, 2022.

[58] Dawa Derksen, Gabriele Meoni, Gurvan Lecuyer, Anne Mergy, Marcus Märtens, and Dario Izzo. Few-shot image classification challenge on-board. In *Workshop-Data Centric AI, NeurIPS*, 2021.

[59] Mingxing Tan and Quoc Le. Efficientnet: Rethinking model scaling for convolutional neural networks. In *International conference on machine learning*, pages 6105–6114. PMLR, 2019.

[60] Pablo Gómez and Gabriele Meoni. Msmatch: Semisupervised multispectral scene classification with few labels. *IEEE Journal of Selected Topics in Applied Earth Observations and Remote Sensing*, 14:11643–11654, 2021.

[61] Dawa Derksen, Gabriele Meoni, Gurvan Lecuyer, Anne Mergy, Marcus Märtens, and Dario Izzo. The ops-sat case dataset. DOI: 10.5281/zenodo.6524750.

[62] Jan Pawel Musial and Jedrzej Stanislaw Bojanowski. Comparison of the novel probabilistic self-optimizing vectorized earth observation retrieval classifier with common machine learning algorithms. *Remote Sensing*, 14(2):378, 2022.

[63] Prasad Deshpande, Anirudh Belwalkar, Onkar Dikshit, and Shivam Tripathi. Historical land cover classification from corona imagery using convolutional neural networks and geometric moments. *International Journal of Remote Sensing*, 42(13):5144–5171, 2021.

[64] David Patterson, Joseph Gonzalez, Quoc Le, Chen Liang, Lluis-Miquel Munguia, Daniel Rothchild, David So, Maud Texier, and Jeff Dean. Carbon emissions and large neural network training. *arXiv preprint arXiv:2104.10350*, 2021.

[65] Mingxing Tan and Quoc Le. Efficientnetv2: Smaller models and faster training. In *International Conference on Machine Learning*, pages 10096–10106. PMLR, 2021.

[66] Brendan McMahan, Eider Moore, Daniel Ramage, Seth Hampson, and Blaise Aguera y Arcas. Communication-efficient learning of deep networks from decentralized data. In *Artificial intelligence and statistics*, pages 1273–1282. PMLR, 2017.

[67] Sanmukh Kaur. Analysis of inter-satellite free-space optical link performance considering different system parameters. *Opto-Electronics Review*, 27(1):10–13, 2019.

[68] Robert Massey, Sara Lucatello, and Piero Benvenuti. The challenge of satellite megaconstellations. *Nature astronomy*, 4(11):1022–1023, 2020.

[69] Unibap spacecloud framework. `https://incubed.esa.int/portfolio/uss/`. *ESA InCubed.*

[70] Nasrin Razmi, Bho Matthiesen, Armin Dekorsy, and Petar Popovski. On-board federated learning for dense leo constellations. In *ICC 2022-IEEE International Conference on Communications*, pages 4715–4720. IEEE, 2022.

[71] Bho Matthiesen, Nasrin Razmi, Israel Leyva-Mayorga, Armin Dekorsy, and Petar Popovski. Federated learning in satellite constellations. *arXiv preprint arXiv:2206.00307*, 2022.

[72] Pablo Gómez, Johan Östman, Vinutha Magal Shreenath, and Gabriele Meoni. Paseos simulates the environment for operating multiple spacecraft. *arXiv preprint arXiv:2302.02659*, 2023.

[73] Wei Yang Bryan Lim, Nguyen Cong Luong, Dinh Thai Hoang, Yutao Jiao, Ying-Chang Liang, Qiang Yang, Dusit Niyato, and Chunyan Miao. Federated learning in mobile edge networks: A comprehensive survey. *IEEE Communications Surveys & Tutorials*, 22(3):2031–2063, 2020.

Space Systems, Quantum Computers, Big Data and Sustainability: New Tools for the United Nations Sustainable Development Goals

Joseph N. Pelton

Chair of the Alliance for Collaboration in the Exploration of Space (ACES WORLD-WIDE); Dean Emeritus, International Space University

Scott Madry

University of North Carolina at Chapel Hill, North Carolina, USA; Professor Emeritus, International Space University

CONTENTS

3.1	Introduction	55
3.2	The Expanding Future to Be Released by Quantum Computing, Big Data, the Internet of Things, Artificial Intelligence and Advanced Space System Capabilities	57
3.3	The Past and the Future of Advanced Computer Processing and Quantum Computing	57
3.4	Developing the Software, the Artificial Intelligence and the Big Data Analysis Capabilities for the Quantum Age	59

DOI: 10.1201/9781003366386-3

3.5	Developing the Space Systems and Sensing Capabilities for the Quantum Age	62
3.6	Evolving Space Systems and Big Data Analytics	63
	3.6.1 Google Earth Engine	63
	3.6.2 Geomatics, GIS, and Decision Support Systems	65
3.7	Advancing the United Nation's Sustainable Development Goals	67
	3.7.1 Goal Number 1: No Poverty	68
	3.7.2 Goal Number 2: No Hunger	79
	3.7.3 Goals Number 3 and 4: Good Health and Well-Being and Quality Education	82
	3.7.4 Goal Number 5: Gender Equality	84
	3.7.5 Goal Number 6: Clean Water	85
	3.7.6 Goal Number 7: Affordable and Clean Energy	85
	3.7.7 Goal Number 8: Decent Work and Economic Growth	88
	3.7.8 Goal Number 9: Industry Innovation and Infrastructure	88
	3.7.9 Goal Number 10: Reduced Inequality	90
	3.7.10 Goal Number 11: Sustainable Cities and Communities	90
	3.7.11 Goal Number 12: Responsible Consumption and Production	92
	3.7.12 Goals Number 13, 14 and 15: Climate Action, Live Below Water, and Live on Land	93
	3.7.13 Goals Number 16 and 17: Peace, Justice & Strong Institutions and Partnerships for the Goals	96
3.8	Current Implementation Efforts	96
3.9	Conclusions	97
Bibliography		102

THE 17 Sustainable Development Goals for 2030 of the United Nations have well defined quantitative objectives in each of these different areas. The ability to note significant progress against these goals, in order to determine if advancement is truly being achieved, is heavily dependent on space sensing systems. Likewise, space systems are also frequently key to detecting problems, enabling progress, and indeed crucial to achieving these goals. New high performance space systems plus enhanced data processing systems, including big data analysis and artificially intelligence, can provide a 'real-time' understanding of global activities—both positive and negative. The petabytes of data generated by advanced multi-spectral Earth Observation Systems are not the essence of these space systems of the future. The prime factor is the advanced intelligence and accelerated data processing techniques that can,

in near real-time, discover environmental or legal infractions, buildings or bridges in danger of collapse, the spread of fires or tree or plant infections, changes in demographics, and more. They can also determine progress toward positive goals. This chapter will explore how space systems linked to Big Data Analysis and AI interpretative systems can support efforts to achieve the United Nations' Sustainable Development Goals, enable enforcement of environmental laws and regulations, and support positive global development in new ways that are much more rapid, accurate, and actionable.

"The authors wish to particularly thank Peter Marshall for his help in reviewing and editing this chapter."

3.1   INTRODUCTION

The 17 Sustainable Development Goals (SDGs) for 2030 of the United Nations (UN), are intended to make the world a better place to live and to provide measurable goals to achieve these. The adverse effects of climate change, rapid population growth, and other forces of change now make these goals increasingly more important, and yet ever more difficult to successfully achieve. Space systems, coupled with big data analytics and advanced computing, can and will making important contributions to achieve these goals. Fortunately, these 17 goals have been formulated and defined in terms of clear quantitative and measurable objectives in virtually all these different areas.

The ability to clearly measure advancement against these goals is critical. Space systems and closely associated data analytics are key to detecting problems and barriers to advancement in these various goals as well as creating tools to allow progress in virtually all the 17 defined areas where rapid progress is needed. This linkage of space systems to achieving the Sustainable Development Goals is sufficiently close that the UN has approved the Space 2030 Agenda.

In short, space systems of many types, coupled to very high-speed data analytics and advanced computing, are key to detecting problems, enabling progress, and achieving these highly diverse goals. New high performance space systems plus enhanced data processing systems, including big data analysis and artificial intelligence, can provide a 'real-time' understanding of global activities—both positive and negative. But how do we do this?

The petabytes of data generated by advanced weather, navigation, and remote sensing satellite systems are then relayed via ultra-fast space networking systems in near-real time for processing on exa-scale computers. These space-based satellite tools and processing systems, are now key to progress against achieving these UN goals. These space-based efforts can become vital global capabilities in eliminating need and hunger, improving health and education, and making strides to supply clean water and energy, and making progress with regard to economic growth and decent work, industrial innovation, more sustainable cities, responsible production and consumption, and a better world environment in terms of the air, land and seas.

The prime enabling factor, the difference that will make a difference, is not the rapidly advancing space systems, but the linking of these space capabilities to the accelerated data collection and processing techniques now becoming available, particularly with the evolving application of Artificial Intelligence (AI) and Machine Learning (ML). These analytical systems can, in near real-time, discover environmental problems or legal infractions, dangers to buildings or bridges in danger of collapse, detect the spread of fires or forest or agricultural infections, grow food more efficiently, detect changes in demographics, aid in rescue situations, provide health and educational systems more widely and efficiently, and much more.

This chapter will explore how Big Data Analysis and AI interpretative systems, when linked to a range of advanced space applications and quantum computing, can make a huge difference to the world of the future. These capabilities can greatly advance achieving the 17 UN SDGs as noted in Annex No. 1 to this chapter.

The truth of the matter is that the original UN Millennium Goals for 2015 were not met, and it was never likely that they would be achieved. These initial Goals, which had to be set, even without much chance of success, have now been replaced by the Sustainable Development Goals for 2030. It now seems likely that these ambitious Goals for 2030 will also not be met. This means that these objectives will thus also likely be replaced by new sustainability goals for 2045. Yet hope can be found in promising new technical capabilities that will allow the gap between the UN goals and performance to be closed by mid-21st century.

Much of this hope is to be found in powerful new space systems technologies, and in the amazing new capabilities of quantum computing. There is tremendous power in the qubits that will replace the digital bits we now use. We will in the decades ahead see new capabilities in AI and ML, the new programming capabilities to be developed for quantum computers, and ever more pervasive big data analytics. It is the power to be found in these new capabilities that suggests the possibility that the "gap" can actually be closed in the coming decades. It is essential for progress in all three of these areas to ultimately reduce poverty and hunger, improve health and education, make our environmental practices greener, and much more. These advances, plus perhaps new advances in the field of family planning, birth control, and environmental sustainability may ultimately unlock a better future for humanity.

3.2 THE EXPANDING FUTURE TO BE RELEASED BY QUANTUM COMPUTING, BIG DATA, THE INTERNET OF THINGS, ARTIFICIAL INTELLIGENCE AND ADVANCED SPACE SYSTEM CAPABILITIES

The future ability to achieve the United Nations' 17 Sustainable Development Goals will be, in many ways, driven by three types of technical advances. These will be:

(a) in computer processing–especially in quantum computing using Qubit processing;

(b) software and programming advances (this will include artificial Intelligence and machine learning, with software optimized for qubit-based operations, and data analytics; and

(c) advances in the development and functional integration of space systems, high-altitude platforms, and in-situ monitoring systems.

These involve a wide range of rapidly evolving capabilities relating to remote sensing, hyperspectral analysis, on-board data processing, Positioning, Navigation and Timing systems, meteorological, hydrological, and climate-related sensing systems, reusable and improved launch systems, data relay systems, and more. In the following section we will examine this unfolding future from a historical and future-trends analysis perspective.

3.3 THE PAST AND THE FUTURE OF ADVANCED COMPUTER PROCESSING AND QUANTUM COMPUTING

The ENIAC gigantic digital processor is often considered the first general purpose electronic computer. It weighed over 30 tons and included almost 18,000 vacuum tubes. This meant ENIAC, as it was developed between 1943 and when it became operation in 1946, cost many millions of dollars and was subject to vacuum tube failure and also used massive amounts of electricity. The use of vacuum tubes was required since it was developed before the age of transistors or integrated circuits. This machine was capable of 5,000 addition problems a second, or operating at 5000 Hz—a remarkable achievement for the mid 1940's Yet today it would be considered an electronic dinosaur. It was originally developed to process artillery ballistics tables, but was used much

Figure 3.1   The ENIAC Computer. (US Army Photo)

more broadly and has been estimated that ENIAC, in its operation from 1946 to 1955, performed more calculations in this one decade than had been made by all humanity up to the time it began operation (See Fig. 3.1).[1]

Today, nearly sixty years later, one might compare the performance of the ENIAC to Intel's Core i9-13900K. This is a compact, mass produced desktop computer chip that operates at 5.8 GHz or 5.8 billion operations a second. This processor can be purchased for about $700, not many millions of dollars, and can operate on a desktop or laptop computer. This computer processor is more than a million times faster than ENIAC, and when inflation is considered, it is probably more than a hundred thousand times lower in cost. And if one looks to the future of computing power as represented by the use of quantum computing, this rapid performance increase seems very likely to continue unabated. In a period of less than sixty years processing capabilities have increased a million times. The leverage to make processing capabilities a million times even more efficient will likely require the abandonment of the world of digital processing that uses only zeros and ones, and to use quantum bits or "qubits" instead. The following explains the enormous advantage in processing power that this allows:

---

[1]ENIAC: Birth of the Computer, Computer History.org, https://www.computerhistory.org/revolution/birth-of-the-computer/4/78 (Last accessed Nov. 20, 2022).

Instead of bits, which conventional computers use, a quantum computer uses quantum bits—known as qubits. "To illustrate the difference, imagine a sphere. A bit can be at either of the two poles of the sphere, but a qubit can exist at any point on the sphere. So, this means that a computer using qubits can store an enormous amount of information and also use less energy doing so than a conventional computer. By entering into this quantum area of computing where the traditional laws of physics no longer apply, we will be able to create processors that are significantly faster (a million or more times) than the ones we use today.[2]

The power of quantum computers that can use "qubits" rather than "binary bits" seems destined to accelerate the power of computers to be more capable, more cost-effective, and use much less electrical power perhaps by orders of magnitude. Google has estimated that its new D Wave quantum computer could become even a 100 million times faster than the most sophisticated supercomputer developed using binary coding. It is possible that such highly capable quantum computers will be able to produce results in four minutes which might have taken a traditional supercomputer 10,000 years to accomplish by using the amazing ability of these new machines to carry out simultaneous calculations.

3.4 DEVELOPING THE SOFTWARE, THE ARTIFICIAL INTELLIGENCE AND THE BIG DATA ANALYSIS CAPABILITIES FOR THE QUANTUM AGE

Of course, the challenge is not only that of the speed of new processors, but the ability to process the data into useful information. New AI and big data analytics must be developed to exploit the powers of quantum computers in order to assess the complex problems the world faces today and that will exist tomorrow. One of the prime challenges to be addressed is finding the answers to true global sustainability for human society and achieving a viable world for all forms of flora and fauna. It must be realized that computer processors have two vital parts. There is the machine that performs the calculations using electrical impulses, or possibly optical light, and then there is the software

[2]Bernard Marr, *What Is Quantum Computing? A Super-Easy Explanation For Anyone*, Forbes Magazine, July 4, 2017, https://www.forbes.com/sites/bernardmarr/2017/07/04/what-is-quantum-computing-a-super-easy-explanation-for-anyone/?sh=26898791d3b5

that allows a problem to be formulated so that the mechanism can calculate an answer.

Ray Kurzweil's book *How to Create a Mind: The Secret of the Human Mind Revealed* essentially says the human mind comes down to: (i) the ability to store and retrieve "memory" in what might be called a database; (ii) the ability to make "calculations" drawing on the stored data, and the ability to create **"pattern recognitions"** to establish connections and create ideas or insight. In the human brain, it is the ability to connect neurons in different ways is seen as the essence of creating ever more complex patterns that represents ways to sense information, understand the world, and think creatively.[3]

Professor Henry Markham headed the "Blue Brain" project in Switzerland, which he first organized in 2005. He began his quest as head of a team of computer scientists dedicated to create the equivalent of a human brain. By 2008 he had created the equivalent of an entire neocortical column of a rat's brain (or 10,000 neurons). By 2011 he had developed a system equivalent to 100 neocortical columns (or 1 million neurons) and which he referred to as a "Meso-circuit." He said at the time that a rat's brain was equivalent to 100 Meso-circuits (or 100 million neurons or the equivalent of a trillion synapses). In terms of computer terminology, he described this intermediate objective as being the equivalent of 10^{14} bytes of memory and 10^{15} flops of computer processing speed. (Note flops = floating point operations per second [1, pp. 124–128].

The funding source of the Blue Brain project has been the École Polytechnique Fédérale de Lausanne (EPFL) in Switzerland. It's website states that the objective was never to try to create an "active consciousness" or to duplicate the functionality of the human brain. Nevertheless, many believe that Professor Markham had often indicated that he and his team might create the equivalent of a human brain, and might do so as early as 2023. Such a capability would have purportedly been comparable to 1000 rat brains. This was, in terms of computer terminology, more or less equivalent to 10^{17} bytes of memory and 10^{18} flops of computer processing speed. Professor Markham has been quoted that all his team had to do was simply to "scale it up." This "Blue Brain" project has yet to produce a human brain equivalence, but it has at least quantified what was a reasonably clear goal and credible means of achieving that goal.

Indeed, another project, known as the Human Brain Project, was officially launched on October 1, 2013. This collaborative project involves hundreds of collaborators across Europe and was envisioned as a decade long undertaking.

---

[3]Ray Kurzweil, *How to Create a Mind: The Secret of Human Thought*, Revealed (2012) Viking Press, N.Y. [1]

The collaborators are using so-called "exa-scale supercomputers" and collaborative information and communication technology (ICT) to create a new type of scientific research infrastructure so that many researchers can participate. These researchers come not only from the field of computer science but also from the medical, neuroscience, and other relevant areas related to the study of the brain.[4]

The problem that noted scientists and computing innovator Ray Kurzweil has identified with regard to the "Blue Brain" initiative was not achieving this very large size of the memory or the exceeding fast processing speed. Rather, he has suggested the ultimate difficulty would arise when it came to what might be called the developing the "complex and learned-over-time software" of the human brain. He notes that this pattern building effort develops as a baby's brain grows and expands in a constantly learning experience. Thus, a baby's brain "learns" through a maze of synapse connections to see, hear, smell, feel, taste, how to behave, speak a language, read and write, understand math, cultural dos and don'ts, and millions of other patterns.

Currently the technical challenge is to develop super-fast processing workstations sufficient to support the develop of AI pattern recognition system capabilities. Today these types of AI ready computer processing chips that are designed to support such sophisticated workstations include NVIDA (DGX Work Stations), Lambda Labs (GPU Work Stations), Lenovo (P Series Work Stations), Edge (XT Work Stations), and 3XS (Data Science Work Stations). Today the Microsoft Azure Cognitive Systems and a host of other companies from Alphabet/Google to Hewlett Packard are focused to developing new processing capabilities in the AI field.

It is these acquired synapse operations that are needed to operate in our human world. The building up of a complex of perhaps quadrillions of synapse collections that shapes pattern recognitions that forms the "software" of the human brain that represents the true crux of the problem for any ultimate "Blue Brain" project. This, in short, according to Kurzweil, is the most difficult of challenges to conquer. The ability to create a software system to upload this kind of capability to a computerized system in Kurzweil's estimation could not be developed until the 2040s. Indeed, Kurzweil has suggested this would need computational power equivalent to 10^{19} calculations per second. At the rate of change due to quantum computing and qubits, this capability seems likely to come sooner than this.[5]

---

[4] Human Brain Project, https://www.humanbrainproject.eu.
[5] The Blue Brain project EPFL, https://www.epfl.ch/research/domains/bluebrain/frequently_asked_questions/.

The challenges to future development of the computer processors and the new software to support them are thus enormous, but the pathway to future success seems relatively clear to those working in these areas. The parallel and rapid development of new space systems that will employ this technology is also unfolding rapidly as well. These new space-based systems will increasingly embrace the ever-faster speeds and software evolving in the field of computer science.

3.5 DEVELOPING THE SPACE SYSTEMS AND SENSING CAPABILITIES FOR THE QUANTUM AGE

If one looks at the progress in satellite technology and systems performance, the same type of rapid increase in space application capability will also be evident. Indeed, some suggest that space systems are increasing such computer processing systems in the sky, with applications specific software and systems needed to carry out their operations. The advent of large-scale satellite constellations in low Earth orbit is one of the more important developments. These types of networks, such as operated today by the company Planet, which has over 200 satellites acquiring data daily over the entire Earth, add important new sensing capabilities. In addition, there are other new capabilities. These include High Altitude Platform Systems (HAPS), Unmanned Aircraft Systems (UAS), and inexpensive aerial drones. These subspace/protospacer/aerial systems create a range of sensing capabilities at an ever-increasing range of altitudes to monitor activities on Earth so that more and more information is available to humanity on an ever-faster time scale.[6]

These space systems are also being matched by exciting new developments in in-situ data collection capabilities, massive sensor nets, and edge computing and data processing, which, together, can provide 'ground verification' capabilities to inform and calibrate our space-based remote sensing data, at local, regional, and eventually, global scales. Merging and integrating in-situ environmental sensor networks with our space and aerial systems and computing capabilities is a key component of this revolution, which is often overlooked, but it will require the same attention and development if we are to achieve our goals.

---

[6]Planet-Our Constellations, https://www.planet.com/our-constellations/ (Last accessed Nov 25, 2022).

3.6 EVOLVING SPACE SYSTEMS AND BIG DATA ANALYTICS

In the field of space-based electro-optical remote sensing there are four types of resolution that define the capabilities of a sensor and its resulting data. These are spatial, temporal, spectral, and radiometric resolution. In the case of spatial resolution, the issue is the desire to produce greater clarity via more pixels of information per square meter (this represent increasing spatial resolution). Then there is often a desire to "see" the same area more frequently (this would represent increased temporal resolution.) Planet now images the entire Earth's surface at least once a day. We also now have hyperspectral remote sensing satellites operating with sensors divided into 100 or even 1,000 different narrow spectrum bands. This sensing capability in discrete bands is useful because various materials, crops, trees, fish, lifeforms, machines, and other things are better revealed at different wavelength (this is an increased spectral resolution). Finally, we are now able to achieve increased radiometric clarity or density, generally represented in bits. Early remote sensing systems had 2^8 or 256 levels of radiometric resolution, or shades of gray if you will, where today some space systems have 2^{12} or 4,096 levels, (this is radiometric resolution). All of the newest existing space systems and evolving high-altitude systems are thus "seeing" more and more precisely the details our entire planet, its oceans and atmosphere and their daily variations, and are streaming massive amounts of data down to us. These data can help us measure and monitor more closely all of types of natural and human activities, global business and governmental operations, weather, natural disasters, and meteorological changes and patterns, as well as thousands of different types of interrelated land, ocean, and atmospheric patterns and changes.

3.6.1   Google Earth Engine

The most advanced system today in this context is the Google Earth Engine.[7] This is a Google Cloud-based system developed for massive-scale processing of satellite imagery and related data. Google has collected and stored on the cloud all available public domain satellite imagery from the US Landsat, ESA Sentinel-2, as well as many other sources of imagery and GIS data. The system then provides the massive computational capability from Google that is needed to conduct rapid analyses over large areas and time-frames. A Python/JavaScript API and GUI are also a part of the system, allowing users to quickly create customized analyses[8].

---

[7]See: https://earthengine.google.com
[8]"Earth Engine Code Editor — Google Earth Engine". *Google Developers.*

This is the first, practical, implementation of such a massive satellite imagery storage and analysis capability, and it has become a state-of-the-art toolkit for many academic and research users. It is provided free, without cost, to academic and research users around the globe, and has been used for hundreds of projects which have resulted in hundreds of publications and papers so far [2]. It does not, however, provide a standard set of tools to conduct the usual suite of image processing routines, such as Normalized Differential Vegetation Indices (NDVI) or unsupervised thematic classifications, processing routines that are provided in standard satellite image processing software systems, and these must be provided by the user through the API or GUI. This is an extraordinary capability, but it still has limitations in our view. Particularly the requirement to understand and write your own code. Most of the world's population are not programmers.

Not everyone with a need for these data and processing capabilities can write JavaScript code, and a major improvement will be the development of natural language capabilities, where one could simply ask "create a time series of land use and land cover from 1980 to 2020 of an area 50 km from Chapel Hill, North Carolina" in the same way that people today ask Siri or Alexa to turn on their favorite radio station. This is not possible now, but such natural language capabilities will surely be available in the near future and will open the utility of these extraordinary tools to many more users in the context of the UN SDGs as discussed below.

So today, current remote sensing systems can conduct massive processing using openly available systems such as the Google Earth Engine, and additional capabilities are in development that will enhance these amazing capabilities that did not exist just a few years ago.

As in the case of "Blue Brain" development project, the greatest challenge for the successors of the Google Earth Engine—or any other such ambitious remote sensing satellite system like it—will be to develop the needed software and processing capability to get useful information into the hands of decision makers in time to actually make a difference, and to make this a routine capability instead of something done by a few thousand highly skilled academic researchers. This would require data processing and analysis, operating in picoseconds, as well as natural language interfaces, to seek near-real-time answers to a huge number of dynamic and complex questions. The challenge of such high-capability space sensing and ground computing systems will be to develop software programs that could process incoming data, perhaps moving at billions of bytes of data per second, and then translate this data into specific, practical actions and concrete and implementable recommendations in near-real time. The objective would be to produce quickly–in time spans represented by minutes and not hours or days–important and specific outputs for not only scientists but for environmental managers, policy makers, decision makers, the press, and, ultimately, the public. These answers might include whether remote potable water systems were leaking or being polluted or poisoned; or whether large tracks of remote farmlands were being infected by pests or disease or were being improperly irrigated or fertilized; and so on to thousands of other practical applications. Each would have to be customized and calibrated to not only the specific geographical and physical regions, but also, in the end, adapted for the cultural context and perspectives of the people. Today, we have large networks of satellite constellations that are populated by highly capable satellites in Polar Sun Synchronous low Earth orbits and out at Geo that are now producing massive amounts of data. These space systems are also being increasingly supplemented in certain areas by unmanned aircraft systems (UAS) and even drone-based systems. All this has to be connected to the ground.

Such comprehensive space networks as described above, will need to be backed by exa-scale super computers, even beyond the current processing of Google's Earth Engine. The greatest need, in the context of the UN SDGs, will be the ability to create currently unavailable, simple to use software geared to produce literally thousands of vital answers to our many efforts to make the world safer, greener, more profitable, and sustainable. In the next decade or two, or even sooner, such types of satellite/data processing systems will be implemented to make our city safer, our farmlands more productive, and the most dangerous of diseases and pandemics detected at the earliest moment. The key is to not only develop the technical tools, but to learn how to use these new technologies wisely and well within the cultural contexts of the end users. The technological challenge of our times is to develop these new capabilities to make our world safer and less endangered by climate change. It is our challenge to make humanity's future less in peril and make the next generations prospects brighter. This will require much more than simply more satellites and faster computers.

3.6.2 Geomatics, GIS, and Decision Support Systems

All of this data from satellites will have to be integrated in near-real time with vast amounts of spatial information generated and managed within Geographic Information Systems (GIS) using Geomatics technologies. In GIS, remote sensing data represent only a fraction of the needed information for appropriate decision making relating to complex human and environmental issues. Data relating to population, economics, public health, crime, stream

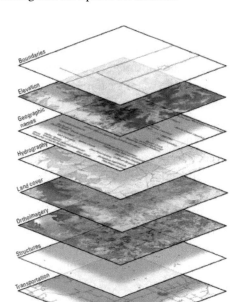

Figure 3.2   A GIS 'layer cake' of co-registered data environmental data, derived from multiple sources. (Image courtesy of USGS)

gauges, and other factors are not accesses through satellites in orbit, but are collected on the ground and are integrated with remote sensing (RS) data and analyzed through GIS systems. In order to effectively address the SDGs, we must go beyond just the types of information generated from satellites and have the ability, using dedicated Decision Support Systems (DSS) within the Geomatics and GIS context, to quickly and accurately analyze and present relevant results to decision makers. This information must then be relayed to decision makers in the formats and contexts that they understand and that can be turned into actions. GIS is yet another domain that will require massive data collection, storage, and processing capabilities using tools not yet available today. (See Fig. 3.2)

The hope is that the newest and most capable remote sensing satellites, as assisted by sophisticated new exa-scale supercomputers, plus advanced communications and networking satellites, Positioning, Navigation, and Timing (PNT) satellites, weather, meteorological, and climate monitoring satellites, as well as scientific satellites and perhaps even solar power satellites (SPS) will allow us to harness these amazing new capabilities. The lower altitude systems and in-situ sensor networks will also serve to supplement and calibrate these space systems as well. One of the most important things these space systems can and will do is to allow the world community to work together to meet the 17 specific sustainable development goals of the United Nations, as will be addressed below.

As stated above, the original goals set by the UN in 2000 were not achieved by 2015. There is every reason to doubt that the 17 SDGs now set for 2030 will be achieved either, but there is also reason to hope that better space tools, integrated with advanced informatics, can assist in closing the gaps between aspired-for-goals and actual achievements. Let's examine the ways that quantum computing, big data analysis, and newly improving space systems, GIS, and Decision Support Systems could be combined to meet the U.N.'s goals for 2030.

3.7 ADVANCING THE UNITED NATION'S SUSTAINABLE DEVELOPMENT GOALS

"If at first you don't succeed, try, try, try, and try again." This is an old saying that reflects the persistence of human initiative. It instills the thought that it takes determined and often difficult effort to make things better even if one must work very hard to succeed. We know the Sustainable Developments Goals currently set for 2030 are almost impossible to achieve, but this means we must try even harder, and smarter, with newly evolving toolkits.[9]

The key is not to give up, the risks are simply too great, and we must continue to develop new and improved tools that will, eventually, allow the most difficult of the SDG objectives to be realized. Certainly, the 17 goals that have been set and represented in the chart below are exacting and difficult (see Fig. 3.3). There is reason to believe that new and improved space systems, big data analytics, in-situ sensors, geomatics, and the improved computing capabilities that will come with quantum computers can assist us in turning the world around to a more sustainable future.

The purpose of the remaining part of the chapter is to go through the UN's 17 sustainable goals for 2030 to see where space systems, particularly when linked to big data and enhanced computer processing and programming capabilities and geomatics, can play an important role in moving forward to achieve those goals (see Table 3.1 at the end of the chapter to find out about the various ways different types of space systems can contribute to these goals).

---

[9]The United Nations Organization, The Sustainable Development Goals Report 2021, https://unstats.un.org > sdgs > report > 2021 (Last accessed Nov. 15, 2022).

Figure 3.3   The United Nation's Seventeen Sustainable Development Goals for 2030. (Graphic courtesy of the United Nations)

3.7.1   Goal Number 1: No Poverty

The worlds of aerospace, digital data and image processing, and information and computer technologies are associated with the capabilities of the developed world and the wealth and technologies of the most powerful and richest nations. Yet space, computers and programming capabilities are very much also linked to raising people and nations out of poverty. Space systems and associate tele-education and tele-health program in India, China, and dozens of other countries have used satellite-based education to provide new opportunities for health services and educating rural and remote populations living in poverty. Author Pelton, while at Intelsat in the 1980s, led the Project SHARE initiative (Satellites for Health and Rural Education). Millions of teachers and well over ten million students and remote residents in the rural areas of China were able to get medical advice, educational lessons, and practical training programs after thousands of remote Earth stations were installed in Chinese villages starting in the mid-1980s. The map below shows the initial locations for television receive only (TVRO) stations. This program started with only about 30 sites, but eventually grew to many thousands over

TABLE 3.1 Examples of Space-Based Services that Aid UN Goals for Sustainable Development. Prepared by Joseph N. Pelton, Ph.D. All Rights Reserved.

Goal	Telecom & Networking Sats	Broadcasting Sats & HAPs	Remote Sensing Sats	Meteoro-logical Sats	Navigation & Timing Sats	Solar Power Sats	Additional Information
No Poverty (Goal 1)	New jobs via telework, opportunity for remote services, training in remote villages	Broad distribution of information on birth control, nutrition, vaccines, etc.	Improved information to support fishing, farming, forestry, mining, etc.	Reduced losses of crops, housing, in-frastructure	Improved farming and fishing via precision geo-location	*In the Future* Lower cost of clean energy to rural and remote locations	Efficient production of mass products thru AI and additive manufactur-ing
Zero Hunger (Goal 2)	More efficient agricultural & fishing processes	Broad distribution of information on nutrition & birth control	More productive farming & lower cost food	Less crop loss due to unpredicted storms, flooding, typhoon, hurricanes	Improved farming and fishing via precision geo-location	TBD	Sea-based farming products for food

(Continued on next page)

TABLE 3.1 (Continued).

Goal	Telecom & Networking Sats	Broadcasting Sats & HAPs	Remote Sensing Sats	Meteorological Sats	Navigation & Timing Sats	Solar Power Sats	Additional Information
Good Health and Well-Being (Goal 3)	Tele-health and remote medical service	Broad distribution of information on birth control, nutrition, vaccines, etc.	Detection of crop or tree disease	Detection of solar flares & ozone holes	Ability to precisely track spread of disease & pandemics	TBD	
Quality Education (Goal 4)	Quality tele-education programs, remote testing programs	Educational radio & television, access to global news	TBD	Less destruction of schools & educational infrastructure	Cost savings on school transportation	*In the Future* Clean energy to remote locations for tele-education	Distribution of information, knowledge sharing in satellite design, operation, etc.

(Continued on next page)

Gender Equality (Goal 5)	Tele-educational programming	Global news and TV broadcasts	TBD	TBD	TBD	TBD	Creation of non-gendered jobs in countries currently without existing industry
Clean Water (Goal 6)	Tele-education on water purification & sanitation	Broadcasts on water purification & sanitation	Detection of polluted water; road access information for water trucks following disasters	Better protection of water reservoirs against storms	Locate polluted waters, storms with acid rain, etc.	TBD	

(Continued on next page)

TABLE 3.1 (Continued).

Goal	Telecom & Networking Sats	Broadcasting Sats & HAPs	Remote Sensing Sats	Meteoro-logical Sats	Navigation & Timing Sats	Solar Power Sats	Additional Information
Affordable and Clean Energy (Goal 7)	Tele-education, Internet access to create solar, wind, tidal, geothermal, etc. energy systems	Broadcasts on energy savings & building clean energy systems	Aid in finding good locations for wind farms, geothermal energy, & tidal energy	Aid in finding good locations for wind farms, geothermal energy, & tidal energy	Assist in location of renewable energy systems	*In the Future* Lower cost clean energy to cities, rural and remote locations	
Decent Work and Economic Growth (Goal 8)	Telework, village training, tele-banking, tele-services	Open university training	Aid to more productive mining, fishing, farming, forestry, & transport	Support new con-struction & design of infrastruc-ture related to climate change	Support for new construction & design of transporta-tion systems	*In the Future* Lower cost clean energy to cities, rural and remote locations	Develop space industry in countries without government or commercial space actors

(Continued on next page)

						In the Future	
Industry, Innovation and Infrastructure (Goal 9)	Tele-education, Internet based innovation, Internet-based technology incubators, protective security for infrastructure	Educational radio & television, access to global news	Aid to more productive mining, fishing, farming, forestry, and transport	Support to new construction & design of infrastructure related to climate change	Support to new construction & design of transportation systems	Lower cost clean energy to cities, rural and remote locations	Develop space industry and associated infrastructure in countries without existing government or commercial space actors
Reduced Inequalities (Goal 10)	Tele-education, Internet-based learning & data bases	Educational radio & television, access to global news	TBD	TBD	TBD	TBD	Small satellites can lead to new industry and produce marketplaces for traditionally less-sophisticated entities

(Continued on next page)

TABLE 3.1 (Continued).

Goal	Telecom & Networking Sats	Broadcasting Sats & HAPs	Remote Sensing Sats	Meteoro-logical Sats	Navigation & Timing Sats	Solar Power Sats	Additional Information
Sustainable Cities and Com-munities (Goal 11)	Substitution of tele-services and tele-work for physical transporta-tion	Educational radio & television, access to global news	Key topographic information for trans-portation, water and sewer planning	Key information related to city infra-structure protection from violent storms	Improved traffic & transporta-tion control	*In the Future* Lower cost clean energy to cities, rural and remote locations	
Responsible Consump-tion and Produc-tion (Goal 12)	Tele-education, tele-work satellite services can be provided worldwide	Broadcasts on tele-work, conservation, energy savings & building clean energy systems	Monitor hazardous waste locations, atmospheric pollution, oil spills, garbage scows, etc.	Note changes in weather & climate due to industrial activities	Accurately pinpoint sources of pollution	*In the Future* Lower cost clean energy to cities, rural and remote locations	

(Continued on next page)

						In the Future
Climate Action (Goal 13)	Tele-education, tele-work, satellite services can be provided worldwide	Broadcasts on tele-work, conservation, energy savings & building clean energy systems	Track ice-cap & glacier melting; measure ocean and air temperatures	Track changes in atmospheric temperatures, intensity of storms, solar activity	Pin point location of atmospheric & oceanic sensors	Lower cost of clean energy
Life below Water (Goal 14)	Tele-education, global internet access; track location of endangered species	Satellite TV & radio can strengthen education, civic activism, knowledge of law	Detection of water & ocean pollution, coral bleaching, fish depletion, etc.	Track ocean storms & hurricanes	Determine exact location of sensor & ocean buoys	TBD
Life on Land (Goal 15)	Tele-education, global internet access; track location of endangered species	Satellite TV & radio can strengthen education, civic activism, knowledge of environmental law	Track animals & endangered species	Monitor violent storms and provide flood & high wind warnings	Determine exact location information of earthquakes, volcanos, coordinate rescue operations	TBD

(Continued on next page)

TABLE 3.1 (Continued).

Goal	Telecom & Networking Sats	Broadcasting Sats & HAPs	Remote Sensing Sats	Meteoro-logical Sats	Navigation & Timing Sats	Solar Power Sats	Additional Information
Peace, Justice and Strong In-stitutions (Goal 16)	Low cost satellite telecommu-nication & Internet access can strengthen education, civic activism, and knowledge of law	Satellite broadcast TV & radio can strengthen education, civic activism, and knowledge of law	Time stamped remote sensing data has been used to prosecute crimes against humanity			TBD	International cooperation in space activities by various nations can lead to the creation of strong institutions; ability to monitor state activities can increase effectiveness of truth and confidence building measures

(Continued on next page)

Partnerships for the Goals (Goal 17)	Satellite manufacturers & service providers can help promote telework, tele-education and tele-health	Satellite manufacturers & satellite broadcasters can help promote tele-work, tele-education and tele-health			TBD	International cooperation in space can lead to mutual respect and ideological alignment that will motivate achieving other goals collectively

Figure 3.4   The black ovals indicate the remote Earth stations initially installed in China for Project SHARE in 1986. (From Intelsat Final Report on Project SHARE)

time. The Central China Television and the Ministry of Education produced most of the programming[10] (See Fig. 3.4).

Today there are a much wider range of satellite opportunities that can help people escape poverty. There are today millions of people in remote villages who are using a combination of Wi-Fi technology and satellite communications networks to communicate and to connect to the Internet. Many of the new so-called mega-constellations such as One Web, Starlink, and (soon) the new satellite networks owned by Amazon (Kuiper), China, and others will provide new levels of Internet, text, telephone, and video connectivity to remote areas around the world. These new systems will make a large impact not only in terms of tele-education and tele-health but also terms of in creating good paying telework and IT support jobs in remote parts of Africa, South and Central America, and Asia.[11]

[10]The Final Report on Project SHARE (1989), Intelsat, Washington, D.C.; see also Pelton, J. (1997). "Project SHARE and the Development of Global Satellite Communications." In Beyond the Ionosphere: Fifty Years of Satellite Communication. A. Butricia, ed. Washington, D. C.: NASA History Office. pp. 257-264. https://ntrs.nasa.gov/api/citations/19970026049/downloads/19970026049.pdf

[11]DNV Technology Outlook 2030, "Mega Constellations on the Horizon", https://www.dnv.com/to2030/technology/mega-constellation-satellites-on-the-horizon.html (Last accessed Nov 18, 2022).

This new form of satellite connectivity can be the first step out of global poverty. Satellites that can provide education and health care do not represent the only benefit. Connectivity can provide vital governmental services, and after training, this can even translate to telework that can even be provided by remote satellite services. In the 1980s American Airlines moved their inventory control operations from Tulsa, Oklahoma to the Dominican Republic after satellite facilities and remote training was offered by satellite connection. In the time of the COVID-19 Pandemic, the advantages of tele-work have been not only understood but very widely adopted, and the world is not going back. Satellite-delivered networking and communications to support education, training and increasing levels of tele-work have enabled a growing number of people to move out of poverty, and this will increase as these new and integrated capabilities are made available, but they must be made available quickly in the less developed economies, and not only in the wealthy and advanced nations. Steps must be taken to ensure that these capabilities and made available equitably.

3.7.2 Goal Number 2: No Hunger

The connection between space systems and combatting hunger is also strong. There are a host of ways satellite remote sensing and weather/meteorological satellites can be used to increase global food production. Early remote sensing projects like NASA's Large Area Crop Inventory Experiment (LACIE) and Agriculture and Resources Inventory Surveys Through Aerospace Remote Sensing (AgriSTARS) clearly demonstrated the power of regional agricultural monitoring and crop estimations over 40 years ago. Author Madry worked on the AgriSTARS project, and its potential for addressing world hunger were clear to us even then with the tech of the 1980's [3] [4, pp. 929–957]. The most significant change in agriculture today, is the development of so-called smart farming or precision agriculture. This involves the analysis of detailed soil and historical crop history data with hyperspectral satellite and local remote sensing data within a GIS environment for individual farm fields to determine exactly how much seed, water, pesticide, and fertilizer should be used for each square meter of a farm throughout a specific growing season for a specific crop. Sophisticated farms around the world today employ hi-tech GPS-equipped tractors and watering systems to automatically distribute seed, fertilizer, and water over their farms, producing equal or improved crop production with significantly lower costs for expensive, and environmentally damaging nitrogen fertilizers and pesticides, while using less precious water for irrigation. This is an excellent example of how space, airborne, IT, and GIS

Figure 3.5   The precision agriculture cycle (Image courtesy of USGS).

can be combined to both increase benefits (crop production) and reduce both the costs and harmful results (nitrogen fertilizers, pesticide, and increased water consumption). (See Fig. 3.5 [5])

In more rural areas, such information can be used with less precise methods for the distribution of water and fertilizers that do not require such advanced automation. Weather and meteorological systems now allow farmers and fishing operations to cope with changes in the weather, rainfall, and even longer-term climate variations to decide on a more scientific way what crops to plant, when to harvest and to take crops to market. Also, communications satellites linked to rural Earth stations and then on to local Wi-Fi systems can allow farmers, fishermen, and food processors and distributors to better understand market conditions for their crops, catches or processed food. This access can inform crop producers not only when to sell their food, but even allow them to sell to markets in other cities, states, or even other countries where demand (and profit) is higher. In short, remote sensing, PNT, and telecommunications satellites can serve to make agricultural, fishing, and processed food markets much more efficient. Producers of foodstuffs, including fishing operations can clearly become more profitable, and global markets can also be more efficient and equitable to both consumers and producers while reducing the environmental impacts of fertilizers and pesticides.

In short, various types of satellite services can truly increase market efficiency, reduce transportation costs for fertilizer or food, or make agricultural or fishing operations much more efficient. The benefits to farming and fishing are now quite long and impressive, but these primarily are benefiting large producers in major, developed markets and have not yet filtered down to smaller producers in the developing world. The technological changes mentioned above could drastically alter this. It should also be noted that exactly the same suite of technologies can be used for global forest management, although this is less fully developed.

There is another way that satellite telecommunications and networking satellites can help to eliminate future hunger and need that is less obvious and clearcut, but nevertheless of future importance. These satellite networks, especially the new mega-LEO constellations that can link rural and remote areas to the Internet, streaming radio and television, as well as via texting and telephone, can provide public education as to the benefits of family planning, ways to access free or low-cost birth control services, condoms, as well as dietary advice as to better ways to maintain healthy and cost-effective diets can make a huge impact on the world of the future. The economic rise of China and the increase in its national health among its citizenry over the past thirty years can be directly linked to improved family planning, birth control, and a reduction in hunger among its 1.2 billion population.

Indeed, in the arena of efforts to eliminate poverty and to reduce hunger, the use of satellite networks to provide information on family planning and birth control can help to forestall exponential population growth and help limit global population to 10 billion people rather than its possible explosion to as many as 12 billion people by the year 2100. The current UNUnited Nations (UN) forecast for the end of the century is 11.2 billion people.[12]

If one reviews the rapid growth of human population going back to 1800, the human experience in terms of rapid and, in fact, exponential expansion is clearly evident. There were only 800 million people on Earth in 1800. This grew to 1.8 billion people by 1900 and to around 6 billion in the 2000 and the day of 8 billion was reached just before years-end in 2022. Today there

---

[12]United Nations, "World population projected to reach 9.8 billion in 2050, and 11.2 billion in 2100", https://www.un.org/en/desa/world-population-projected-reach-98-billion-2050-and-112-billion-2100.

are 25 countries in Africa with annual birth rates that range between 4.5% per annum (i.e., Equatorial Guinea) and 7.15% (i.e., Niger). This type of exponential growth is unsustainable in the longer term, not only in terms of meeting goals for reduced poverty and hunger but many other environmental, educational, and health goals as well. There is an urgent need for a sustained global program to encourage family planning and to publicize this initiative globally through expanded health and educational programs is critical. These new programs need to utilize global space communications networks, linked in rural communications systems, and other relevant programming and do so on an urgent basis. The advantages of slowing population growth need to be explained globally via satellite, wi-fi, broadcasting systems and internet networks. If this is not done, the result by century end will very likely be a human population well above 11.2 billion and possibly as many as 12 billion. A new population goal as low as 10 billion versus an uncontrolled growth of humanity to 12 billion people represents a tremendous difference in terms of sustainability. The impact to 2 billion extra people in terms of expanded poverty and hunger, in terms of unmet health services and educational needs, and perhaps most impactful unsustainable environmental damages to Planet Earth will be staggering. A review of all 17 of the UN sustainable development goals will likely reveal adverse impacts that will come from overly rapid population growth in every walk of life for all of humanity. Zero population growth, or even better, some reduced numbers, would best support the U.N.s desire to achieve a more sustainable world. Satellite networks could provide a powerful tool to address these global overpopulation concerns.[13]

3.7.3   Goals Number 3 and 4: Good Health and Well-Being and Quality Education

These two goals are combined because the need to use space technologies to provide these vital tele-services are quite parallel. The power of tele-health, tele-education and expanded governmental services that global satellite telecommunications systems and especially the new networking capabilities of the new large-scale, low latency satellite constellations in low Earth orbit (LEO) and medium Earth orbit (MEO) can provide networking services with high efficiency to the unserved or under-served areas of the world. These networks, when combined with local village Wi-Fi type networks, can revolutionize telecommunications and networking services to billions of people in rural and remote areas. It is no accident that one of the first one of the satellite constellations was known as O3b. This stood for the Other Three Billion

---

[13]See Chapter 9 on Overpopulation in [6, pp. 153–163].

people on planet Earth that did not have reliable and easily accessible links to telecommunications and networking services that were available to the other four to five billion people on our planet.

Earlier it was noted how the Intelsat "Project SHARE" experiment in China opened the door to millions of rural and remote villages in that country. This was largely a one-way broadcast satellite broadcast system that provided health, nutritional, and educational services through broadcasted programming that was augmented in many cases with on-site instruction. Since the first attempts at these types of programs in the 1980s, great strides have been made to create interactive tele-health, tele-medicine, tele-education, tele-training and tele-work systems provided both by communications satellites and Internet-based health and educational services also provided by wireless systems and conventional terrestrial systems.

The provision of satellite-based tele-health, tele-medicine and tele-education has proved to be of particular relevance in the time of the COVID-19 pandemic, and in emergency disaster management and rescue operations. The obvious first need is to provide the necessary satellite and terrestrial links between the impacted area and those providing the medical treatment or educations services, but much more is also involved. There is the need to "harmonize protocols" associated with video, telephone, text, or internet sites on a technical level. There is also a need to find common staff procedures or practices between doctors, nurses, or educators as well as in some cases there is a need for translators or interpreters or simply someone who knows and understands local customs [7].

Human spaceflight has been a guiding force and technology demonstrator in tele-medicine, with NASA and other agencies providing tele-med services to astronauts from the earliest days of human spaceflight to the routine operations on the International Space Station and now moving forward for the future Artemis programs. Telehealth and telemedicine services to the Pacific Islands using L-band communications on the GOES-2 satellite via the PEACESAT operations from Hawaii, (see Fig. 3.6[14]), the University of the West Indies network to serve the West Indies Island, and the telemedicine projects providing services to remote areas of Australia and Canada, were among the first practical relayed telemedicine and telehealth operations. Other satellite functions such as Lifesat related medical journal articles and health and medical research articles to doctors and nurses in remote areas in the 1980s and 1990s, but today, satellites are often used to provide Internet connectivity to support both classroom education and remote medical and health-related

---

[14]The graphic is accessible on PEACESAT archives, on http://peacesat.hawaii.edu/ 4ORESOURCES/Library/Papers/images/FIG-02.JPG.

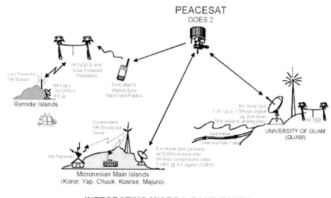

Figure 3.6   The PEACESAT Remote Island Service to South Pacific Islands (Graphic from PEACESAT archives).

services. Millions of people and students around the world depend on satellite connectivity to receive these services. Breakthroughs in satellite system design, low latency connectivity, and low-cost satellite receivers connected to Wi-Fi local coverage continues to expand these services globally. One of the challenges with regard to SDG Goals 3 and 4 is finding a way to locate and quantify all of the hybrid satellite and terrestrial/Wi-Fi networks that are allowing the rapid expansion of educational and telemedicine/telehealth services being provided globally, and especially in to rural and remote areas previously underserved or totally unserved at all.

3.7.4   Goal Number 5: Gender Equality

The contribution of technology and satellites, computer sciences and artificial intelligence is of course quite limited here, but it is still relevant. In fact, the lack of women in these fields are one of the key areas which have been a target for improvement for some time. The main contribution that satellites can make, of course, is with regard to global education to spread the word that a modern world must provide equal opportunities to all people. This means that satellite education to currently underserved remote and rural areas should be structured to advance a message of social equity. This would mean to not restrict the education, the careers, or participation of any person, and especially to open the door to all women and their advancement on merit and with equal access to all educational systems and all careers. Providing other SDGs such as improved education, health care, and employment will all benefit women around the world disproportionately.

3.7.5 Goal Number 6: Clean Water

This goal is closely linked to the environmental goals number 13, 14, and 15 that are related to climate-change mediation and additional protection to the Earth's land, the atmosphere, and the oceans. Acid rain, pollution of all types, the depletion of water in aquifers, the rapid melting of glaciers to create saltwater, leakage of water from storage facilities and pipelines, and more are all major global problems related to the lack of clean water in many parts of our planet. It is exactly all of these challenges to which remote sensing satellites with ever more sophisticated sensors with improved resolution (of all four types), in conjunction with ground-based sensor networks, GIS systems, and Decision Support Systems are best able to address. In short, well-designed applications satellites are critical to progress in this area. Today there is legitimate concern about climate change and danger to our atmosphere and oceans, global warming and the melting of the thousands of tons of ice per week. People are worried about climate change from many perspectives, but the shortage of potable water with an exponentially increasing global population may be the shortest fuse that lights the war of human survival across our small planet.

As noted earlier under Goal 2—with regard to hunger–the greatest hunger in coming decades may be for potable water. The most important solution may be to stem the tides of global overpopulation. It is here that wide-spread, low-cost, and carefully prepared satellite-based education—especially to rural and remote areas may provide an important tool to teach the world the dangers ahead and the need for responsible family planning. It is no small irony that on a planet that is bright blue when seen from space, with water covering over 70% of its surface, only 2.5% of that water is fresh, and over half of that is locked up in ice or otherwise unavailable to us in the ground. Only 1.2% of the Earth's water is available for human use. Perhaps a new SDG should be the practical, environmental, and cost-effective desalinization of sea water.

3.7.6 Goal Number 7: Affordable and Clean Energy

There is ambiguity in this goal in that "affordable and "clean" do not always go together. Hydro-electric and increasingly solar energy are both largely "green" and today both are largely affordable. Coal remains reasonably plentiful in supply, but is becoming more expensive and increasingly under fire for the pollution created, both in terms of gasses and poisonous slag refuse. It represents the dirtiest source of energy and there are mounting efforts to eliminate coal-fired power plants in many countries, but this must be countered by the plentiful supply and low-cost alternative and clean fuel sources. Atomic energy is, of course, very controversial in that it involves the as yet unsolved environmental problem of the disposal of radioactive waste. Indeed, at least three atomic power plants around the world (in Japan, the U.S. and Ukraine) have become environmental disasters that have involved over a trillion dollars in cleanup costs. In times of war, these facilities can also become prime targets of attack, and, as demonstrated in Japan, natural disasters can trigger nuclear disasters as well. Solar cell-derived energy represents a constant of renewable source energy which has become cheaper and cheaper. Fortunately, the Sun constantly keeps supplying more energy than all of humanity consumes–and by wide margins of orders of magnitude. Many thus hail energy taken directly from the Sun's rays as the most logical source of the energy of the future. Ocean and tidal turbines, wind turbines, geothermal, and ocean thermal energy conversion are perhaps alternative "green" choices, but they are not practical in many areas and their costs are much more fixed at this stage and are not declining as much as solar cell technology. They are also not predictable in output. The problem remains to find a timely and cost-effective way to make the transition from dirty energy–especially from outmoded coal plants and gasoline–to new "greener" sources. But the good news is that this conversion is underway.

Space systems can aid meeting this goal in several ways. One of the possibilities, that has been explored in depth for over a half century, is the creation of space-based solar power satellites. New developments in lower cost, easier to manufacture, and lighter weight solar cells are being developed all the time. Some produce higher energy efficiency conversion due to improved multiple PhotoVoltaic (PV) junction solar cells that can capture light in the higher energy ultra-violet end of the spectrum. Researchers in Australia announced in 2021 the first 30% efficiency solar panels, using perovskite and silicon tandem solar cells.[15]

Future quantum dot solar cells of improved efficiency may prove to offer even greater efficiencies in time. Lower cost reusable launch vehicles that can lift heavier payloads at lower cost are also becoming increasingly available. The idea is that solar power satellites can operate 24 hours a day and for most days of the year. This is because satellite located appropriately, such

---

[15]See   https://www.pv-magazine.com/2023/02/08/australian-scientists-achieve-30-3-efficiency-for-tandem-perovskite-silicon-solar-cell/.

as in geosynchronous orbit, are very rarely blocked from solar radiation. This type of space operation would allow energy to be continuously beamed down to Earth, and even to areas that are not very sunny. These space-based solar power satellite would be designed to convert solar light (photons) to radio power transmission down to ground or sea-based rectennas. A rectenna is a type of receiving antenna that is used for converting electromagnetic energy into direct current (DC). There are issues as to where such space-based solar power satellites would be placed in orbit and also there is controversy as to where the relatively large rectennas would be located. These round-shaped rectennas would need to be very large in area, because an overly concentrated microwave beam could potentially "fry" birds and even passengers on aircraft. Many such systems have been proposed and the technology to make it viable has increased in clarity and costs have come down, but none are yet operational.[16] The old space joke is that solar power satellites are the space technology of the future, and always will be. The old space joke is that solar power satellites are the space technology of the future, and always will be.

There are indirect ways to conserve energy so that it is never needed to be produced, stored or used. This would be accomplished by developing lower cost quantum computers, or by designing 'smarter cities' designed for telework so that many fewer people will need to commute to work. Further cars, trucks, buses and other vehicles will in time become all electric or use other energy efficient propellants. AI, robotics, and other computer related innovations could make many systems more efficient in terms of work requirements and energy use.

Of course, the role of space systems might not be to supply the energy, but to make energy systems "smarter' or more secure, or efficient. Power networks today are timed-coordinated and managed using PNT (positioning, navigation and timing) satellite systems such as the US GPS system, and dedicated power GIS systems also are key technologies. Power grids around the world are synchronized, and alternative current (AC) power networks are timed, using GPS signals. There are ways to install satellite interconnected internet of things (IoT) units that can constantly update the security of every element of an energy system through generators, transmission lines, transformers, atomic energy power plants and more. Further, there are cosmic threats from space such as solar storms and in particular coronal mass ejections (CMEs). Satellites in space are constantly monitoring our Sun, watching for dangerous solar

---

[16]Daniel Clery "Space-based solar power is getting serious—Can it solve Earth's energy woes?", Science, Oct. 19, 2022. doi: 10.1126/science.adf4118 (https://www.science.org/content/article/space-based-solar-power-getting-serious-can-it-solve-earth-s-energy-woes)

events that could disrupt our power grids and damage satellites in orbit. In this case "blocking capacitors" need to be installed to protect vulnerable transistors. In this scenario space system trackers could be installed to prevent a cascade of power surges and outages.[17]

3.7.7 Goal Number 8: Decent Work and Economic Growth

Satellites, computer improvements, automation, robotics, and artificial intelligence are serving on a global basis to reduce the number of employees needed for both service jobs and manufacturing. Even in agriculture and mining, significant automation has lowered employment needs across the world. Jobs in the satellite and computer world are well compensated and provide good health and retirement benefits, bur require education and skills. Satellite, computer, automation and AI technologies can become a road to environmentally cleaner, less demanding, and more beneficial employment conditions around the world, but it can also threaten global employment at the same time. These trends are hard to measure and are much too long in development to see fundamental change even by mid-century. There will clearly be economic growth fueled by technology and efficiency gains, but the nature of work will take a significant amount of time to take hold in the poorest economies with the highest unemployment and lowest wages. This is, in short, not the area where satellites and computers, in the shorter term, will have a major impact on jobs in developing economies within the next seven years—and indeed for decades to come, but the potential is there in the long term.

3.7.8 Goal Number 9: Industry Innovation and Infrastructure

This is another area in which space systems, computer advances such as in quantum systems and use of qubit encoding, robotics, and artificial intelligence is making significant strides forward and introducing innovation is of great importance. The important contribution in the space sector is the valuable new infrastructure that satellites are now adding to global capabilities. Decades ago, telephone and text messaging depended on the stringing of wire on telephone and telegraph poles. But today it is vital wireless systems involving satellites, wireless cellular systems and Wi-Fi capabilities that allows more and more people to communicate and receive broadcast media. Today there are over 10,000 television channels broadcast by satellites. Soon there

---

[17]NOAA, "Space Weather Conditions Space Weather Prediction Center", https://www.swpc.noaa.gov/communities/space-weather-enthuiasts (Last accessed Nov. 18, 2022).

will be tens of thousands of satellites that will support global connectivity to provide access to the broadband Internet and dozens of applications in the 6G wireless world. These wireless apps will support driverless cars, broadband wireless voice and texting, television and video streaming, and vital services such as search and rescue, police and military communications, emergency alert systems.

In the 19th and 20th centuries, infrastructure was rooted on the ground. In the 21st century more and more infrastructure related to communications, computer networking, Internet of Things (IoT) connectivity, security systems, broadcasting and entertainment will operate wirelessly and via satellite networks as an alternative to terrestrial wire and fiber systems. The important role of PNT systems for timing and navigation cannot be understated.

The challenge of the future will be to seamlessly interconnect vital infrastructure on the ground and those in the sky. In the future we will see a host of new space systems for improved communications, computer networking, and monitoring of storms and weather and climate change. We will also see enhanced capabilities with regard to space systems for positioning, navigation, and timing as well as global monitoring of frequencies for policing of communities and the high seas. One of the most powerful space technologies will be systems to provide connectivity for the control of transportation systems, pipelines, water and sewer conduits, and more. Thus, billions of "smart" IoT connected units can be linked by space systems to provide control and security for infrastructure within a GIS environment. In many cases the cost of creating new capabilities for wireless or satellite networks will be faster, less costly, and more flexible than ground-based systems that must be mounted on poles or buried in the ground. These new types of infrastructure are key to creating the world of tomorrow and space systems, supplemented by high altitude platforms, and unmanned aircraft systems are key tools to create the world of tomorrow in all parts of the world.

A fairly recent study by CITI bank of the space industry has quantified current annual space-related revenues at $424 billion (U.S.) for the year 2020, which is up some 70% from 2010. This report and other studies by Merrill Lynch have suggested that this growth of space systems will continue and produce annual revenues of $1 trillion (U.S.) as soon as 2040. It hard to identify which space assets can accurately be identified as infrastructure and quantified accurately in financial terms, but some would suggest that current satellite facilities in Earth orbit represent many hundreds of billions of dollars in economic value and that they indeed represent vital infrastructure for the global economy.[18]

3.7.9 Goal Number 10: Reduced Inequality

In this area, the strength of satellite communications in terms of tele-education can be used to create new awareness of global needs and the reasons for changes to the structure of the global economy to make the world more equitable. Satellite based television and radio networks also makes the world more interconnected and global economic goals more achievable. Satellite communications systems have made island nations able to sell their products for more money and also negotiate prices down to lower levels by seeking needed imports through international channels rather buying what ships bring in and thus can set the price for incoming goods as well as export items. Around the world, global satellite communications have let developing countries talk and collaborate directly with each other. No longer do global communications lines follow the pathways of former colonial rule. This is especially important in the global south.

A world where satellite telecommunications, networking and broadcasting provides direct links from all countries to all other countries, rather than connecting developing countries only through colonial capitals is a world that is more equal and where lingering inequalities are reduced.

3.7.10 Goal Number 11: Sustainable Cities and Communities

One of the tantalizing possibilities of the future is to integrate a wide nexus of space, computer, robotics, AI technology, smart energy, and transport systems into well planned, more responsive, and "greener" cities. Geomatics will play a major role here. Infrastructure management is much more a GIS problem that a remote sensing problem, and there are dedicated and capable systems today for the management of complex, urban infrastructures. The best way to start is not with the technology, but rather with citizen, government and business needs. This would be a city designed around how to make health care, education, transportation, and a host of other services more readily available, efficient and hopefully, lower in cost. In the future, cities need to be designed so that basic services, as well as social and communal needs, all work well together as a whole rather than a collection of parts.

---

[18]Michael Sheetz, "The space industry is on its way to reach $1 trillion in revenue by 2040 Citi says", CNBC, May 5, 2022. https://www.cnbc.com/2022/05/21/space-industry-is-on-its-way-to-1-trillion-in-revenue-by-2040-citi.html

The challenge is to find a way for urban communities to work economically, environmentally, peacefully together. Better communications can lower crime, and, in a social and cultural context, improved connectivity can bring a sense of contentment and even occasional joy to its inhabitants. Today the world, for the first time in history, is predominantly urban. Due to rapid population growth and job centralization, it may be 80% urban at the end of the century. For millennia, since the first agricultural revolution some 5,000 years ago, humans lived in rural settings and tended to be self-sufficient, and towns and large cities were the exception.

The reality of cities today is far from the ideal described above. Cities are often seen as being overcrowded, filled with smog, foul air, traffic jams, significant limitations when it comes to adequate education and health care, significant rates of crime, cultural and racial strife, poverty, and city governments struggling to try to meet the needs of the city. In many cases, the root cause is a surge of migration to the city to find jobs. Population projections for cities forecast that more than a half billion people will be moving to and inhabiting cities in the next twenty years. This means that perhaps a total of 2 billion more people with inundate the world's cities while net global population growth will be somewhere around 1.5 billion. If these projections are correct, the hope for sustainable cities will be very, very difficult to achieve [8].

New sustainable cities need to be created to relieve the congestion of the growing number of megacities of 10 million and more, as well as the even faster growing number of cities in the developing world, primarily located along the world's coastlines. These higher rates of growth are now apparently in all types of cities. This is true for cities in the 1-3 million range, in the 3-5 million range and in the 5-10 million range. The continuing influx of people in to the cities will undoubtedly overload the physical infrastructure and the key social and governmental services of these cities—particularly those in the economically developing world with annual birth rates in the 3% to 7% per annum range growth. The compounded birth rates will have staggering consequences that will frustrate efforts to create sustainable cities and a sustainable world.

Satellite systems will at least be able to monitor rates of urban growth and develop, vectors of growth, and produce a significant amount of data about areas of food production and rates of growth, levels of environmental pollution, increasing levels of traffic congestion, other critical factors such as changes to water supply in reservoirs, etc. Unfortunately, remote sensing satellites, positioning, navigation, and timing satellites, and other space networks cannot provide answers to the problems cited above. They can, however, identify the rates of change and help to ameliorate the most serious of these difficulties. The related technologies of Geomatics and Geographic Information Systems (GIS) can play vital roles in this regard, and imagery and the related AI and other tools, integrated with dedicated city management GIS systems, can improve city services and provide more efficient and robust infrastructures with the required redundancy needed for such rapid growth.

3.7.11 Goal Number 12: Responsible Consumption and Production

The keys to responsible consumption and production are different. Intelligent and improved consumption seeks to understand the implications and results that come from wasteful, single use, and wilful overconsumption. Environmentally enlightened economists have outlined the virtues of a circular economy that is based on recycling and reuse of consumable resources and satellite based tele-education can serve to instill these ideas in students and responsible citizenry. The rapidly emerging Internet of Things and its networks of enabled devices, linked to computer networks, can be used in a multitude of ways to further smart inventory of products and enable systems to promote appropriate recycling and reuse.

Production is the other side of the circle of production and consumption. Production must allow a circular economy to exist and indeed prosper. This means producers of soft drinks and water need to create reusable plastic or metal containers and a workable system for their recycling. This means, at the very least, that shoppers have to increase shop by means of re-usable tote bags for goods they buy. Every product, in fact, needs to be re-examined to see if it can be made fully or partially reusable, or made more permanently durable. It is here where big data analytics can be hugely productive to see how millions of products can be redesigned or reconstituted in terms of parts and ingredients to help create greener and recyclable products that meet consumer needs, and perhaps also can be designed to be more profitable in the longer term.

Finally, remote sensing satellites can help to inventory the world and its various sources of metal ores and mineral deposits, forests, rare earth minerals, chemicals, and other raw resources to develop a much better understanding of how many resources are being consumed and at what rate. This process, aided by big data analytics, can also monitor the world's atmosphere, waters, lands, mountains, and icecaps to create accurate models of increasing rates of pollution, loss of green space and rates of urbanization and dozens of other data points to measure the consequences of poorly conceived and harmful production on one hand and wasteful or dangerous consumption on the other hand.

3.7.12 Goals Number 13, 14 and 15: Climate Action, Live Below Water, and Live on Land

These three climate and environmental goals are very closely related and the space systems to address these three efforts to determine new and ongoing information about increasing levels of pollution, human impact on the world's flora and fauna, problems with fire, crop or tree disease, insect infestation, tidal waves, flooding, tsunamis, and much more will largely be addressed by complementary Earth sensing satellites, meteorological and weather satellites designed to track storms, lightning strikes, climate changes over time, and solar storms, radiation flares and coronal mass ejections. Various types of sensors in different orbits, on the ground and in the oceans, will be able to provide more precise data that can be processed in near-real time. Improvements in big data analytics will be able to detect various types of environmental problems, pollution, dangers to various species due to temperature changes and radiation, and other dangerous trends—and with greater precision. The ability of telecommunications satellites and data relay systems will provide information for data analytics at increasing rates and generate warnings and new "actionable information" more quickly. The addition of high-altitude platforms, unattended aircraft systems, and in-situ sensor networks will augment information in critical areas of particular concern.

The general trend is that there will be more and more capable satellite remote sensing systems linked to faster data systems, onboard data processing, higher resolution sensors and imaging equipment gathering data across more spectrum ranges, as well as improved and more cost-effective launch vehicle systems to keep the constellations populated. These new and improved space systems will churn out greater amounts of data to be interpreted and translated in a myriad of actions to make the world's environment safer and more protected. This will become one vital aspect of an integrated Geomatics capability for regional and global monitoring and the multiple applications involved in the SDGs.

The largest unknown is how this ever-increasing space-based torrent of data from the skies, as delivered by ever more versatile and precise space systems, will be translated from raw data to actionable information. Here the options are manifold and the pathways forward not well defined. Today, much "raw data" is never even processed, or is crunched by super computers on the ground to seek basic patterns and unseen relationships. Raw data, telemetered to the ground, must be quickly preprocessed (georectification and sensor calibration, for example), processed into the appropriate data such as land use or land cover. It must then be rapidly and automatically provided for integration with other data in the GIS environment, where various change detection or other analyses are conducted, including calibration by in-situ sensors. AI will then be harnessed to compare data over time, spot important trends, trigger alerts and alarms when needed, and generate actionable intelligence, in the form of maps, briefings, statistical summaries, and visualizations, all tailored for the specific situation and the specific customers, including governmental decision makers, academics, disaster responders, and more. We have a lot of work to do to get us to this point, but this is our task.

The development of key algorithms and signatures of various gases, pollutants, and crop or tree diseases over time may help to process data more precisely to identify environmental impacts. There remain several key questions. How will onboard data processing, AI and ML allow big data analytics to become more fully automated to be not only accurate, but also provide specific advice as to how to respond to this information? What activities will require human intervention for deeper understanding before actions are taken and what others can be automated? How will this be determined? Traditionally, there has been a need for so-called "ground truthing" for field verification of satellite data categorization, and also for detailed georectification to match exactly the coordinate reference system being used with appropriate precision depending on the scale of the study area. In the future, there will be so much information that machine-intelligence will tend to play an ever-greater role in decision-making. This leads to a host of moral, legal, ethical, policy, and regulatory issues yet to be completely understood, let alone resolved.

For years there have been arguments as to what sensing data can be sufficiently automated through pre-processing of data on board satellites so that data can be compressed, so that less data has to be transmitted to the ground or data relay satellite to reach processing centers and, ultimately, decision makers. What is clear is that in the age of satellite sensors where large scale satellite constellations are perhaps acquiring exabytes of data per minute–or ultimate perhaps per second—only quantum super computers and AI-based programs will be capable of performing the big data analytics need to process such vast quantities of raw and unprocessed data into useful and actionable information. Again, it is not only the imagery or image-derived data, but these must be integrated into a much larger Geomatics and decision support environment. This will be where actionable intelligence will be derived for decision makers. Who will decide who can decide? What do we do when the machines make a mistake? HAL[19] are you listening?

---

[19]The rogue computer from the movie "2001: A Space Odyssey"

Back some thirty years ago, NASA started a vast satellite remote sensing project to gather data about the environment and the planet with a network of remote sensing satellites. They called this effort "Mission to Planet Earth." After the satellites and their sensing capabilities were initially designed, the group that was assigned to create this new space system moved on to figure out how the data would be relayed to the ground and then processed to interpret the raw data. They soon realized that they would have to redesign the Earth Observation satellite constellation they had first conceived, because there was not the capacity to relay the data to processing centers at the needed speeds, nor the processing power to interpret the massive amounts of data received, without their system being completely overloaded.

Today the satellites are truly much more capable and the prospect of quantum computers and new qubit-based programming can conceivably cope with this Niagara of raw data. Nevertheless, questions about what level of pre-processing of data on board satellites are appropriate. The need for some level of ground-truthing and human level of analysis and debugging of automated big data analytical systems will remain. It is in these areas where designers will need to start with the most vital questions. They will need to ask what is the most critical data to be acquired and where and how can the best and most actionable data be determined so as to allow the most important remedial actions to be undertaken. AI will play a central role in this process.

The latest GOES Satellites operated by the US National Oceanic and Atmospheric Administration (NOAA) are equipped with precision lightning sensors that can allow detailed tracking of violent storms, but they also now confirm that due to climate change, and due to more energy in the atmosphere, there are now about 40% more lightning strike now occurring with violent storms. This has many implications about the increase in forest fires, the increasing intensity of storms, the rate of increase of climate change, and dozens of other important insights into our changing world. We are looking at the dawn of a new and exciting era of integrated space and ground capabilities, that has the potential for making progress in each of these three goal areas defined in the United Nations sustainable goal areas. What is clear is that in less than a decade's time we will know much more about the world's atmosphere, our seas, oceans, and aquifers, and about out landmasses than in recorded history. We can only hope that this knowledge will allow us to act to save planet Earth, or at least an Earth that can support humans.

We humans have been much better at designing, building, launching, and operating sensors, and computers than in using these to measurably impact political, economic, and social issues. We have created complex data systems to move, store, and analyze these petabytes—and soon exabytes–of data, but we still have a terrible time convincing our elected political leaders to act on the information that our sensors tell us requires urgent response. This problem takes us out of the engineering and technical world and into the complex and difficult domain of anthropology, sociology, and politics. We must improve our ability to influence in practical ways the public and our decision makers around the world with the data that we collect and process. These are not engineering questions, but human social questions, and, if we are to meet the SDGs, we must put more effort into understanding the 'last mile' of how political decisions are actually made, and how we can better learn to influence these processes with the data that we collect and process. They are very different domains, and the space community has largely failed in this to date. The move from ever more capable technology to cogent political and social action is the most critical link of all.

3.7.13   Goals Number 16 and 17: Peace, Justice & Strong Institutions and Partnerships for the Goals

These are important goals, vital to our collective future, but Space Systems, Big Data and Data Analytics are now key to enabling the achievement of these goals and thus no further comment on these two goals are offered here, beyond a statement that success in providing food, water, education, and jobs plus providing improved communications can play a significant supporting role. Pervasive remote sensing can play a vital role in monitoring environmental crimes, monitor peace accords, document war atrocities, and provide strong support for these goals.

3.8   CURRENT IMPLEMENTATION EFFORTS

The above discussion may have given the impression that success in using space systems, AI, quantum computing and big data analysis is ultimately dependent on future applications. This is not true. There are currently efforts underway to use space systems in new and impactful ways to meet the UN SDGs right now. The Alliance for Collaboration in the Exploration of Space (ACES Worldwide) is embarked on efforts to develop "Compact Agreements" that outline new efforts to use a variety of space systems and services to meet one or more of the UN sustainability goals.

In a pilot program for South Asia there are three Compact agreements in the process of being implemented. One Compact Agreement involves the use of remote sensing systems to provide for safe and sustainable fishing off the South coast of India and with the possibility that this will extend to Sri

Lankan applications as well. Another Compact Agreement involves expanded use of satellite telecommunications for expanded primary and post primary education in India. Yet another involves the use of satellite-generated data to support the growth of ocean-based agriculture to process "sea bio-growth" into inexpensive protein-rich food.

A paper is being presented on behalf of ACES Worldwide at the International Astronautical Congress in Baku, Azerbaijan in October 2023 in panel E-3 on Space, Policy, Regulations and Economics on these initiatives. This paper "Space & Sustainability Using Compact Agreements-A New Initiative to Realize the Space 2030 Agenda." provides the latest updates on this program. It is hoped that by that time we have expanded the scope of this initiative to other parts of the world. The plan is to expand this effort organically as additional partner organizations are added.

Another important thing that can be addressed immediately is the broader adoption and promotion of Open Source tools. In order to address these vital goals, we need the best and most capable minds around the world, and this includes the majority of the world's population who do not live in the affluent places such as North America and Western Europe. And most of the implementation of the SDGs will be done where English may not be the main language, where commercial space and IT software are not affordable, and where cost, upgradability, and regional customization are vital to success. The adoption and promotion of Free and Open Source Software (FOSS) in support of the SDGs is something that can be done now and that can lay the foundation for success. Open Source GIS, remote sensing, drone flight management, and in-situ sensor nets are now available, robust. These Open Source tools have multiple advantages, including the ability to see and alter the source code, the ability to customize the code and the user interface, the capability to construct user interfaces in local languages rather than only in English, and more. Perhaps most importantly, these new openly available tools do not require the expensive end user licenses that often prevent their use in developing nations. It lays a foundation for international collaboration and speeds the development curve. Adopt Open Source tools.

3.9 CONCLUSIONS

Space Systems, Big Data Analytics, Quantum Computing, Significant New Use of Qubit Programming, AI, ML, GIS, PNT, Decision Support Systems, and related capabilities, when combined together in creative ways, can be strong forces for changing our world for the better and actually achieving the SDGs. These types of technologies, like most intellectual tools, can be used to improve the quality of life, help save the world's environment for future generations, and sustain the long-term future of humanity. Or perhaps not. Experience has shown us that scientific data about our world—despite great detail and supporting information—has still not resulted in needed action to respond quickly enough to the global climate crisis we face. Scientists are better at generating data and conclusions than they are at convincing decision makers to act. Carbon dioxide, methane, and other greenhouse gases continue to mount to alarming levels. There are shrinking amounts of potable water in the world as human global population continues to rise at an alarming rate. There is no current way to alter irresponsible industrial decisions or reduce wasteful consumption. Many efforts aimed at coping with climate change dangers are being negated simply by a global population that may swell to as much as 12 billion by 2100.

Despite the breakthroughs in space systems, computers, AI and geomatics which are producing a new data about climate change, the response to this information has been dangerously slow. Our patterns of wasteful consumption, our ways of living with coal-fired electrical plants, gas-burning cars, and the ever-larger footprint of human habitation, at times seems irreversibly out of control. Ultimately, the most important thing is not more data on the environment. It is not more UN goals that are in some ways being disregarded and pushed down the road for others to deal with. It is not better data about what is going wrong with human overpopulation and the hazards of human-produced wastes, pollution, and dangerous products.

What is needed is a global understanding that the world's scientists are telling us that the time for serious action is now. Not in ten years, not in twenty years, not in fifty years, but now. Our space sensing satellites and our computers on the ground are telling us one simple, profound message. Most of our actions around the world are undercutting the longer-term sustainability of humanity, and we will never meet these or future sustainable goals, if we continue to act as we have since the beginning of the industrial age. New ways forward are needed.

Ultimately, gathering more and better data will be irrelevant if we cannot find ways to translate these into meaningful actions and better decision making. But better data and better ways of analyzing the data are a start, and the improvements presented in this chapter will an important part this process.

ACRONYM

AI Artificial Intelligence

ACES Alliance for Collaboration in the Exploration of Space

API Application Programming Interface

CME Coronal Mass Ejection

DC Direct Current

DSS Decision Support Systems

ENIAC Electronic Numerical Integrator and Computer

EPFL École Polytechnique Fédérale de Lausanne

HAPS High Altitude Platform Systems

ICT Information and Communication Technology

IoT Internet of Things

GIS Geographic Information Systems

GPS Global Positioning System

GUI Graphical User Interface

FOSS Free and Open Source Software

LACIE Large Area Crop Inventory Experiment

LEO Low Earth Orbit

MEO Medium Earth Orbit

NASA National Aeronautics and Space Administration

NDVI Normalized Differential Vegetation Indices

NOAA National Oceanic and Atmospheric Administration

RS Remote Sensing

SDGs Sustainable Development Goals

SHARE Satellites for Health and Rural Education

SPS Solar Power Satellites

TBD To be defined

TVRO Television Receive Only

O3b Other Three Billion

PNT Positioning, Navigation and Timing

PV PhotoVoltaic

UAS Unmanned Aircraft Systems

UN United Nations

US United States

USGS United States Geological Survey

GLOSSARY

6G: in telecommunications, 6G is the sixth generation mobile system standard currently under development for wireless communications technologies supporting cellular data networks.

AgriSTARS: NASA's remote sensing crop monitoring projects mission, Agriculture and Resources Inventory Surveys Through Aerospace Remote Sensing

Big Data: primarily refers to data sets that are too large or complex to be dealt with by traditional data-processing application software

Blue Brain: the Blue Brain Project is a Swiss brain research initiative that aims to create a digital reconstruction of the mouse brain. The project was founded in May 2005 by the Brain and Mind Institute of École Polytechnique Fédérale de Lausanne (EPFL) in Switzerland.

COVID-19 Pandemic: the COVID-19 pandemic, also known as the coronavirus pandemic, is an ongoing (–2023) global pandemic of coronavirus disease 2019 (COVID-19) caused by severe acute respiratory syndrome coronavirus 2 (SARS-CoV-2).

EINAC: the Electronic Numerical Integrator and Computer or EINAC was the first programmable, electronic, general-purpose digital computer, completed in 1945.

Intelsat: a multinational satellite services provider, owning and managing a constellation of communications satellites providing international telecommunications and broadcast services.

Internet of Things (IoT): the emerging technology where billions of objects, 'things' have embedded sensors, software, and connectivity, creating a globally interconnected web of sensors and data.

megaLEO: a new generation of satellite constellations in Low Earth Orbit (LEO) consisting of thousands of interconnected small satellites. The Starlink system by SpaceX is planned to consist of some 42,000 satellites, and there are several others planned. These create new issues relating to tracking of space objects, space debris, and problems of light pollution for astronomical observation.

SHARE project: the Intelsat System to celebrate its 20th anniversary donated satellite service capacity to countries and organizations seeking to test the use of satellites for rural tele-education or tele-health services. SHARE stood for "Satellites for Health and Rural Education" Many dozens of demonstrations were undertaken involving over 100 countries.

Starlink: a LEO satellite internet constellation operated by SpaceX, providing satellite Internet access coverage globally. It also aims for global mobile phone service.

Space 2030 Agenda: the UN. Committee on the Peaceful Uses of Outer Space (COPUOS) and the UN Office of Outer Space (UN OSSA) has provided support to the UN Organization in the development of the U.N. Sustainable Development Goals and explored ways that space systems could support the attainment of many of the 17 goals set for 2030. The U.N. General Assembly approved the Space 2030 Agenda that supported these objectives. The UN OOSA is required to report progress that has been made in using space systems to achieve this goals starting in 2025.

Other Three Billion O3b: an ambitious new satellite network that was created by entrepreneur Greg Wyler in partnership with the SES satellite corporation of Luxembourg. The medium Earth orbit (MEO) constellation was successfully deployed and operated by SES. SES then bought out and attained full ownership of this satellite network. It is now deploying an even more powerful global satellite networking system, also in MEO, which is known as mPower.

UN Millennium Goals: the United Nations established 17 a number of far-reaching development goals at the start of the 21st century for 2015. These objectives were known as the Millennium Goals These ambitious goals were not met by this deadline. The General Assembly in 2014-15 accordingly established new goals for 2030 now know as the Sustainable Development Goals.

Qubit processing: Quantum Information Processing focuses on information processing and computing based on quantum mechanics. While current digital computers encode data in binary digits (bits), quantum computers aren't limited to two states. They encode information as quantum bits, or qubits, which can exist in superposition. This means that one qubit can represent 2 bits, two qubits can represent four bits, three qubits can represent 8 bits and thus increase by or two to the nth power or 2^n

FURTHER READING

Becskei, A. and Serrano, L. (2000) "Engineering stability in gene networks by autoregulation". *Nature*, 405: 590–593.

Madry, S. (2019) "Disruptive Space Technologies and Innovations: The Next Chapter". *Springer Nature*, Geneva. 252 pages.

Rosenfeld, N., Elowitz, M.B., and Alon, U. (2002) "Negative auto-regulation speeds the response time of transcription networks". *J. Mol. Biol.*, 323: 785–793.

Savageau, M.A. (1976) "Biochemical Systems Analysis: A study of Function and Design in Molecular Biology". *Addison-Wesley*. Chap. 16.

Savageau, M.A. (1974) "Comparison of classical and auto-genous systems of regulation in inducible operons". *Nature*, 252: 546–549.

Viggiano, G. (2023) "Convergence: Artificial Intelligence and Quantum Computing". *Wiley Press*, N.Y.

BIBLIOGRAPHY

[1] Kurzweil Ray. How to create a mind–the secret of human thought revealed. *Viking, New York [USA]*, 2012.

[2] Meisam Amani, Arsalan Ghorbanian, Seyed Ali Ahmadi, Mohammad Kakooei, Armin Moghimi, S Mohammad Mirmazloumi, Sayyed Hamed Alizadeh Moghaddam, Sahel Mahdavi, Masoud Ghahremanloo, Saeid Parsian, et al. Google earth engine cloud computing platform for

remote sensing big data applications: A comprehensive review. *IEEE Journal of Selected Topics in Applied Earth Observations and Remote Sensing*, 13:5326–5350, 2020.

[3] Scott Madry. Precision agriculture and forestry: Bytes for bites. *Handbook of Small Satellites: Technology, Design, Manufacture, Applications, Economics and Regulation*, pages 1–29, 2020.

[4] Joseph N Pelton and Scott Madry. *Handbook of Small Satellites: Technology, Design, Manufacture, Applications, Economics and Regulation*. Springer, 2020.

[5] A Comparetti et al. Precision agriculture: past, present and future. *Agroinzinerija ir energetika*, 16:216–230, 2011.

[6] Joseph N Pelton. *Space Systems and Sustainability: From Asteroids and Solar Storms to Pandemics and Climate Change*. Springer Nature, 2021.

[7] Shashi Bhushan Gogia. *Fundamentals of telemedicine and telehealth*. Academic Press, 2019.

[8] Joseph Pelton, Indu Singh, et al. Smart cities of today and tomorrow. *Cham: Springer International Publishing*, 2019.

III

Applications & Use-Cases

CHAPTER 4

Neuromorphic Computing and Sensing in Space

Dario Izzo

Advanced Concepts Team (ACT), European Space Research & Technology Centre (ESTEC), Keplerlaan 1, 51014 AG Noordwijk, The Netherlands

Alexander Hadjiivanov

Advanced Concepts Team (ACT), European Space Research & Technology Centre (ESTEC), Keplerlaan 1, 51014 AG Noordwijk, The Netherlands

Dominik Dold

Advanced Concepts Team (ACT), European Space Research & Technology Centre (ESTEC), Keplerlaan 1, 51014 AG Noordwijk, The Netherlands

Gabriele Meoni

Advanced Concepts Team (ACT), European Space Research & Technology Centre (ESTEC), Keplerlaan 1, 51014 AG Noordwijk, The Netherlands
Φ-lab, European Space Research Institute (ESRIN), Via Galileo Galilei, 1, 00044 Frascati RM, Italy

Emmanuel Blazquez

Advanced Concepts Team (ACT), European Space Research & Technology Centre (ESTEC), Keplerlaan 1, 51014 AG Noordwijk, The Netherlands

CONTENTS

4.1   Introduction 108
4.2   Spiking neural networks 110
4.3   Learning algorithms for SNNs 114
4.4   Existing hardware 116
4.5   Application to onboard processing for Earth observation 119
        4.5.1   Rate-based SNNs for onboard Earth observation 121

DOI: 10.1201/9781003366386-4

	4.5.2	Spike-time-based SNNs for onboard Earth observation	124
4.6		Neuromorphic sensing	125
	4.6.1	Structural and functional organisation of the retina	126
	4.6.2	Adaptation and homeostasis in the retina	128
	4.6.3	Retinomorphic models	129
	4.6.4	Sparse convolution	130
	4.6.5	Existing silicon retinas	137
	4.6.6	The dynamic vision sensor (DVS)	137
	4.6.7	Why in space?	140
	4.6.8	Landing with events	140
	4.6.9	From events to egomotion	142
Bibliography		....	145

T HE recent success of AI comes at the price of high data and power demands. However, this success also rekindled a lot of interest in revisiting biology to explore and develop novel bio-inspired sensor and computing architectures, promising to not only reduce the power and data footprints of current technologies, but also uncover novel paradigms and technologies altogether. In this chapter, we review the potential opportunities of emerging neuromorphic technologies for edge computing and learning in space, focusing on preliminary results in the areas of event-based sensing, spiking neural networks and neuromorphic hardware for onboard applications.

4.1 INTRODUCTION

The term "neuromorphic" refers to systems that are closely resembling the architecture and/or the dynamics of biological neural networks [1,2]. Typical examples would be novel computer chips designed to mimic the architecture of a biological brain, or sensors that get inspiration from, e.g., the visual or olfactory systems in insects and mammals to acquire information about the environment. This approach is not without ambition as it promises to enable engineered devices able to reproduce the level of performance observed in biological organisms – the main immediate advantage being the efficient use of scarce resources, which translates into low power requirements. Nowadays, the neuromorphic approach is mostly investigated at two levels (i) algorithmic and (ii) hardware. On the algorithmic level, it leverages spike-based processing and training [2] to build novel machine learning pipelines able to process data efficiently. At the hardware level, the neuromorphic approach is pursued in designing novel analog and digital circuits and computer chips inspired by biological neural systems. This results in novel sensing devices believed to produce particularly good candidates to emulate biological vision, as well as in the design of computer chips dedicated to efficiently implement the spike-based systems just introduced. In fact, due to the discontinuous nature of spike-based communication and the temporal dynamics of spiking neurons, simulating the behaviour of a whole network of spiking neurons on conventional computer hardware is computationally – and thus energy-wise – very inefficient. This has now also a close relative in the field of artificial intelligence (AI) where Geoffrey Hinton has recently introduced the concept of "Mortal Computation" [3]: a form of computing where no separation between software and hardware exists. In "Mortal Computation", neural network solutions are uniquely tied to their underlying (analogue) hardware substrate, which Hinton argues (according to us correctly) might be the only way of obtaining large-scale neural networks that are energy-efficient – a concept that is closely following the neuromorphic paradigm.

The emphasis on low power and energy efficiency of neuromorphic devices is a perfect match for space applications. Spacecraft – especially miniaturized ones – have strict energy constraints as they need to operate in an environment which is scarce with resources and extremely hostile [4]. Numerous works have been investigating different energy-efficient solutions, especially leveraging commercial off-the-shelf (COTS) devices, aiming at optimizing model performance and energy usage trade-offs [4–8]. Much less have investigated neuromorphic devices. Early work [9, 10] performed in 2010 at the Advanced Concepts Team (ACT) suggested considering a neuromorphic approach for onboard spacecraft applications. Focusing on optic flow detection [11, 12], these preliminary works showed the possibility to safely land a spacecraft on an unknown planetary surface, assuming a neuromorphic approach to sensing based on the Elementary Motion Detector (EMD) [13], a device inspired by the visual system in flying insects. More recently, with the availability of new neuromorphic sensors such as the Dynamic Vision Sensor [14, 15] and chips such as Loihi [16], TrueNorth [17], Akida and others, the interest on neuromorphic architectures for spacecraft missions grew considerably.

Another significant potential advantage of neuromorphic hardware for space applications concerns radiation. Earth's atmosphere and magnetic field protects us from a lot of the cosmic radiation, but this poses a considerable problem even in relatively low orbits. While in many instances it can damage the actual hardware, radiation can also interfere with its operation (for instance, by flipping bits in memory), leading to software failure. Neuromorphic hardware can potentially mitigate these issues since, apart from intermittent spikes, it is in fact mostly silent. [18].

In addition to processors, also event-based cameras have contributed to the growing interest in the field of neuromorphic engineering. Event-based vision sensors are well equipped for operation in space: they have a very high dynamic range (on the order of $120dB$), respond only to moving segments in the visual field with very low latency and, most importantly, consume very little power due to their sparse output (on the order of mW). Naturally, these advantages come at a price – for instance, very high noise in very dark environments and proportionally low fidelity for slow-moving objects in the visual field.

In Section 4.6, we present several research lines that aim to harness these advantages and mitigate the downsides of event-based vision. Before diving into neuromorphic sensing though, we will first give a brief introduction to neuromorphic algorithms (Sections 4.2 and 4.3) and neuromorphic hardware platforms (Section 4.4) to then discuss past and current research mainly conducted by the ACT on evaluating the feasibility of a neuromorphic approach for onboard AI applications (Section 4.5). We hope that this chapter will stimulate further research pursuing a neuromorphic approach to spacecraft onboard computation and sensing.

4.2 SPIKING NEURAL NETWORKS

Arguably, the feature which is found most often in modern neuromorphic algorithms is spike-based communication. In the mammalian brain, neurons communicate with each other using action potentials ('spikes' or 'events') – electrical pulses with a stereotypical shape where only the time at which the spike occurs carries information (Fig. 4.1A). This realizes a highly sparse, and hence energy efficient, computing paradigm, as neurons only actively change their internal state when excited by an incoming spike, remaining passive otherwise.

Mathematical models of spiking neurons differ substantially in terms of computational complexity and biological realism. A widely used model (both in terms of algorithms and neuromorphic hardware) is the Leaky Integrate-and-Fire (LIF) neuron model, which adequately balances complexity and realism. The LIF model represents a biological neuron as an RC-circuit[1] with capacitance C and resistance R. The capacitor maintains an electric voltage u (the 'membrane potential') and is charged by an electric current I on a

---

[1]I.e., an electrical circuit with a resistor and capacitor coupled in parallel.

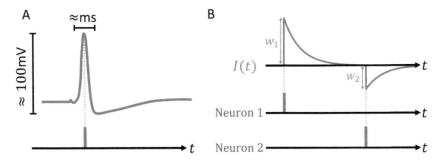

Figure 4.1 **(A)** Illustration of an action potential or sodium spike [19], an electrical pulse that is sent through the axon, which acts as the 'output cable' of a neuron, as a signal to other neurons. For computational purposes, it can be reduced to simply the time of its occurrence (bottom). **(B)** Presynaptic spikes trigger currents $I(t)$ at the postsynaptic neuron, with strength and direction depending on the interaction strength w_i (commonly referred to as synaptic *weight* in ANN research).

characteristic time scale $\tau = R \cdot C$ (the 'membrane time constant'):

$$\tau \frac{\mathrm{d}}{\mathrm{d}t} u(t) = -u(t) + R \cdot I(t). \qquad (4.1)$$

More specifically, $u(t)$ represents the potential difference between the inside and outside of the neural cell caused by different ion concentrations. Via R, ions can 'leak' through the membrane until an equilibrium potential is reached (here, the equilibrium potential is set to 0 for simplicity). The current $I(t)$ is caused by the spikes of source ('presynaptic') neurons (labelled with indices $i \in \mathbb{N}$ here), which connect to the target ('postsynaptic') neuron through synapses with interaction strengths $w_i \in \mathbb{R}$. In the absence of spikes, $I(t)$ decays to 0 and increases or decreases only whenever a presynaptic spike arrives, with strength and polarity depending on the value and sign of w_i (Fig. 4.1B). The postsynaptic neuron emits a spike at time t_s when the membrane potential crosses a threshold value[2] ϑ, $u(t_s) = \vartheta$ (Fig. 4.2A). In biological neurons, the ability to spike is diminished for a short period of time (known as the 'refractory period') immediately after spiking, as the mechanism responsible for creating action potentials has to recover first. This is often modelled by clamping the membrane potential to a reset value for the refractory period (Fig. 4.2A, yellow shaded areas). Without the leak term $-u(t)$, the model is reduced to the Integrate-and-Fire (IF) neuron model.

---

[2]This threshold is not physically manifested in biological neurons, but it is a reasonably good approximation for modelling their response behaviour [20].

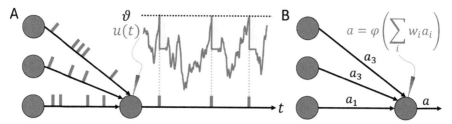

Figure 4.2 Comparison of SNNs and ANNs. **(A)** In a LIF-based SNN, pre-synaptic spikes (left, blue) lead to fluctuations in the membrane potential of the postsynaptic neuron (right, orange). When the membrane u reaches a threshold value ϑ, a spike is emitted and the membrane potential is clamped to a reset value for a period of time during which subsequent spiking is impossible (area highlighted in yellow). **(B)** In an ANN, real-valued activations a_i are multiplied by weights w_i and summed up at the postsynaptic neuron. The output a of the postsynaptic neuron is a non-linear function $\varphi(\cdot)$ of this sum.

The realism of the LIF model can be further increased by extending Eq. (4.1) with additional terms. For instance, the Adaptive Exponential LIF model (AdEx) [21] further adds action potential shapes to $u(t)$ as well as spike-rate adaptation, where prolonged tonic spiking (i.e., permanent activity) of a neuron is preceded by a period of increased or decreased activity. The temporal dynamics of AdEx neurons is capable of replicating a broad range of dynamics observed in biological neurons. In particular, it has been found that adaptation is employed by neurons for solving temporal tasks such as sequence prediction (see, e.g., [22, 23]).

In principle, every architecture that exists for ANNs can be converted into a spiking neural network (SNN) by replacing artificial neurons with spiking neurons, e.g., LIF neurons. However, ANNs and SNNs are different in two key aspects: (i) as discussed above, spiking neurons are characterised by a dynamic internal state, while artificial neurons possess no intrinsic state or dynamics, and therefore have no intrinsic 'awareness' of time; and (ii) spiking neurons interact by triggering currents at specific times through spikes, while artificial neurons communicate directly with each other via real-valued activations (Fig. 4.2B). Even though the event-based nature of SNNs makes them potentially more energy efficient than ANNs, it is still an open question whether it also provides other advantages (e.g., in terms of performance, robustness to noise or training time). A key challenge hereby is identifying how information can be efficiently encoded in the temporal domain of spikes [24].

TABLE 4.1 Different ways of encoding information using spikes.

Encoding scheme	Description
Time-To-First-Spike (TTFS) [25]	Given a stimulus onset at time t_0, information is encoded in the time required for *any* neuron in a population of neurons to emit its first spike (Fig. 4.3A). This encoding scheme allows fast processing (i.e. low latency) and is highly energy efficient since each involved neuron spikes only once at the most in order to solve a given task.
Rank order [25]	Information is encoded by the order in which spikes occur in a population of neurons given the onset of a stimulus (Fig. 4.3B). Thus, in contrast to TTFS, the exact spike time becomes irrelevant, but information can still be processed using only a low number of spikes depending on how many of the earliest spikes are used for encoding the input.
Spike patterns	Instead of only using single spikes to represent information, whole spike patterns of individual neurons or populations can be used [26, 27]. For instance, in [28], symbolic information (e.g., abstract concepts) is represented by spike trains, and relatedness between concepts is encoded in the similarity of spike trains (Fig. 4.3C).
Sampling	In neural sampling, neurons represent sampled values of binary random variables (refractory after spiking $= 1$, non-refractory $= 0$), allowing them to encode probability distributions (Fig. 4.3D). This is suitable for approximating, e.g., sampling-based deep learning architectures such as restricted Boltzmann machines [29, 30].
Bursts	Information can be encoded in bursts, i.e., short periods of high spike activity (Fig. 4.3E). For instance, in [31], bursts have been used to propagate information through SNNs that guides learning.
Rates	An instantaneous spike rate obtained by averaging over spiking activity (either over time or populations) carries the information (Fig. 4.3F). Since the spiking rate is a continuous variable, it encodes information in a similar way to ANNs.

A selection of proposed spike-based encoding schemes addressing this question is summarized in Table 4.1 and illustrated in Fig. 4.3.

In general, for machine learning applications it would be preferred to have a learning algorithm that automatically finds the optimal (combination of) encoding schemes. Although training SNNs has been a daunting task for a long time, recent progress in terms of both theory and software infrastructure has enabled exactly such end-to-end learning for SNNs, opening novel opportunities for building highly efficient and powerful spike-based AI systems.

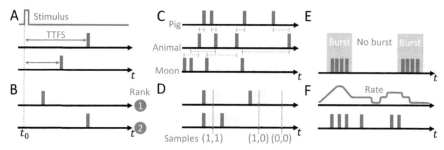

Figure 4.3   An illustration of spike-based encoding schemes. **(A)** TTFS encoding. **(B)** Rank order encoding with a population of two neurons. **(C)** Example of how concepts can be encoded and decoded (via similarity) in spike trains. Here, the spike train representing 'Pig' and 'Animal' are similar, while 'Moon' and 'Animal' are dissimilar. For clarity, the differences between spike times are shown in gray. **(D)** Encoding as random samples. At every point in time, a sample can be read out from the network, with neurons being in state '1' while refractory (yellow area) and in state '0' otherwise. **(E)** Encoding with bursts, short and intense periods of spike activity. **(F)** Rate-based encoding.

4.3   LEARNING ALGORITHMS FOR SNNs

One of the main enablers of the incredible success of deep learning in recent years is the error backpropagation algorithm ('backprop'). However, for a long time, this success did not pass over to SNNs – mostly due to the threshold mechanism of spiking neurons that leads to vanishing gradients at all times except at the time of threshold crossing [32] (cf. Fig. 4.4 for an illustration of the concept of surrogate gradients, which present one solution to this problem). Instead, to apply SNNs to a variety of machine learning applications, it was customary to convert the parameters of a trained ANN to a SNN [33]. SNNs 'trained' this way typically show a significant degradation in performance compared to the original ANN. In addition, the mapping promotes purely rate-based encoding in the SNN which has been found to reduce the energy efficiency of SNNs [34, 35]. An alternative approach is to use biologically inspired learning rules, such as Spike-Time-Dependent Plasticity (STDP) [36] or variants thereof, although they do not scale well beyond shallow networks[3].

---

[3]As of late, several biologically plausible learning rules that are applicable to deep neural architectures have been proposed [37]. However, they lack the flexibility and theoretical guarantees of backprop, i.e., they are not proven to minimize a task-dependent (and customizable) loss function. Nevertheless, they present intriguing alternatives for training SNNs.

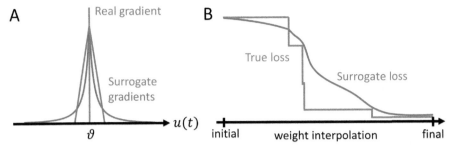

Figure 4.4   Illustration of the surrogate gradient method. **(A)** The gradient of the threshold function is a Dirac delta peak, which is infinite at $u = \theta$ and vanishes otherwise (gray). In the surrogate gradient approach, this gradient is replaced by a well-behaved function, such as a mirrored Lorentz function (orange) or a triangular function (blue). **(B)** In a SNN, the landscape of the loss function has discontinuities due to the spiking mechanism (gray). Especially long plateaus with constant loss inhibit learning. Using surrogate gradients 'smoothens' the loss landscape (blue trace). Here, the loss landscape is illustrated by interpolating between two sets of weights of a SNN. Reproduced based on [32].

Recently, several approaches have been found for successfully unifying SNNs and backprop, which have become the *de facto* state of the art for training SNNs on machine learning tasks and are currently being investigated at the ACT. For instance, one approach is based on SNNs using neuron models where the TTFS is both analytically calculable and differentiable, allowing gradients to be calculated exactly without having to deal with the discrete nature of the threshold mechanism [38–41], although this method is limited to neurons that only spike once.

At the time of writing, the most widely used approach is the so-called "Surrogate Gradient" method, which is applicable to all kinds of spiking neuron models [32, 42]. Here, the gradient of the threshold function is replaced with a surrogate function that has non-zero parts for membrane potential values away from the threshold (Fig. 4.4A). This 'softening' of the spike threshold gradient function allows gradient information to flow continuously through the network (Fig. 4.4B), enabling end-to-end training of SNNs capable of utilizing the temporal domain of spikes to find, e.g., highly sparse and energy efficient solutions. For instance, in [43] SNNs are trained that reach a competitive classification accuracy ($\approx$ 1.7% test error) on the MNIST handwritten digits dataset [44] with, on average, only 10 to 20 spikes per inference. Surrogate gradients can be used to optimise not only the weights but all parameters that influence the dynamic behaviour of spiking neurons, such as time constants and spike thresholds of individual neurons – with potential benefits for the robustness and expressiveness of SNNs [23, 45]. A major downside of the surrogate gradient method is that currently there is no theoretical framework for choosing the shape of the surrogate gradient function. Nevertheless, initial evidence suggests that the approach is relatively robust with respect to this choice [43]. In the future, this downside could be alleviated through exact derivations of gradient-based learning rules for SNNs (see, e.g., [46]). Additional information on the current standards for training SNNs can be found in [32, 47].

To summarize, the surrogate gradient method enables end-to-end training of SNNs using error backpropagation. Several open-source packages are available that standardize and ease the implementation and training of SNNs by utilising existing libraries for automatic differentiation. Currently, four general SNN libraries based on pyTorch are being developed: Norse [48], spikingjelly [49], snnTorch [47] and BindsNET [50]. The first three adopt both the workflow and class structure of pyTorch, effectively extending it to support SNNs. The last one is geared towards developing machine learning algorithms that take inspiration from biology. Even though it does not support gradient-based learning, it contains a larger variety of neuron models and biologically inspired learning rules than the other three packages. In general, the emerging landscape of SNN libraries greatly reduces the development time of SNN-based algorithms, painting a promising picture for further exploration of the capabilities and potential benefits of SNNs. In addition, it increases the accessibility of SNNs to researchers outside the field of neuromorphic computing and computational neuroscience by making it possible to seamlessly exchange or interweave ANNs and SNNs in a common framework. Thus, in the coming years we can expect to see an increasing number of contributions from the aerospace field attempting to incorporate spike-based algorithms and hardware onboard spacecraft.

4.4 EXISTING HARDWARE

Although end-to-end training of SNNs is possible nowadays, exploring their capabilities remains challenging due to the computationally demanding nature of simulating the internal neuron dynamics of spiking neurons. In contrast to artificial neurons, the dynamics of spiking neurons have to be solved using numerical solvers for ordinary differential equations, introducing a significant overhead. Thus, on conventional hardware systems, the benefit of spike-based coding (in the absence of spikes, no active computations are performed) is not immediately apparent.

An emerging technology that is capable of harnessing the potential of spike-based information processing is neuromorphic hardware, which explores novel computing architectures and paradigms that closely emulate how the brain processes information. Standard computer architectures separate processing units and memory (storing data and instructions for the processor), introducing a bottleneck due to the constant flow of data between memory and processor (the so-called 'von Neumann bottleneck'). Instead, neuromorphic hardware follows several design philosophies that can be found in the mammalian brain.

- No separation between processing and memory. Thus, a processing unit (e.g., a neuron) can only access information that is locally available (e.g., synaptic weights or activity of other neurons that it is connected to);

- Large-scale parallel computing;

- Asynchronous instead of clocked computations;

- Event-based information processing (i.e., using spikes instead of continuous values);

- (Re)programming the chip consists of (re)mapping networks or adapting network parameters through learning;

- Time-continuous and locally constrained learning rules.

Broadly speaking, neuromorphic hardware comes in two flavours: digital and analogue. Both designs are transistor-based (i.e., CMOS), but the transistors are used in different operating regimes. In simple terms, digital means calculating with discrete-valued bits (i.e., transistors take the states 'on' and 'off'), while analogue means calculating with continuous-valued currents and voltages. Digital chips provide stable solutions that are closer to commonly used hardware chips. Analogue chips have to deal with noise (e.g., manufacturing noise of components, cross-talk between electrical components, temperature dependence) and have a longer development cycle, but provide intriguing advantages such as high energy efficiency and potentially accelerated (compared to biology) emulation of neuron and network dynamics.

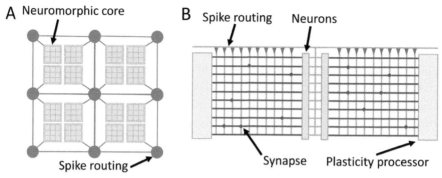

Figure 4.5 **(A)** Simplified schematic of the Loihi chip [60]. **(B)** Simplified schematic of the BrainScaleS-2 chip. Spikes emitted from the neurons enter the synaptic crossbar through synapse drivers (orange triangles) and, in essence, arrive at the target neuron when a connection is set (red dot). Each horizontal line in the crossbar is an input line for a neuron [55].

In addition to CMOS, memristor-based circuits[4] are being developed that open novel opportunities for energy-efficient and adaptive neuromorphic devices.

Examples of digital neuromorphic systems are Loihi (1 and 2) (developed by Intel) [16], TrueNorth (developed by IBM) [17], Akida (developed by BrainChip), SpiNNaker(1 and 2) [51, 52], Darwin [53] and Tianjic [54]. Examples of analogue chips are BrainScaleS (1 and 2) [55] and Spikey [56], DYNAPs [57], ROLLS [58] and Neurogrid [59]. A thorough review of state-of-the-art neuromorphic platforms at the time of writing can be found in [1]. Two of these platforms (Loihi by Intel and BrainScaleS-2 by the University of Heidelberg) are discussed in more detail below to further illustrate the difference between digital and analogue platforms.

First, Loihi's (Fig. 4.5A) main components are neuromorphic cores, which are processing units with custom circuitry and small amount of memory necessary to simulate a population of LIF neurons and their plasticity (i.e., learning rules). Spikes are exchanged between neuromorphic cores using a routing grid. In contrast, the BrainScaleS-2 chip (Fig. 4.5B) realizes physical AdEx neurons and synapses to emulate biology – in other words, neurons and synapses are not *simulated* but rather *implemented* directly as analogue circuits. Hence, no simulation time step exists and the system

---

[4]A memristor is a fundamental electrical element that displays a hysteretic resistance profile where the resistance depends on the recent history of current that has passed through the element.

evolves continuously. Although neurons and synapses are implemented using analogue circuits, spikes are transmitted digitally (also known as "mixed-signal"). Through a synaptic crossbar, connections can be set flexibly and are not pre-wired.

The main potential of neuromorphic hardware lies in enabling the deployment of energy-efficient low-latency AI systems suitable for edge applications. First glimpses into this potential can already be obtained nowadays: in Table 4.2, we list a few neuromorphic platforms that provide both performance and energy benchmarks for MNIST classification, with most systems reaching an energy footprint on the order of μJ per inference. We chose MNIST here as it is the most commonly used benchmark and allows for easy comparison across different (non-)neuromorphic platforms. However, it has been noted that SNNs perform best on data with temporal structure, meaning that MNIST is less suited as a benchmark for showing the advantages of SNNs as well as neuromorphic hardware [43, 69]. In fact, we argue that in order to better assess the potential of neuromorphic devices, standardized and application-specific benchmarks are required to fairly evaluate both algorithms and hardware platforms.

A recent feature of neuromorphic chips that is especially interesting for space applications is on-chip learning, found in chips like Loihi, BrainScaleS-2 and SPOON [66]. On-chip learning enables continuous learning directly integrated in the neuromorphic chip with a low energy footprint, suitable for the distributed design philosophy of neuromorphic devices. In the future, such 'intelligent' neuromorphic devices could be especially useful for retraining or fine-tuning onboard models during deep space missions toautonomously adjust and adapt to previously unknown or unexpected circumstances.

4.5 APPLICATION TO ONBOARD PROCESSING FOR EARTH OBSERVATION

As previously mentioned, the potential advantages of SNNs (e.g., in terms of energy efficiency) are promising for data processing, especially onboard miniaturized satellites. However, when applied to static data, as is the case in the classification of Earth observation images, it is not clear how SNNs compare to ANNs in terms of energy, latency and performance trade-offs [35]. For instance, some previous works [70, 71] showcase that when rate coding is used, the advantages of SNNs in terms of energy/performance decrease for classification datasets which contain complex features. This makes satellite data a challenging task for SNNs because of the complexity of their features [35]. For that reason, ESA's ACT, in collaboration with ESA's Φ-lab, is currently

TABLE 4.2 A list of neuromorphic platforms benchmarked on MNIST. As a baseline, [41] provides values for a convolutional ANN implemented on a nVidia Tesla P100, reaching 99.2% accuracy with an average energy / image of $852\mu J$. It should be noted that all neuromorphic platforms here process images sequentially, while the GPU uses batching. For sequential processing, [61] reports an accuracy of 98.90% with an average energy / image of $37mJ$ (Intel i7 8700) and $16mJ$ (nVidia RTX 5000). Of the results shown above, only TrueNorth, Akida and SPOON implement a convolutional SNN. Table adapted from [41].

Platform	Type	Tech	Model	MNIST	Energy / Image
Unnamed (Intel) [62]	digital	10nm	LIF	97.70%	$1.7\mu J$
Intel Loihi [16]	digital	14nm	LIF	96.40%	n.a.* [63]
				94.70%	2.47mJ [61]
IBM TrueNorth [17]	digital	28nm	LIF[1]	92.70%	$0.268\mu J$
				99.42%	$108\mu J$
Brainchip Akida [64]	digital	28nm	IF	99.20%	n.a.*,2
SpiNNaker-2 [52]	digital	28nm	_3	96.60%	$23\mu J$ [65]
SPOON [66]	digital	28nm	_4	97.50%	$0.3\mu J^5$
BrainScales-2 [55]	analogue	65nm	AdEx	96.90%	$8.4\mu J$ [41]
SpiNNaker [51]	digital	130nm	_3	95.00%	3.3mJ [67]

* Not available.

[1] [17] states that TrueNorth implements "(...) a dual stochastic and deterministic neuron based on an augmented integrate-and-fire (IF) neuron model [68]".

[2] https://doc.brainchipinc.com/zoo_performances.html. Accessed: 2022-11-07.

[3] SpiNNaker is an ARM-based processor platform and can therefore support arbitrary neuron models as long as a software implementation for SpiNNaker is available.

[4] SPOON is an "event-driven convolutional neural network (eCNN) for adaptive edge computing" where "TTFS encoding" is used "in the convolutional layers" [66].

[5] Pre-silicon simulation result.

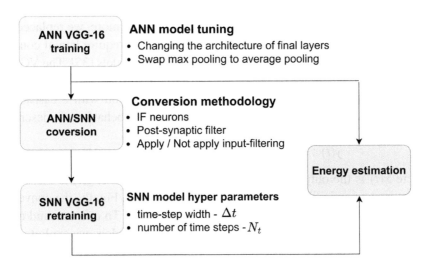

Figure 4.6   Rate-based SNN vs ANN performance/energy benchmark methodology used in our previous work [35].

benchmarking SNNs and ANNs for onboard scene classification. The latter was chosen as a target application due to the abundance of benchmark datasets such as EuroSAT [72, 73] which was also the dataset chosen by the ACT for this project. EuroSAT is a 10-classes land-user and land-cover classification dataset consisting of images captured by the Sentinel-2 satellite. For the study, the RGB version of the images was used. We are mostly investigating different information encoding solutions (including rate encoding and spike-time encoding) and their impact on the energy and accuracy that could be expected in performing onboard inference using an end-to-end neuromorphic approach. The results of past and ongoing efforts are presented in Sections 4.5.1 and 4.5.2, respectively.

4.5.1   Rate-based SNNs for onboard Earth observation

Rate-based models were explored in a previous work by the ACT [35], where the SNN model was trained by exploiting the approximate equivalence between spike rates of IF neurons and ReLU activations of artificial neurons, fundamentally using a weight conversion technique as discussed in Section 4.3 with limited loss in accuracy [35]. The methodology is summarized in Fig. 4.6.

As a first step, we started from a VGG16 ANN model [74] pretrained on ILSVRC-2012 (ImageNet) [75]. We replaced the last three dense layers with a cascade of average pooling and dense layers. Furthermore, we replaced the max pooling with average pooling layers that do not require lateral connections and offer an easier implementation for spiking models [35]. The VGG16 network was then trained on the EuroSAT RGB dataset. To convert the trained ANN model, we replaced the ReLu activation function with a cascade of IF neurons and post-synaptic filters, whose dynamic behaviour is described in [35]

$$y(t) := \left(1 - e^{-\Delta t/\tau}\right) \cdot x(t) + e^{-\Delta t/\tau} \cdot x(t-1), \tag{4.2}$$

where $x(t)$ is the output spike train of a neuron at time t, $y(t)$ is the output of the postsynaptic filter, and Δt is the timestep width. Finally, the converted SNN model is retrained to optimize the performance. To estimate and compare the energy consumption of the SNN and ANN models, we exploited the methodology implemented in *KerasSpiking*. In particular, more than performing an accurate estimation of the energy consumption, the methodology used aimed at allowing a relative comparison [35]. By considering a single spiking layer L, the used methodology assumes that the energy consumption during the inference is mostly due to two contributions: energy dissipated due to the synaptic activity E_s and energy usage due to neuron updates E_n. E_s can be calculated as

$$E_s = E_o \cdot \sum_N S_N \cdot f_{\text{in}} \cdot N_t \cdot \Delta t, \tag{4.3}$$

where E_o is the energy for a single synaptic operation and depends on the hardware used, S_N is the number of synapses for every neuron, f_{in} is the average neuron spiking rate, N_t is the number of timesteps in the simulation and Δt is the timestep width. For artificial models, $f_{\text{in}} = \frac{1}{\Delta t}$ and $N_t = 1$. Furthermore, E_n can be calculated as

$$E_n = E_u \cdot N_n \cdot N_t, \tag{4.4}$$

where E_u is the energy for a neuron update and depends on the hardware (similarly to E_o), and N_n is the number of neurons in a layer. The total energy consumption for a model is then given by the sum of the contributions of the different layers. Since for the SNN models, the energy depends on N_t and Δt, different SNN models were trained to explore various energy/accuracy trade-offs. To that end, we trained different models by setting the number of timesteps N_t ranging from 1 to 32 and making Δt a learnable parameter. The constant τ of every neuron post-synaptic filter was also trained to optimize the dynamic response of each neuron. In addition, since E_s is proportional to f_{in}. Therefore, in order to reduce the average rate for each neuron, the input

TABLE 4.3 Summary of the results shown in Table 2 of [35].

Model	Accuracy [%]	T	Δt [s]	Energy [J] on GPU	Energy [J] on Loihi
ANN	95.07	1	-	0.06996	0.00636
ANN + Prewitt	90.19	1	-	0.06996	0.00636
SNN (Best accuracy)	85.11	4	0.0381	-	0.00444
SNN + Prewitt (Lowest energy)	87.89	4	0.0403	-	0.00205
SNN + Prewitt (Highest accuracy)	85.07	1	0.0813	-	0.00476

was processed with a Prewitt filter, whose dynamics is given by

$$\mathbf{X}' := \max\left(c\sqrt{(\mathbf{G} \circledast \mathbf{X})^2 + (\mathbf{G}^\mathsf{T} \circledast \mathbf{X})^2}, \, 0.0078 \cdot \mathbb{1} \right), \qquad (4.5)$$

where

$$\mathbf{G} := \begin{pmatrix} 1 & 1 & 1 \end{pmatrix}^\mathsf{T} \begin{pmatrix} 1 & 0 & -1 \end{pmatrix},$$

and $\circledast$ represents the two-dimensional convolution operator, $\mathbb{1}$ is a unitary tensor having the same shape as the input $\mathbf{X}$, and c is a normalization constant. The effect of the Prewitt filter is to concentrate the input spikes mostly around the boundaries of different crops while keeping the input color information. Training was performed with and without applying the Prewitt filter to test its effects. Since the energy depends on the hardware used through the constants E_o and E_u, to estimate the energy consumption we tested various models on different hardware, including on desktop CPU (Intel i7-4960X), one embedded processor (ARM Cortex-A), one GPU (nVidia GTX Titan Black) for ANN models and SpiNNaker [51], SpiNNaker 2 [52], and Loihi [16] neuromorphic processors for both the ANN and SNN networks. The test results are summarized in Table 4.3.

The maximum accuracy of the SNN model was 85.11% and 87.89% with and without Prewitt filtering, respectively [35]. The ANN models reached 95.07% and 90.19% with and without Prewitt filter, respectively. In addition, the input filter had a strong effect on the tradeoff between energy consumption and accuracy for SNN models, whereby the energy consumption dropped by half while maintaining similar accuracy or the accuracy increased by over 2% with a similar energy consumption. The drop in accuracy of the SNN + Prewitt (highest accuracy) and SNN + Prewitt (lowest energy) was 1.18% and 10%, respectively, which was significant. However, when compared to the GPU implementation of the ANN, inference with SNN + Prewitt (highest accuracy) and SNN + Prewitt (lowest energy) on Loihi required 14.70 and 34.13 times less energy, respectively. Since this improvement was partially because of the higher efficiency of Loihi compared to the GPU, we also tested the inference performance of the ANN model on Loihi to eliminate the effects of the hardware. Even in this case, the energy consumption was 3.1 and 1.3 times lower compared to the SNN + Prewitt (lowest energy) and SNN + Prewitt (highest accuracy) models. However, to obtain more realistic estimates, future work will aim at more hardware-aware proxies that also take into account other effects, such as on-chip saturation effects and power-consumption due to off-chip accesses, which might affect the estimates presented here. In addition, the results seem to confirm trends described in [70,71], according to which longer simulation leads to better accuracy at the expense of energy consumption when rate coding is used. Because of that, as described in Section 4.5.2, different information encoding solutions shall be explored.

4.5.2 Spike-time-based SNNs for onboard Earth observation

Additional encoding schemes, such as temporal encoding, and surrogate gradient training approaches are currently being explored that could yield better trade-offs for Earth observation. By encoding information in spike times instead of rates, information can be processed with a much lower number of spikes, leading to higher sparsity in the network activity and consequently lower energy consumption. We are focusing on two approaches: (i) training SNNs that solely use TTFS encoding, either trained with exact gradients or surrogate gradients, and (ii) training SNNs end-to-end using surrogate gradients without imposing any specific encoding scheme. Our evaluation of models is two-fold: performance and network metrics (for instance, the average number of spikes per inference) are investigated in simulation using SNN frameworks such as Norse. In addition, we are mapping parts of these models to the neuromorphic chips Loihi and BrainScaleS-2 to obtain an estimate for the energy efficiency of a fully neuromorphic implementation. As this is still an ongoing study, only a few preliminary results will be discussed here.

Performance-wise, we found that convolutional architectures trained end-to-end show the highest accuracy. We implemented a spike-based convolutional neural network (CNN) with 4 layers (3 convolutional, 1 dense layer) that takes the pixel values of images (rescaled to be between 0 and 1) as input currents. In this architecture, the output layer consists of LIF neurons without a threshold mechanism, i.e., without spiking. All other layers consist of LIF neurons. A classification result is given by the index of the neuron with the maximum membrane potential value during the whole simulation time – in simpler terms, the label neuron that obtained most votes from the last hidden layer provides the classification outcome. Trained end-to-end, this model achieves competitive results ($\approx$ 91 $-$ 92% test accuracy on EuroSAT RGB) with low spike activity (on average $\approx$ 2 $-$ 3 spikes per neuron during inference).

Moreover, we are investigating SNNs based on TTFS encoding for all layers (input, hidden and output). Thus, in the label layer, the classification outcome is given by the label neuron that spikes first. We are currently exploring both models that can be trained using exact gradients and ones that can be trained using surrogate gradients. Preliminary results show that these models reach slightly lower accuracy, although they also require fewer spikes per inference than models without pre-imposed encoding. In general, these initial results already hint at a confirmation that direct training of SNNs using surrogate gradients leads to better results than (rate-based) weight conversion for Earth observation image processing tasks.

4.6 NEUROMORPHIC SENSING

In addition to *processing*, neuromorphic computing has also entered the domain of *perception*, most notably in the area of computer vision. Biological systems provide a rich source of inspiration due to their energy efficiency, strong data compression and feature extraction properties as well as remarkable adaptability. A widely used example of biological computations is the visual system in flying insects, which has been deeply researched in the last decades (e.g. [76, 77]). Research on the neuronal systems in insects allows for single-neuron analysis and thus provides very detailed insight on the internal working mechanisms of insect vision. The primary source of motion information in insects is the so-called Elementary Motion Detector (EMD) (e.g. [13]). EMDs are a minimalistic neural architecture able to extract important motion information from only two photoreceptors. Preliminary studies performed at the European Space Agency on their applications to spacecraft landing [9–12] have suggested the possibility to use EMDs as the only sensor necessary for landing a spacecraft in specific scenarios. Another example of a highly efficient and adaptive biological sensor is the mammalian retina, which is composed of several layers of neurons specialised in processing visual information. The mammalian visual system is important for understanding the biological inspiration behind the concept of event-based vision, which we focus on later in this chapter. Thus, the following section provides a brief overview of the structure and function.

4.6.1 Structural and functional organisation of the retina

The structural and functional organisation of the mammalian retina has been explored in great detail over the past few decades. The elaborate 1965 study by Hubel and Wiesel [78] was the first to present a collective summary of many fundamental properties of the retina, such as the hierarchy of retinal neurons, their receptive fields (RFs) and orientation selectivity, and so forth. The mammalian retina consists of several layers containing nerve cells and receptors that perform highly specialised functions [79]. The following is a brief overview of the main types of cells found in the mammalian retina, together with their main distinctive features.

Receptors   The outer nuclear layer (ONL) contains the actual photosensitive receptors: rods, which are responsible for scotopic vision (low-light conditions), and cones, which are responsible for photopic vision (in well-lit conditions). Furthermore, rods function in low-light conditions but do not distinguish colour, whereas cones are active in display colour specialisation (e.g., responding to light in the red, green or blue part of the visible spectrum). Receptors are excitatory cells.

Horizontal cells   The signal produced by receptors in the ONL is modulated by horizontal cells, which provide local inhibitory signal informed by the immediate neighbourhood of the receptor. Horizontal cells are responsible for local brightness adaptation (especially in the foveal region) as well as the formation of ON/OFF RFs in ganglion cells [80]).

Bipolar cells   The raw visual input normalised through feedback from horizontal cells is then fed into bipolar cells (BCs), which effectively split the visual channel into two pathways (ON and OFF) [80]. This means that ON BCs respond to an increase in brightness, whereas OFF BCs respond to a *decrease* in brightness. The separate pathways are preserved all the way to the visual cortex and display asymmetric behaviour – for instance, the OFF pathway reacts faster and sends stronger signals to the cortex compared to the ON pathway [81]. In addition, while horizontal cells provide the retina with spatial brightness adaptation capabilities, BCs are responsible for temporal adaptation and filtering [82, 83].

Amacrine cells   The excitatory signals produced by BCs are combined with inhibitory feedback from amacrine cells in many different configurations. There are more than 30 types of amacrine cells, whose primary function is to combine the signals arriving from BCs into many different circuit configurations that define, modulate and control the RFs of ganglion cells via inhibitory synapses [83] (however, there is a possibility that they also provide excitatory signals via gap junctions [84]).

Ganglion cells   Amacrine and BCs form different wiring patterns providing input to the ganglion cell layer (GCL). Different retinal ganglion cells (RGCs) have different RFs, which determine their *selectivity* to certain patterns (both static and dynamic, such as orientation, motion or combinations of both) in the visual field [80]. For instance, some RGCs respond to edges, others are highly sensitive to motion relative to the background in the visual field, and yet others respond to approaching motion while remaining silent in response to lateral motion [83]. RGCs are considered to be the only neurons in the retina that produce spikes [85] (the other cells have graded responses), although there are indications that at least BCs also produce a limited number of spikes [86] that are phase-locked to visual stimuli.

Counterintuitively, the ONL is the *innermost* layer of the retina, which means that light has to travel through all the other layers before it reaches the receptors. This and other structural properties of cellular layers in the retina have several functional implications.

- The GCL contains a lot of cell bodies as well as axons which relay spikes via the optic nerve to the visual cortex. Since the GCL is on top of the receptor layer, the part of the retina where RGC axons interface with optic nerve is devoid of receptors, giving rise to the infamous *blind spot*.

- A small area of the retina known as the *fovea*, which is located at the centre of the retina directly behind the lens, contains the highest concentration of receptors. To maximise the resolution in the fovea, the axons of RGCs in the fovea are highly stretched out, forming a dense circular bunch of axon bodies around the fovea, where the resolution starts to drop sharply.

- At least in humans (and many other mammals), most of the colour-sensitive receptors are in the foveal region. Away from the fovea, vision is dominated by rods, which are much more sensitive (down to a single photon [87]) but not colour-specific. This is easily observed in low-light conditions, where our visual perception shifts towards detecting only silhouettes – in other words, only contrast without colour information.).

- RGCs in the foveal region have RFs composed of very few receptors (or even a single receptor for the central region of centre-surround RFs). The RF size increases steadily further away from the fovea. This eccentricity-dependent RF size has been found to have a number of interesting consequences for scene recognition, including cortical magnification, scale invariance and object recognition [88, 89].

4.6.2 Adaptation and homeostasis in the retina

dAdaptation is a hallmark of most types of neurons and circuits found in the retina. Considerable research effort has been dedicated to unveiling the underlying principles of visual adaptation [90, 91]. A number of different homeostatic processes have been observed in practically all layers in the retina. For instance, cones have a very large dynamic range and are also extremely sensitive to small changes in contrast, both in terms of response time and contrast range. In comparison, rods can detect very low levels of illumination (as low as individual quanta of light [92]), with the trade-off that they adapt very slowly and become saturated by strong and sudden changes in light intensity [92]. ON-type (OFF-type) BCs have a light center/dark surround (dark center/light surround) RF and respond strongly to positive (negative) contrast. The ON/OFF differentiation is maintained in RGCs, which receive input from one of the two types of BCs. The RFs of both ON and OFF RGCs independently cover the entire retina [93]. The retina uses adaptation mechanisms in BCs and RGCs to adapt to both spatial and temporal variations in illumination [94, 95]. Both BCs and RGCs homeostatically regulate the intensity of their output depending on the strength of the visual stimulus, allowing the retina to adapt to changes in luminance that span 9 to 12 orders of magnitude over a 24-hour period [96, 97].

Both ON and OFF BCs also maintain a fast 'push-pull' mechanism mechanism that balances the activation of the center relative to the surround for the purpose of locally enhancing the contrast within the RF [98, 99]. In addition, RGCs adapt their activity to both the mean and the variance of the visual stimulus [94, 96, 100] and even employ a gain control mechanism to change the effective *size* of the RF based on the illumination [101].

Finally, the eye is not a static sensor – it constantly performs a combination of movements (ocular tremors, drift, microsaccades and saccades [102]) that span various spatial and temporal scales. These different types of movements serve different purposes: for instance, small random movements (such as drift and microsaccades) are outside voluntary control and occur even while fixating on a target. They prevent the phenomenon of visual fading (also known as retinal fatigue), in which the rapid adaptation of RGCs to a fixed visual scene leads to the complete elimination of neural response (the eye effectively becomes blind to anything that is not moving) [102, 103].

Thus, the concept of a vision sensor modelled on the retina emerged more than three decades ago by building upon detailed research on the neuroscience of the mammalian retina. The following section presents a brief overview of the major milestones since the conception of the earliest retinomorphic circuits to present-day commercial event-based cameras.

4.6.3   Retinomorphic models

Neuromorphic systems that specifically deal with modelling the mammalian visual system (mostly the retina) are known as 'retinomorphic' systems [104]. While some the operating principles of the circuits differ substantially from that of biological retinal cells, they are in many respects *functionally* identical (or at least very similar) to biological retinal cells.

To optimize the power consumption and response speed of sensors, most recent research on event-based vision is heavily geared towards hardware-based solutions, which are difficult to customise for the purpose of implementing novel computer vision algorithms and processing paradigms. In addition, existing software libraries are geared towards processing and emulating the existing hardware rather than providing a generic base for implementing novel algorithms.

The ACT is thus pursuing research on retinomorphic models to explore the question of whether sophisticated visual processing can be performed in the sensor itself, rather than being *learned* by a downstream algorithm. A prime example is the specialised RGCs which are sensitive to certain types of motion in their RF, to the exclusion of all other types of motion. For instance, there are RGCs that are sensitive to approaching motion while being insensitive to lateral or receding motion [83, 105]. Such selectivity can be useful for applications such as spacecraft landing, where the landscape is perceived as approaching from the perspective of the camera.

To maximise the efficiency and speed of onboard processing, it would be beneficial to design sensors that provide insight about various types of motion and features in the visual field, such as edges, depth or optical flow. Therefore, it would be beneficial to be able to create models of such cells and their functions in order to inform the design of novel dedicated hardware sensors.

In this context, the ACT is developing a simulation framework that can efficiently model many features of the mammalian retina in order to facilitate rapid prototyping of retinomorphic architectures. In particular, our research and development efforts have been geared towards implementing support for all of the following features:

- All major types of cells in the retina (cf. 4.6.1);

- Spatial and temporal adaptation and homeostasis (cf. 4.6.2);

- Complex RFs (e.g., centre/surround);

- Foveation and eccentricity-dependent RFs;

- Sparsification of cells towards the edges of the retina (i.e., cell distribution becomes increasingly sparse towards the edges of the retina, in proportion to increasing RF size).

- Saccadic movements (microsaccades, drift, etc.);

As outlined above in Section 4.6.1, the retina consists of several layers, whereby cells in each layer (except the photoreceptor layer) receive input from a compact region of cells in the preceding layer(s). This mechanism is behind the inspiration for CNNs, which have occupied a central role in computer vision domain of AI research [106, 107]. The successful application of CNNs to a wide range of tasks, combined with the availability of software packages that offer highly optimised CNN implementations out of the box, has largely dampened any incentive to explore alternative architectures or mechanisms for implementing convolution. However, incorporating all of the above features into a coherent framework is not straightforward, specifically with respect to saccadic movements and eccentricity-dependent RFs. This requires rethinking how convolution is implemented.

4.6.4 Sparse convolution

Convolution is a costly operation to implement, and state-of-the-art CNNs rely on a number of tricks to speed up the process. N-dimensional convolution is usually implemented using of a number of convolutional kernels (represented as a single $(N + 2)$-dimensional tensor, where N is the number of dimensions of the input and two extra dimensions representing the number of kernels and the batch size). Therefore, without loss of generality, we focus on $2D$ convolution, unless explicitly stated otherwise.

The generic expression for convolving a batch of N $2D$ inputs (each with dimensions H and W) using C_{in} convolutional kernels can be expressed as follows:

$$Conv2d(N, C_{out}, H, W) = \sum_{i=0}^{C_{in}} W_{(C_{out}, i)} \star Input(N, i) \qquad (4.6)$$

The dimensions H_{out} and W_{out} of the output channels are computed as

$$H_{out} = \left\lfloor \frac{H_{in} + 2p_H - d_H(k_H - 1) - 1}{s_H} + 1 \right\rfloor \qquad (4.7)$$

$$W_{out} = \left\lfloor \frac{W_{in} + 2p_W - d_W(k_W - 1) - 1}{s_W} + 1 \right\rfloor, \qquad (4.8)$$

Here, p, d, k and s (with indices H and W) represent the padding, dilation, kernel size and stride of the convolution operation (in the direction of H and W, respectively). These four parameters impose constraints on each other depending on the input since H_{out} and W_{out} in Eqs. (4.7) and (4.8) must be integer. For instance, convolving an image of size 32×32 with a kernel of size 5×5, a stride of 2 and dilation and padding of 0 means that the convolution would not be symmetrical at the edges of the image. This is a direct consequence of the constraint that H_{out} and W_{out} be integer enforces the convolution operation to stop short of covering the last column and row of the image. Adding a padding of 2 around the image mitigates this problem and ensures that the convolution is symmetric. We refer to the relevant literature for an in-depth overview of the meaning of these parameters [108].

Software implementations of convolution rely on a crucial but often understated preprocessing operation, known as 'image-to-column' (or *im2col* for short) [109]. This operation involves stretching out the convolutional kernels and arranging them into rows of a matrix, as well as stretching *all* kernel-sized subsections of the input into one-dimensional column vectors and arranging them as columns of another matrix (hence the term *im2col*, since $2D$ convolutions are most commonly applied to images) (Fig. 4.7). In this way, the convolution operation is reduced to a single dense matrix-matrix multiplication.

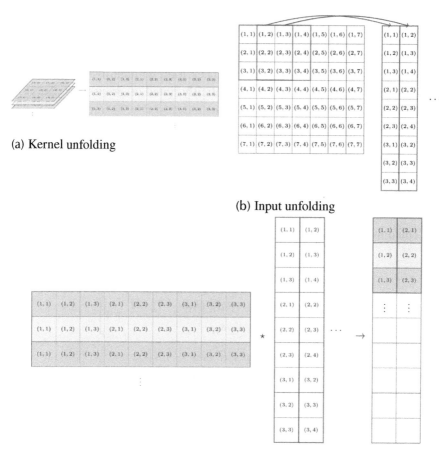

(a) Kernel unfolding

(b) Input unfolding

(c) Convolution operation.

Figure 4.7   Default Im2col operation.

It is immediately clear that while it improves the efficiency of convolution, this approach introduces a number of cascading limitations that impede the effort to implement the desired features outlined in 4.6.3:

- All kernels must be the same size, which precludes foveation and eccentricity-dependent RFs.

- A fixed kernel size also limits the type of RFs that can be implemented; specifically, centre/surround RFs are difficult to implement with a fixed kernel size, especially in combination with foveation.

- Sparsification is impossible to model since the operation is by definition **dense** as it applies the kernel to all segments of the input as defined in Eq. (4.6).

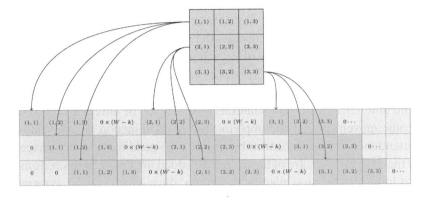

Figure 4.8   Unfolding a kernel into a sparse matrix. The redundancy in terms of space introduced in this operation is offset by the elimination of the *im2col* operation.

- It is not trivial to implement saccadic movements as this would entail performing the *im2col* preprocessing step for each movement. Furthermore, as some saccadic movements are an attention-shifting mechanism, they are much more useful in combination with foveation.

- Without saccades, adaptation (specifically temporal adaptation) as observed in the retina quickly erases features in stationary parts of the image.

To address these limitations, we have developed a straightforward alternative to the default convolution operation. Rather than unfolding the input, our method relies on a *kernel* unfolding step (tentatively named *kernel2row*). As the name suggests, this operation consists of unfolding the kernels, splitting them into continuous chunks and arranging those chunks into a *sparse* matrix, while the input is unfolded into a single large column vector. The kernel chunks are arranged in such a way that the result of multiplying the sparse kernel matrix and the stretched input vector is identical to that obtained with the above dense convolution method. The *kernel2row* operation is illustrated in Fig. 4.8.

The most notable advantage of performing convolution in this way is that the kernel size can be decoupled from the implementation constraints, allowing for kernels of different sizes to be used within the same convolution matrix.

Most of the desired features for simulating a retinal model as outlined in 4.6.3 can be implemented trivially with our sparse convolution method.

For instance, foveation can be implemented by using smaller kernels towards the centre of the image than towards the edges. Similarly, a separate kernel 'instance' can be used to convolve each pixel of a $2D$ image, which allows us to implement complex RFs (such as ones relying on the centre-surround mechanism) as a difference of Gaussians ('Mexican hat' filter) represented by the difference of two convolutions with kernels of different size. Finally, saccadic movements can be implemented in combination with foveation as an attention shifting mechanism by simply zero-padding the input image in a way that offsets the centre of the resulting padded image by the desired amount equal to the distance travelled by the 'fovea' during the saccadic movement.

To demonstrate the flexibility of this method, we have implemented a simple pipeline emulating all retinal layers (see §4.6.1). In this model, the input to excitatory layers (receptors, BCs and RGCs) is modulated by feedback from inhibitory layers (horizontal and amacrine cells). Cells in the excitatory layers have eccentricity-dependent RFs (smaller towards the centre of the image and gradually growing in size towards the edges). A local spatial normalisation map is implemented using the horizontal layer, whereas the bipolar layer implements a temporal filter as well as the split into separate ON and OFF pathways. The temporal filter is implemented as an exponential running mean which is updated iteratively as follows at every frame:

$$\mu_{t+1} = \alpha x + (1 - \alpha)\mu_t, \tag{4.9}$$

where μ_t is the mean at time step t, α is the 'forgetting rate' that determines how much of the preceding input is remembered, and x is the input frame. The mean is computed in this way for each pixel.

The ganglion layer preserves the separation of the ON and OFF paths and implements two types of RGCs with ON-centre / OFF-surround and OFF-centre / ON-surround RFs, respectively. For the sake of brevity, we have omitted the demonstration of saccadic movements, which can be used for inducing events from an otherwise static input. The parameters for this model are given in Table 4.4. For the input, we use part of the Perseverance landing sequence as recorded with the rover's onboard camera minutes before touchdown [110].

Several interesting observations can be made about the results in Fig. 4.9. First, despite the simplicity of the model, it can clearly perform certain types of preprocessing, such as local brightness normalisation at the horizontal layer, edge extraction (e.g., at the bipolar layer) and event generation at the ganglion layer. It is also noteworthy that ON/OFF and OFF/ON RGCs highlight the two sides (brighter and darker) of edges in the visual scene.

It is noteworthy that the RF organisation in this case successfully suppresses noise in the output of the ganglion layer. This demonstration of effects

TABLE 4.4 Parameters for the retinomorphic model in 4.6.4

Cell type	Parameter	Value
Receptors	Input type	Grayscale
Horizontal, bi-polar, amacrine, ganglion	RF size distribution	Gaussian[1]
	RF size distribution: mean	$(H/2, W/2)$[2]
	RF size distribution: SD	$(H/3, W/3)$
	RF type	Proportional[3]
Horizontal, bi-polar, amacrine	Min. / max. RF size	$1 \times 1 / 4 \times 4$
Ganglion	Min. / max. RF size (centre)	$1 \times 1 / 4 \times 4$
	Min. / max. RF size (surround)	$4 \times 4 / 9 \times 9$
BCs	α[4]	0.95
Ganglion cells	Threshold[5]	15

[1] For all cell types except horizontal cells, the Gaussian determining the RF size is *inverted*, meaning that the RF size *increases* away from the fovea. For horizontal cells, the RF size is *larger* at the centre and decreases towards the edges. This is consistent with the fact that horizontal cells are concentrated around the foveal region [80], where they facilitate the mechanism of adaptation to the local brightness level. A larger RF size for horizontal cells in the foveal region leads to a sharper and more contrasted image in the fovea.

[2] In this table, H and W denote the height and width of the input image, respectively.

[3] Here, 'proportional' means that the value of each element of the kernel is the inverse of the kernel size (e.g., $1/9$ for a 3×3 kernel).

[4] α is the 'forgetting rate' of the exponential running average used as a temporal filter in the bipolar layer.

[5] The threshold indicates the difference between the intensity at the centre vs. the surround of the receptive field. An event is produced if this threshold is crossed.

observable in biological retinas is encouraging as we are working towards implementing more sophisticated processing of the incoming visual information, such as optical flow reconstruction and detection of approaching motion for applications such as obstacle avoidance and landing.

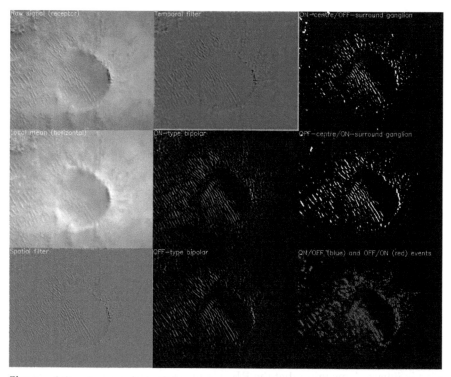

Figure 4.9 A snapshot of the state of each layer and the combination of ON/OFF and OFF/ON events. (a) Raw input (converted to grayscale). (b) Simulated horizontal cell layer computing the local mean illumination. (c) Raw signal with the mean illumination from (b) subtracted. (d) Temporal filter (implemented as a running average of the normalised signal in (c). (e) ON and (f) OFF BCs computing positive and negative deviation from the temporal mean, respectively (i.e., parts that are brighter and darker than the mean). RGCs with (g) ON/OFF and (h) OFF/ON centre/surround RFs. In ON/OFF RGCs, the centre is driven by ON BCs and the surround by OFF BCs, and vice versa for OFF/ON RGCs. An RGC produces a binary spike (event) when the difference between the centre and surround crosses a predefined threshold (a model parameter). Finally, (i) shows the combined activity of ON/OFF (white) and OFF/ON (black) RGCs.

The retinal simulation code is part of a library named *pyrception*, which in turn is part of the ACT open-source ecosystem [5]. The ultimate goal of the *pyrception* library is to provide convenient interfaces for converting input from multiple sensory modalities (such as visual, auditory and olfactory)

---

[5]https://gitlab.com/EuropeanSpaceAgency/pyrception

into spike trains, which can be used downstream for spike-based multi-modal learning tasks.

4.6.5 Existing silicon retinas

The understanding that the retina deals with the inherent unreliability of 'wetware' by responding to relative intensity changes (rather than the absolute magnitude) in the visual field [111] has been central to the development of silicon analogues to the retina. One of the earliest circuits that modelled the adaptive behaviour of retinal cells was a photoreceptor that mimicked some homeostatic processes in the retina. Specifically, the receptor could detect *changes* in illumination rather than the absolute illumination level [112,113]. The idea behind this circuit was inspired by the operation of receptors and BCs (see §4.6.1). It employed a feedback mechanism that compared the response to the changes in intensity to a filtered version of the output of the receptor, with an added hysteretic element that provided a 'memory' of recent illumination. Importantly, the circuitry emulated the way that biological photoreceptors adapt to changes in light intensity on a logarithmic rather than linear scale (a phenomenon known as Weber's law). In other words, if the output of the receptor is given as $V \propto log(I)$, then the change in that response would be logarithmic as well ($\delta V \propto \delta I / I$).

A prominent milestone along this line of research was the development of a 'silicon retina' [114], where each photoreceptor could adapt to the local light intensity through feedback from simulated horizontal cells and the adapted signal was further amplified by silicon analogues of BCs. The silicon retina demonstrated how a globally connected matrix of elements can still adapt to local fluctuations in light intensity, producing a familiar output comparable to that of an actual biological retina. Ultimately, a circuit was demonstrated that could model the behaviour of all five types of retinal cells, including RGCs [115].

Subsequently, the Address Event Representation (AER) [116, 117] was developed with a slightly different objective, namely the efficient communication of a continuous signal in the form of multiplexed spike trains. The generic nature of AER was used to demonstrate how a circuit could perform spatial convolution (using, for instance, Gabor filters), allowing for feature extraction and tracking with high temporal resolution.

4.6.6 The dynamic vision sensor (DVS)

One of the main problems with the early circuits described above was that they were built primarily as proof-of-concept devices – they were never meant to be used as actual camera sensors. This changed with the development of the dynamic vision sensor (DVS) [14, 15].

In many ways, DVS is a simplification on the original approach of modelling the retina holistically (i.e., all the layers separately). For instance, pixels in the DVS camera are entirely independent of all other pixels, enabling asynchronous operation that employs AER to represent events using x and y coordinates, a timestamp t and a polarity p. DVS pixels are relatively simple and implement the entire pipeline from receptors through BCs and RGCs within a single circuit. The main principle of event generation and adaptation remains similar to that in earlier research: each pixel detects changes in the brightness (log intensity) in either positive or negative direction relative to a baseline. Once the derivative of the logarithmic intensity crosses a fixed threshold Θ, an event is produced and the brightness level of the threshold is set as the new baseline. In this way, rapid changes in brightness result in a large number of events, whereas slow fluctuations produce few events. Importantly, since the baseline is *reset independently for each pixel* to the level of the threshold each time that pixel produces an event, the sensor as a whole has an exceedingly large dynamic range – on the order of $140dB$. This means that strong contrast does not saturate the sensor because the response of a pixel on the dark side of an edge does not depend on the response on the bright side of the same edge; it only depends on *relative* changes in brightness (Fig. 4.10).

The DAVIS sensor [14] adds active pixel sensor circuitry to the core DVS logic in order to be able to keep information about the absolute brightness in the scene while retaining the advantages of the DVS sensor (such as high dynamic range, low latency and sparse activity). This allows a conventional frame-based readout to be obtained simultaneously with the fast asynchronous event-based output of the DVS sensor.

DVS pixels also show a characteristic junction leakage current, which depends strongly on the temperature [118]. This leakage current is the source of sporadic (noisy) ON events, with the noise level becoming more prominent at high temperature and in dark environments.

The DVS pixel circuit enabled event-based sensors to bridge the gap between proof of concept and usability, leading to several commercial implementations (e.g., by Inivation[6] and Prophesee[7]).

There are also some notable points where the DVS departs from the operating principles of the mammalian retina. For instance, the DVS does not fully preserve the ON/OFF pathway separation since a single pixel can

---

[6]https://inivation.com/
[7]https://www.prophesee.ai/

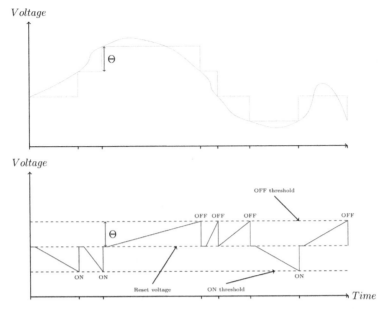

Figure 4.10   Operating principle of a DVS pixel. Every time the brightness of the area of the visual scene covered by the pixel increases (decreases) by a fixed amount (quantified by the threshold Θ), the pixel generates an ON (OFF) event. The AER representation is commonly used for encoding the output of an event sensor, whereby an AER event consists of a timestamp, the x and y coordinates of the pixel that generated the event and the polarity of the output (+1 for ON and −1 for OFF events).

produce both ON and OFF events (whereas in biological systems these pathways are in fact kept separate all the way to the visual cortex [81]). Also, due to the lack of cross-talk between pixels, it is impossible for DVS cameras to implement RFs (although at the time of writing there are efforts to introduce RFs [119], mainly with the goal of suppressing noise). Other features of the retina and the eye in general, such as microsaccades [120] and foveation [121], have also been attracting the attention of event-based vision researchers in recent years. In parallel, there are also efforts to develop software that can emulate, simulate or otherwise complement the capabilities offered by hardware event-based sensors. For instance, the development of the video2events (v2e) framework [122] and the ESIM library [123] was driven by the objective to harness existing conventional video datasets for event-based vision research.

4.6.7 Why in space?

In 2021, the DAVIS240 sensor became the first neuromorphic device to be launched in space and is, at the time of writing, also onboard the ISS launched as part of the ISS resupply payload in the Falcon Neuro Project [124]. This shows that neuromorphic sensing is attracting the attention of aerospace experts. However, we are only just beginning to understand and quantify the advantages of event-based cameras in the context of space missions [125–129]. In addition to the discussed benefits of pursuing a neuromorphic approach whenever low-resource requirements are driving the design of a mission, event-based cameras can offer additional advantages over conventional ones due to their higher dynamic range of the former, thus achieving superior performance in high-contrast scenes as well as allowing a relaxation on the constrains imposed by light entering the field of view of the sensor. The high temporal resolution and sparse output of an event sensor are also of interest, as well as its recently demonstrated resilience to different types of radiation (such as wide-spectrum neutrons) that are often present in space operating environments [130]. Promising scenarios where event-based cameras could make a difference in a space context are landing [125, 126], pose estimation [128], planetary surface mapping, astronomy [127, 129], monitoring of specific events (e.g., explosive events, pebble dynamics around asteroids, etc.). In all these cases it can be argued that the neuromorphic nature of the camera (combining considerable power efficiency and other beneficial properties), could introduce advantages with respect to conventional systems, especially combined with a neuromorphic approach from the algorithmic side to tackle the interpretation of the produced events [125]. However, an open question that remains in this field is the quantification of the domain gap introduced by the use of synthetic data produced by video-to-event converters [122]. A recent work [128] addresses this issue by proposing and testing an algorithm (trained on a synthetic dataset) on event streams created by a realistic mock-up experiment performed in a lab.

4.6.8 Landing with events

The ACT has recently demonstrated the suitability of event-based vision for autonomous planetary landing operations by inferring and processing optical flow measurements to predict time-to-contact (TTC) and divergence. TTC is defined as:

$$\text{TTC} = -\frac{z}{v_z} \tag{4.10}$$

where z is the altitude of the spacecraft with respect to the planetary surface and v_z its the vertical component of the velocity vector, so that a descending spacecraft will always satisfy $z > 0$ and $v_z < 0$. Previous work has already demonstrated the suitability of TTC feedback for controlling a spacecraft during flight operations and landing scenarios by using visual information from traditional frame-based sensors onboard micro air vehicles [131, 132]. Our recent studies have shown that event data could also be effective for reconstructing on-board TTC, therefore bridging the gap from on-board event data streams to real-time control [125, 126].

In these studies, event streams representative of relevant ventral landing scenarios were constructed synthetically by converting sequences of frames similar to those provided by traditional frame-based sensors to events. The PANGU (Planet and Asteroid Natural Scene Simulation Utility) [133] software was used to generate proxy models of the Moon surface and render frames representative of the output of onboard vision instrumentation. From the generated synthetic events, TTC was estimated by identifying features in the event stream corresponding to static structures in the planetary body (such as craters and boulders) and computing their rate of perceived expansion or optical flow divergence. The reconstructed TTC from the event stream was fed back into a closed-loop control system to simulate real-time autonomous ventral landing by enforcing constantly decreasing TTC. Simulations were performed for different lunar terrains characterized by diverse crater and boulder distributions. Results showcased the suitability of the solution as a proof of concept for real-time onboard processing of events in planetary landing scenarios, where the main areas with room for improvement that could benefit from future research are the inference of divergence and the inclusion of more accurate noise models for event modelling based on known properties of DVS hardware. [125]

In recent work, the divergence estimation procedure was further improved by using a contrast maximisation [134, 135] formulation for event-based divergence estimation, demonstrating its applicability on both synthetic and real event datasets [126]. Spacecraft position and velocity vectors were obtained for several planetary ventral landing profiles computed via indirect optimisation methods. A UR5 robotic arm with a Prophesee GEN 4 event sensor was employed to replicate these ventral trajectories using 2D and 3D printed planetary surfaces. Some of the generated trajectories were used to render 3D landing reconstructions in PANGU, which were then passed through the v2e [122] pipeline to generate synthetic event streams. A mathematical formulation for the event-based radial flow was proposed, and a GPU-accelerated optimisation procedure was used to maximise the contrast of the resulting flow-compensated event images and estimate the divergence of event batches. Comparisons to other state-of-the-art divergence estimation methods showcased the accuracy and stability of the proposed procedure across a wide variety of event streams, with competitive run times achieved owing to the GPU-accelerated implementation.

4.6.9 From events to egomotion

Events as produced by the DVS or some other device based on retinomorphic models contain information on the perceived scene motion as projected in the camera plane. But what can be learned about the spacecraft egomotion (or pose) using only this information? It is conceivable that events can be used for reconstructing the motion field, whereby the problem of motion field interpretation has already been discussed in the literature [136]. A number of researchers have commented on this issue, starting with the work by Longuet Higgins [137], which addressed the question for the case of retinal vision, and subsequent works that lay down the mathematical structure of the problem, mapping it into a linear problem and thus marking it as "solved" [136]. More recently, substantial work in the area of drones and robotic vision has been produced based on these fundamental results established in the 80s and 90s.

Let us consider the familiar equations [136, 137]:

$$\begin{cases} u = (v_{xc} + xv_{zc})h(x, y) + q + ry - pxy + qx^2 \\ v = (v_{yc} + yv_{zc})h(x, y) + rx - p - py^2 + qxy \end{cases}$$

where u, v are the velocities of the feature in x, y on the camera plane, p, q, r the angular velocity components of the camera, v_{xc}, v_{yc}, v_{zc} the velocity of the camera center of mass and $h(x, y)$ the inverse of the depth map.

From these equations, if $h, v_{xc}, v_{yc}, v_{zc}$ constitutes a solution, so does $\frac{h}{k}, kv_{xc}, kv_{yc}, kv_{zc}$; it is thus impossible to invert the motion field to absolute values of all motion parameters, as only relative depth can be distinguished. Approximating the surface as an infinite plane, one can show that necessarily:

$$h(x, y) = \alpha x + \beta y + \gamma$$

where $H^2(\alpha^2 + \beta^2 + \gamma^2) = 1$, having denoted the spacecraft altitude with H. Let us analyse the structure of the resulting equations. If we assume that we know the motion field u_i, v_i, $i = 1..n$ at n different points, we can write a system of $2n$ equations in the unknowns $v_{xc}, v_{yc}, v_{zc}, p, q, r, \alpha, \beta, \gamma$. As before, only relative depth can be estimated, hence one can set $H = 1$ and consider $\gamma^2 = 1 - \alpha^2 - \beta^2$ to conclude that in theory $n = 4$ points of the motion field

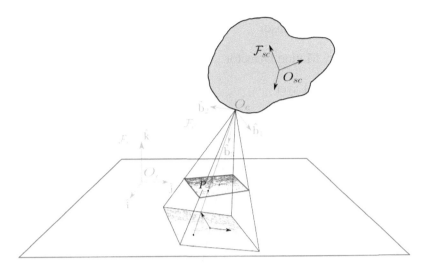

Figure 4.11   Camera geometry during planetary landing. The three reference systems, namely planet (i), camera (c) and spacecraft (sc), are shown as well as the camera plane projection of a generic surface feature assumed to be stationary.

are necessary and sufficient to fully determine the spacecraft egomotion in relative terms to the altitude H. This conclusion does not consider the effect of measurement noise on the resulting estimates, a problem that is highly dependent on the sensor used (i.e., the quality of the events produced and their translation into a motion field). To conclude, it is worth mentioning the recent work from Jawaid *et al.* [128] on pose estimation from events. In a docking scenario (both simulated and reproduced in a lab mockup), the approach proposed in that work is able to determine from event data also the absolute values of mutual distances. This is not in contradiction to what is claimed above, as the algorithm is not exploiting depth cues from a reconstructed motion field but rather cues learned directly from a predefined dataset of known poses where the absolute dimensions appear explicitly.

ACRONYM

AER Address Event Representation

ANN Artificial Neural Network

BC Bipolar cell

CMOS Complementary metal-oxide-semiconductor

DVS Dynamic Vision Sensor

EMD Elementary Motion Detector

GCL Ganglion Cell Layer

LIF Leaky Integrate-and-Fire

ONL Outer Nuclear Layer

RF Receptive Field

RGC Retinal Ganglion Cell

SNN Spiking Neural Network

STDP Spike-Time-Dependent Plasticity

TTC Time To Contact

TTFS Time-To-First-Spike

GLOSSARY

Neuromorphic: The concept of designing hardware systems and computational methods that take inspiration from neuronal systems in nature.

Biological neural network: By biological neural network, we denote neural architectures as found in nature, e.g., in the human neocortex.

Spiking neural network: Neural network type where neurons communicate with each other using discrete temporal events, similarly to how most biological neurons communicate with each other using electrical pulses.

(Leaky) Integrate-and-Fire neuron: Abstract mathematical model of biological neurons.

Synapse: Connection point between two neurons.

Presynaptic: Quantities (e.g., spikes) coming before a synapse.

Postsynaptic: Quantities coming after a synapse.

Refractory period: Time period after a neuron spiked, during which the ability of the neuron to spike again is inhibited.

Rate-based coding: Information is encoded in the population or time-averaged spike activity of neurons.

Spike-based coding: Information is encoded in the exact spike timings of neurons.

Time-To-First-Spike: Spike-based coding where information is encoded in the latency between stimulus onset and subsequent spiking.

FURTHER READING

Kucik, A. S., and Meoni, G. (2021) "Investigating spiking neural networks for energy-efficient on-board ai applications. a case study in land cover and land use classification". *Proceedings of the IEEE/CVF Conference on Computer Vision and Pattern Recognition*, pp. 2020-2030.

Sikorski, O., Izzo, D., & Meoni, G. (2021) "Event-based spacecraft landing using time-to-contact". In *Proceedings of the IEEE/CVF Conference on Computer Vision and Pattern Recognition* (pp. 1941-1950).

Open Neuromorphic. `https://github.com/open-neuromorphic/open-neuromorphic`. Accessed: 04/2023.

BIBLIOGRAPHY

[1] Charlotte Frenkel, David Bol, and Giacomo Indiveri. Bottom-up and top-down neural processing systems design: Neuromorphic intelligence as the convergence of natural and artificial intelligence. *arXiv preprint arXiv:2106.01288*, 2021.

[2] Kaushık Roy, Akhilesh Jaiswal, and Priyadarshini Panda. Towards spike-based machine intelligence with neuromorphic computing. *Nature*, 575(7784):607–617, 2019.

[3] Geoffrey Hinton. The forward-forward algorithm: Some preliminary investigations. *arXiv preprint arXiv:2212.13345*, 2022.

[4] Gianluca Furano, Gabriele Meoni, Aubrey Dunne, David Moloney, Veronique Ferlet-Cavrois, Antonis Tavoularis, Jonathan Byrne, Léonie Buckley, Mihalis Psarakis, Kay-Obbe Voss, et al. Towards the use of artificial intelligence on the edge in space systems: Challenges and opportunities. *IEEE Aerospace and Electronic Systems Magazine*, 35(12):44–56, 2020.

[5] Fredrik C Bruhn, Nandinbaatar Tsog, Fabian Kunkel, Oskar Flordal, and Ian Troxel. Enabling radiation tolerant heterogeneous gpu-based onboard data processing in space. *CEAS Space Journal*, 12(4):551–564, 2020.

[6] Gianluca Giuffrida, Luca Fanucci, Gabriele Meoni, Matej Batič, Léonie Buckley, Aubrey Dunne, Chris van Dijk, Marco Esposito, John Hefele, Nathan Vercruyssen, Gianluca Furano, Massimiliano Pastena, and Josef Aschbacher. The phi-sat-1 mission: The first on-board deep neural network demonstrator for satellite earth observation. *IEEE Transactions on Geoscience and Remote Sensing*, 60:1–14, 2022.

[7] Georges Labrèche, David Evans, Dominik Marszk, Tom Mladenov, Vasundhara Shiradhonkar, Tanguy Soto, and Vladimir Zelenevskiy. Opssat spacecraft autonomy with tensorflow lite, unsupervised learning, and online machine learning. In *2022 IEEE Aerospace Conference*, 2022.

[8] Aksel S Danielsen, Tor Arne Johansen, and Joseph L Garrett. Self-organizing maps for clustering hyperspectral images on-board a cubesat. *Remote Sensing*, 13(20):4174, 2021.

[9] V. Medici, G. Orchard, S. Ammann, G. Indiveri, and S. Fry. Neuromorphic Computation of Optic Flow Data. techreport 08/6303, European Space Agency, the Advanced Concepts Team, 2010. Available online at www.esa.int/act.

[10] F. Valette, F. Ruffier, and S. Viollet. Neuromorphic Computation of Optic Flow Data. techreport 08/6303b, European Space Agency, the Advanced Concepts Team, 2009. Available online at www.esa.int/act.

[11] Florent Valette, Franck Ruffier, Stéphane Viollet, and Tobias Seidl. Biomimetic optic flow sensing applied to a lunar landing scenario. In *2010 IEEE International Conference on Robotics and Automation*, pages 2253–2260. IEEE, 2010.

[12] Dario Izzo, Nicolás Weiss, and Tobias Seidl. Constant-optic-flow lunar landing: Optimality and guidance. *Journal of Guidance, Control, and Dynamics*, 34(5):1383–1395, 2011.

[13] Mark Frye. Elementary motion detectors. *Current Biology*, 25(6):R215–R217, 2015.

[14] Christian Brandli, Raphael Berner, Minhao Yang, Shih-Chii Liu, and Tobi Delbruck. A 240× 180 130 db 3 μs latency global shutter spatiotemporal vision sensor. *IEEE Journal of Solid-State Circuits*, 49(10):2333–2341, 2014.

[15] Gemma Taverni, Diederik Paul Moeys, Chenghan Li, Celso Cavaco, Vasyl Motsnyi, David San Segundo Bello, and Tobi Delbruck. Front and back illuminated dynamic and active pixel vision sensors comparison. *IEEE Transactions on Circuits and Systems II: Express Briefs*, 65(5):677–681, 2018.

[16] Mike Davies, Narayan Srinivasa, Tsung-Han Lin, Gautham Chinya, Yongqiang Cao, Sri Harsha Choday, Georgios Dimou, Prasad Joshi, Nabil Imam, Shweta Jain, et al. Loihi: A neuromorphic manycore processor with on-chip learning. *IEEE Micro*, 38(1):82–99, 2018.

[17] Filipp Akopyan, Jun Sawada, Andrew Cassidy, Rodrigo Alvarez-Icaza, John Arthur, Paul Merolla, Nabil Imam, Yutaka Nakamura, Pallab Datta, Gi-Joon Nam, et al. Truenorth: Design and tool flow of a 65 mw 1 million neuron programmable neurosynaptic chip. *IEEE transactions on computer-aided design of integrated circuits and systems*, 34(10):1537–1557, 2015.

[18] Zhilu Ye, Rui Liu, Jennifer L Taggart, Hugh J Barnaby, and Shimeng Yu. Evaluation of radiation effects in rram-based neuromorphic computing system for inference. *IEEE Transactions on Nuclear Science*, 66(1):97–103, 2018.

[19] Alan L Hodgkin and Andrew F Huxley. Action potentials recorded from inside a nerve fibre. *Nature*, 144(3651):710–711, 1939.

[20] Wulfram Gerstner and Richard Naud. How good are neuron models? *Science*, 326(5951):379–380, 2009.

[21] Romain Brette and Wulfram Gerstner. Adaptive exponential integrate-and-fire model as an effective description of neuronal activity. *Journal of neurophysiology*, 94(5):3637–3642, 2005.

[22] Guillaume Bellec, Darjan Salaj, Anand Subramoney, Robert Legenstein, and Wolfgang Maass. Long short-term memory and learning-to-learn in networks of spiking neurons. *Advances in neural information processing systems*, 31, 2018.

[23] Bojian Yin, Federico Corradi, and Sander M Bohté. Accurate and efficient time-domain classification with adaptive spiking recurrent neural networks. *Nature Machine Intelligence*, 3(10):905–913, 2021.

[24] Friedemann Zenke, Sander M Bohté, Claudia Clopath, Iulia M Comşa, Julian Göltz, Wolfgang Maass, Timothée Masquelier, Richard Naud, Emre O Neftci, Mihai A Petrovici, et al. Visualizing a joint future of neuroscience and neuromorphic engineering. *Neuron*, 109(4):571–575, 2021.

[25] Simon Thorpe, Denis Fize, and Catherine Marlot. Speed of processing in the human visual system. *nature*, 381(6582):520–522, 1996.

[26] Robert Gütig and Haim Sompolinsky. The tempotron: a neuron that learns spike timing–based decisions. *Nature neuroscience*, 9(3):420–428, 2006.

[27] Dominik Dold, Josep Soler Garrido, Victor Caceres Chian, Marcel Hildebrandt, and Thomas Runkler. Neuro-symbolic computing with spiking neural networks. In *Proceedings of the International Conference on Neuromorphic Systems 2022*, pages 1–4, 2022.

[28] Dominik Dold. Relational representation learning with spike trains. In *2022 International Joint Conference on Neural Networks (IJCNN)*, pages 1–8, 2022.

[29] Lars Buesing, Johannes Bill, Bernhard Nessler, and Wolfgang Maass. Neural dynamics as sampling: a model for stochastic computation in recurrent networks of spiking neurons. *PLoS computational biology*, 7(11):e1002211, 2011.

[30] Mihai A Petrovici, Johannes Bill, Ilja Bytschok, Johannes Schemmel, and Karlheinz Meier. Stochastic inference with spiking neurons in the high-conductance state. *Physical Review E*, 94(4):042312, 2016.

[31] Alexandre Payeur, Jordan Guerguiev, Friedemann Zenke, Blake A Richards, and Richard Naud. Burst-dependent synaptic plasticity can coordinate learning in hierarchical circuits. *Nature neuroscience*, 24(7):1010–1019, 2021.

[32] Emre O Neftci, Hesham Mostafa, and Friedemann Zenke. Surrogate gradient learning in spiking neural networks: Bringing the power of gradient-based optimization to spiking neural networks. *IEEE Signal Processing Magazine*, 36(6):51–63, 2019.

[33] Michael Pfeiffer and Thomas Pfeil. Deep learning with spiking neurons: opportunities and challenges. *Frontiers in neuroscience*, page 774, 2018.

[34] Simon Davidson and Steve B Furber. Comparison of artificial and spiking neural networks on digital hardware. *Frontiers in Neuroscience*, 15:651141, 2021.

[35] Andrzej S Kucik and Gabriele Meoni. Investigating spiking neural networks for energy-efficient on-board ai applications. a case study in land cover and land use classification. In *Proceedings of the IEEE/CVF Conference on Computer Vision and Pattern Recognition*, pages 2020–2030, 2021.

[36] Guo-qiang Bi and Mu-ming Poo. Synaptic modifications in cultured hippocampal neurons: dependence on spike timing, synaptic strength, and postsynaptic cell type. *Journal of neuroscience*, 18(24):10464–10472, 1998.

[37] Bernd Illing, Jean Ventura, Guillaume Bellec, and Wulfram Gerstner. Local plasticity rules can learn deep representations using self-supervised contrastive predictions. *Advances in Neural Information Processing Systems*, 34:30365–30379, 2021.

[38] Hesham Mostafa. Supervised learning based on temporal coding in spiking neural networks. *IEEE transactions on neural networks and learning systems*, 29(7):3227–3235, 2017.

[39] Iulia M Comsa, Krzysztof Potempa, Luca Versari, Thomas Fischbacher, Andrea Gesmundo, and Jyrkı Alakuijala. Temporal coding in spiking neural networks with alpha synaptic function. In *ICASSP 2020-2020 IEEE International Conference on Acoustics, Speech and Signal Processing (ICASSP)*, pages 8529–8533. IEEE, 2020.

[40] Saeed Reza Kheradpisheh and Timothée Masquelier. Temporal backpropagation for spiking neural networks with one spike per neuron. *International Journal of Neural Systems*, 30(06):2050027, 2020.

[41] Julian Göltz, Laura Kriener, Andreas Baumbach, Sebastian Billaudelle, Oliver Breitwieser, Benjamin Cramer, Dominik Dold, Akos Ferenc Kungl, Walter Senn, Johannes Schemmel, et al. Fast and energy-efficient neuromorphic deep learning with first-spike times. *Nature machine intelligence*, 3(9):823–835, 2021.

[42] Friedemann Zenke and Surya Ganguli. Superspike: Supervised learning in multilayer spiking neural networks. *Neural computation*, 30(6):1514–1541, 2018.

[43] Friedemann Zenke and Tim P Vogels. The remarkable robustness of surrogate gradient learning for instilling complex function in spiking neural networks. *Neural computation*, 33(4):899–925, 2021.

[44] Yann LeCun and Corinna Cortes. MNIST handwritten digit database. 2010.

[45] Nicolas Perez-Nieves, Vincent CH Leung, Pier Luigi Dragotti, and Dan FM Goodman. Neural heterogeneity promotes robust learning. *Nature communications*, 12(1):1–9, 2021.

[46] Timo C Wunderlich and Christian Pehle. Event-based backpropagation can compute exact gradients for spiking neural networks. *Scientific Reports*, 11(1):1–17, 2021.

[47] Jason K Eshraghian, Max Ward, Emre Neftci, Xinxin Wang, Gregor Lenz, Girish Dwivedi, Mohammed Bennamoun, Doo Seok Jeong, and Wei D Lu. Training spiking neural networks using lessons from deep learning. *arXiv preprint arXiv:2109.12894*, 2021.

[48] Christian Pehle and Jens Egholm Pedersen. Norse - A deep learning library for spiking neural networks, January 2021. Documentation: https://norse.ai/docs/.

[49] Wei Fang, Yanqi Chen, Jianhao Ding, Ding Chen, Zhaofei Yu, Huihui Zhou, Timothée Masquelier, Yonghong Tian, and other contributors. Spikingjelly, 2020.

[50] Hananel Hazan, Daniel J. Saunders, Hassaan Khan, Devdhar Patel, Darpan T. Sanghavi, Hava T. Siegelmann, and Robert Kozma. Bindsnet: A machine learning-oriented spiking neural networks library in python. *Frontiers in Neuroinformatics*, 12:89, 2018.

[51] Steve B Furber, David R Lester, Luis A Plana, Jim D Garside, Eustace Painkras, Steve Temple, and Andrew D Brown. Overview of the spinnaker system architecture. *IEEE transactions on computers*, 62(12):2454–2467, 2012.

[52] Christian Mayr, Sebastian Hoeppner, and Steve Furber. Spinnaker 2: A 10 million core processor system for brain simulation and machine learning. *arXiv preprint arXiv:1911.02385*, 2019.

[53] Juncheng Shen, De Ma, Zonghua Gu, Ming Zhang, Xiaolei Zhu, Xiaoqiang Xu, Qi Xu, Yangjing Shen, and Gang Pan. Darwin: a neuromorphic hardware co-processor based on spiking neural networks. *Science China Information Sciences*, 59(2):1–5, 2016.

[54] Jing Pei, Lei Deng, Sen Song, Mingguo Zhao, Youhui Zhang, Shuang Wu, Guanrui Wang, Zhe Zou, Zhenzhi Wu, Wei He, et al. Towards artificial general intelligence with hybrid tianjic chip architecture. *Nature*, 572(7767):106–111, 2019.

[55] Christian Pehle, Sebastian Billaudelle, Benjamin Cramer, Jakob Kaiser, Korbinian Schreiber, Yannik Stradmann, Johannes Weis, Aron Leibfried, Eric Müller, and Johannes Schemmel. The brainscales-2 accelerated neuromorphic system with hybrid plasticity. *Frontiers in Neuroscience*, 16, 2022.

[56] Thomas Pfeil, Andreas Grübl, Sebastian Jeltsch, Eric Müller, Paul Müller, Mihai A Petrovici, Michael Schmuker, Daniel Brüderle, Johannes Schemmel, and Karlheinz Meier. Six networks on a universal neuromorphic computing substrate. *Frontiers in neuroscience*, 7:11, 2013.

[57] Saber Moradi, Ning Qiao, Fabio Stefanini, and Giacomo Indiveri. A scalable multicore architecture with heterogeneous memory structures for dynamic neuromorphic asynchronous processors (dynaps). *IEEE transactions on biomedical circuits and systems*, 12(1):106–122, 2017.

[58] Ning Qiao, Hesham Mostafa, Federico Corradi, Marc Osswald, Fabio Stefanini, Dora Sumislawska, and Giacomo Indiveri. A reconfigurable on-line learning spiking neuromorphic processor comprising 256 neurons and 128k synapses. *Frontiers in neuroscience*, 9:141, 2015.

[59] Ben Varkey Benjamin, Peiran Gao, Emmett McQuinn, Swadesh Choudhary, Anand R Chandrasekaran, Jean-Marie Bussat, Rodrigo Alvarez-Icaza, John V Arthur, Paul A Merolla, and Kwabena Boahen. Neurogrid: A mixed-analog-digital multichip system for large-scale neural simulations. *Proceedings of the IEEE*, 102(5):699–716, 2014.

[60] Intel. `https://en.wikichip.org/wiki/intel/loihi`. Accessed: 2022/11/14.

[61] Amar Shrestha, Haowen Fang, Daniel Patrick Rider, Zaidao Mei, and Qinru Qiu. In-hardware learning of multilayer spiking neural networks on a neuromorphic processor. In *2021 58th ACM/IEEE Design Automation Conference (DAC)*, pages 367–372. IEEE, 2021.

[62] Gregory K Chen, Raghavan Kumar, H Ekin Sumbul, Phil C Knag, and Ram K Krishnamurthy. A 4096-neuron 1m-synapse 3.8-pj/sop spiking neural network with on-chip stdp learning and sparse weights in 10-nm finfet cmos. *IEEE Journal of Solid-State Circuits*, 54(4):992–1002, 2018.

[63] Chit-Kwan Lin, Andreas Wild, Gautham N Chinya, Yongqiang Cao, Mike Davies, Daniel M Lavery, and Hong Wang. Programming spiking neural networks on intel's loihi. *Computer*, 51(3):52–61, 2018.

[64] Anup Vanarse, Adam Osseiran, Alexander Rassau, and Peter van der Made. Application of neuromorphic olfactory approach for high-accuracy classification of malts. *Sensors*, 22(2):440, 2022.

[65] Chen Liu, Guillaume Bellec, Bernhard Vogginger, David Kappel, Johannes Partzsch, Felix Neumärker, Sebastian Höppner, Wolfgang Maass, Steve B Furber, Robert Legenstein, et al. Memory-efficient deep learning on a spinnaker 2 prototype. *Frontiers in neuroscience*, 12:840, 2018.

[66] Charlotte Frenkel, Jean-Didier Legat, and David Bol. A 28-nm convolutional neuromorphic processor enabling online learning with spike-based retinas. In *2020 IEEE International Symposium on Circuits and Systems (ISCAS)*, pages 1–5. IEEE, 2020.

[67] Evangelos Stromatias, Daniel Neil, Francesco Galluppi, Michael Pfeiffer, Shih-Chii Liu, and Steve Furber. Scalable energy-efficient, low-latency implementations of trained spiking deep belief networks on spinnaker. In *2015 International Joint Conference on Neural Networks (IJCNN)*, pages 1–8. IEEE, 2015.

[68] Andrew S Cassidy, Paul Merolla, John V Arthur, Steve K Esser, Bryan Jackson, Rodrigo Alvarez-Icaza, Pallab Datta, Jun Sawada,

Theodore M Wong, Vitaly Feldman, et al. Cognitive computing building block: A versatile and efficient digital neuron model for neurosynaptic cores. In *The 2013 International Joint Conference on Neural Networks (IJCNN)*, pages 1–10. IEEE, 2013.

[69] Mike Davies. Benchmarks for progress in neuromorphic computing. *Nature Machine Intelligence*, 1(9):386–388, 2019.

[70] Maxence Bouvier, Alexandre Valentian, Thomas Mesquida, Francois Rummens, Marina Reyboz, Elisa Vianello, and Edith Beigne. Spiking neural networks hardware implementations and challenges: A survey. *ACM Journal on Emerging Technologies in Computing Systems (JETC)*, 15(2):1–35, 2019.

[71] Bing Han, Aayush Ankit, Abhronil Sengupta, and Kaushik Roy. Cross-layer design exploration for energy-quality tradeoffs in spiking and non-spiking deep artificial neural networks. *IEEE Transactions on Multi-Scale Computing Systems*, 4(4):613–623, 2017.

[72] Patrick Helber, Benjamin Bischke, Andreas Dengel, and Damian Borth. Eurosat: A novel dataset and deep learning benchmark for land use and land cover classification. *IEEE Journal of Selected Topics in Applied Earth Observations and Remote Sensing*, 2019.

[73] Patrick Helber, Benjamin Bischke, Andreas Dengel, and Damian Borth. Introducing eurosat: A novel dataset and deep learning benchmark for land use and land cover classification. In *IGARSS 2018-2018 IEEE International Geoscience and Remote Sensing Symposium*, pages 204–207. IEEE, 2018.

[74] Karen Simonyan and Andrew Zisserman. Very deep convolutional networks for large-scale image recognition. *arXiv preprint arXiv:1409.1556*, 2014.

[75] Jia Deng, Wei Dong, Richard Socher, Li-Jia Li, Kai Li, and Li Fei-Fei. Imagenet: A large-scale hierarchical image database. In *2009 IEEE conference on computer vision and pattern recognition*, pages 248–255. Ieee, 2009.

[76] Nicolas Franceschini, Jean-Marc Pichon, and Christian Blanes. From insect vision to robot vision. *Philosophical Transactions of The Royal Society Of London. Series B: Biological Sciences*, 337(1281):283–294, 1992.

[77] Young Min Song, Yizhu Xie, Viktor Malyarchuk, Jianliang Xiao, In-hwa Jung, Ki-Joong Choi, Zhuangjian Liu, Hyunsung Park, Chaofeng Lu, Rak-Hwan Kim, et al. Digital cameras with designs inspired by the arthropod eye. *Nature*, 497(7447):95–99, 2013.

[78] David H. Hubel and Torsten N. Wiesel. Receptive Fields and Functional Architecture in Two Nonstriate Visual Areas (18 and 19) of the Cat. *Journal of Neurophysiology*, 28(2):229–289.

[79] Eleonora N. Grigoryan. Self-Organization of the Retina during Eye Development, Retinal Regeneration In Vivo, and in Retinal 3D Organoids In Vitro. *Biomedicines*, 10(6):1458.

[80] Richard H. Masland. Cell Populations of the Retina: The Proctor Lecture. *Investigative Opthalmology & Visual Science*, 52(7):4581–4591.

[81] Reece Mazade, Jianzhong Jin, Carmen Pons, and Jose-Manuel Alonso. Functional Specialization of ON and OFF Cortical Pathways for Global-Slow and Local-Fast Vision. *Cell Reports*, 27(10):2881–2894.e5.

[82] Thomas Euler, Silke Haverkamp, Timm Schubert, and Tom Baden. Retinal bipolar cells: Elementary building blocks of vision. *Nature Reviews Neuroscience*, 15(8):507–519.

[83] Tim Gollisch and Markus Meister. Eye Smarter than Scientists Believed: Neural Computations in Circuits of the Retina. *Neuron*, 65(2):150–164.

[84] Richard H. Masland. The tasks of amacrine cells. *Visual Neuroscience*, 29(1):3–9.

[85] E. J. Chichilnisky and R. S. Kalmar. Temporal Resolution of Ensemble Visual Motion Signals in Primate Retina. *Journal of Neuroscience*, 23(17):6681–6689.

[86] Tom Baden, Federico Esposti, Anton Nikolaev, and Leon Lagnado. Spikes in Retinal Bipolar Cells Phase-Lock to Visual Stimuli with Millisecond Precision. *Current Biology*, 21(22):1859–1869.

[87] Jonathan N. Tinsley, Maxim I. Molodtsov, Robert Prevedel, David Wartmann, Jofre Espigulé-Pons, Mattias Lauwers, and Alipasha Vaziri. Direct detection of a single photon by humans. *Nature Communications*, 7(1):12172.

[88] Yena Han, Gemma Roig, Gad Geiger, and Tomaso Poggio. Scale and translation-invariance for novel objects in human vision. *Scientific Reports*, 10(1):1411.

[89] R.T. Pramod, Harish Katti, and S.P. Arun. Human peripheral blur is optimal for object recognition. *Vision Research*, 200:108083.

[90] Horace B. Barlow and Peter Földiák. Adaptation and Decorrelation in the Cortex. In R. Durbin, C. Miall, and G. Mitchison, editors, *The Computing Neuron*, pages 54–72. Addison-Wesley.

[91] Robert Shapley and Christina Enroth-Cugell. Visual adaptation and retinal gain controls. *Progress in Retinal Research*, 3:263–346.

[92] Marie E Burns and Trevor D Lamb. Visual Transduction by Rod and Cone Photoreceptors. In John S. Werner and Leo M. Chalupa, editors, *The Visual Neurosciences*, page 19. MIT Press.

[93] C. P. Ratliff, B. G. Borghuis, Y.-H. Kao, P. Sterling, and V. Balasubramanian. Retina is structured to process an excess of darkness in natural scenes. *Proceedings of the National Academy of Sciences*, 107(40):17368–17373.

[94] Daniel K. Freeman, Gilberto Graña, and Christopher L. Passaglia. Retinal Ganglion Cell Adaptation to Small Luminance Fluctuations. *Journal of Neurophysiology*, 104(2):704–712.

[95] Fred Rieke. Temporal Contrast Adaptation in Salamander Bipolar Cells. *The Journal of Neuroscience*, 21(23):9445–9454.

[96] Andrew T. Rider, G. Bruce Henning, and Andrew Stockman. Light adaptation controls visual sensitivity by adjusting the speed and gain of the response to light. *PLOS ONE*, 14(8):e0220358.

[97] James T. Pearson and Daniel Kerschensteiner. Ambient illumination switches contrast preference of specific retinal processing streams. *Journal of Neurophysiology*, 114(1):540–550.

[98] Markus Meister and Michael J. Berry. The Neural Code of the Retina. *Neuron*, 22(3):435–450.

[99] Michiel Van Wyk, Heinz Wässle, and W. Rowland Taylor. Receptive field properties of ON- and OFF-ganglion cells in the mouse retina. *Visual Neuroscience*, 26(3):297–308.

[100] Fred Rieke and Michael E. Rudd. The Challenges Natural Images Pose for Visual Adaptation. *Neuron*, 64(5):605–616.

[101] H. B. Barlow, R. Fitzhugh, and S. W. Kuffler. Change of organization in the receptive fields of the cat's retina during dark adaptation. *The Journal of Physiology*, 137(3):338–354.

[102] Susana Martinez-Conde and Stephen L. Macknik. Unchanging visions: The effects and limitations of ocular stillness. *Philosophical Transactions of the Royal Society B: Biological Sciences*, 372(1718):20160204.

[103] Robert G. Alexander and Susana Martinez-Conde. Fixational Eye Movements. In Christoph Klein and Ulrich Ettinger, editors, *Eye Movement Research*, Studies in Neuroscience, Psychology and Behavioral Economics, pages 73–115. Springer International Publishing.

[104] K. Boahen. Retinomorphic vision systems. In *Proceedings of Fifth International Conference on Microelectronics for Neural Networks*, pages 2–14.

[105] Todd R Appleby and Michael B Manookin. Selectivity to approaching motion in retinal inputs to the dorsal visual pathway. *eLife*, 9:e51144.

[106] Y. LeCun, B. Boser, J. S. Denker, D. Henderson, R. E. Howard, W. Hubbard, and L. D. Jackel. Backpropagation Applied to Handwritten Zip Code Recognition. *Neural Computation*, 1(4):541–551.

[107] Alex Krizhevsky, Ilya Sutskever, and Geoffrey E Hinton. Imagenet classification with deep convolutional neural networks. In F. Pereira, C.J. Burges, L. Bottou, and K.Q. Weinberger, editors, *Advances in Neural Information Processing Systems*, volume 25. Curran Associates, Inc., 2012.

[108] Ian Goodfellow, Yoshua Bengio, and Aaron Courville. *Deep Learning*. MIT Press.

[109] Kumar Chellapilla, Sidd Puri, and Patrice Simard. High performance convolutional neural networks for document processing. In *Tenth international workshop on frontiers in handwriting recognition*, 2006.

[110] Perseverance rover landing on mars recorded with onboard camera. https://mars.nasa.gov/mars2020/multimedia/videos/?v=461. Accessed: 2022/11/23.

[111] Carver Mead. Adaptive Retina. In Carver Mead and Mohammed Ismail, editors, *Analog VLSI Implementation of Neural Systems*, volume 80 of *The Kluwer International Series in Engineering and Computer Science*, pages 239–246. Springer US.

[112] Carver Mead. A Sensitive Electronic Photoreceptor. In *1985 Chapel Hill Conference on Very Large Scale Integration*, pages 463–471.

[113] T. Delbrück and C. A. Mead. An electronic photoreceptor sensitive to small changes in intensity. In *Proceedings of the 1st International Conference on Neural Information Processing Systems*, NIPS'88, pages 720–727. MIT Press.

[114] Misha A Mahowald and Carver Mead. The Silicon Retina. *SCIENTIFIC AMERICAN*, page 8.

[115] Kareem A Zaghloul and Kwabena Boahen. A silicon retina that reproduces signals in the optic nerve. *Journal of Neural Engineering*, 3(4):257–267.

[116] T. Serrano-Gotarredona, A.G. Andreou, and B. Linares-Barranco. AER image filtering architecture for vision-processing systems. *IEEE Transactions on Circuits and Systems I: Fundamental Theory and Applications*, 46(9):1064–1071.

[117] Rafael Serrano-Gotarredona, Teresa Serrano-Gotarredona, Antonio Acosta-Jimenez, and Bernab Linares-Barranco. A Neuromorphic Cortical-Layer Microchip for Spike-Based Event Processing Vision Systems. *IEEE Transactions on Circuits and Systems I: Regular Papers*, 53(12):2548 2566.

[118] Yuji Nozaki and Tobi Delbruck. Temperature and Parasitic Photocurrent Effects in Dynamic Vision Sensors. *IEEE Transactions on Electron Devices*, 64(8):3239–3245.

[119] Tobi Delbruck, Chenghan Li, Rui Graca, and Brian Mcreynolds. Utility and feasibility of a center surround event camera. In *2022 IEEE International Conference on Image Processing (ICIP)*, pages 381–385. IEEE, 2022.

[120] Elvin Hajizada, Patrick Berggold, Massimiliano Iacono, Arren Glover, and Yulia Sandamirskaya. Interactive continual learning for robots: A neuromorphic approach. In *Proceedings of the International Conference on Neuromorphic Systems 2022*, pages 1–10. ACM.

[121] Amélie Gruel, Dalia Hareb, Antoine Grimaldi, Jean Martinet, Laurent Perrinet, Bernabé Linares-Barranco, and Teresa Serrano-Gotarredona. Stakes of Neuromorphic Foveation: A promising future for embedded event cameras.

[122] Yuhuang Hu, Shih-Chii Liu, and Tobi Delbruck. v2e: From video frames to realistic dvs events. In *2021 IEEE/CVF Conference on Computer Vision and Pattern Recognition Workshops (CVPRW)*, pages 1312–1321, 2021.

[123] Henri Rebecq, Daniel Gehrig, and Davide Scaramuzza. Esim: an open event camera simulator. In Aude Billard, Anca Dragan, Jan Peters, and Jun Morimoto, editors, *Proceedings of The 2nd Conference on Robot Learning*, volume 87 of *Proceedings of Machine Learning Research*, pages 969–982. PMLR, 29–31 Oct 2018.

[124] Matthew G McHarg, Richard L Balthazor, Brian J McReynolds, David H Howe, Colin J Maloney, Daniel O'Keefe, Rayomand Bam, Gabriel Wilson, Paras Karki, Alexandre Marcireau, et al. Falcon neuro: an event-based sensor on the international space station. *Optical Engineering*, 61(8):085105, 2022.

[125] Olaf Sikorski, Dario Izzo, and Gabriele Meoni. Event-based spacecraft landing using time-to-contact. In *2021 IEEE/CVF Conference on Computer Vision and Pattern Recognition Workshops (CVPRW)*, pages 1941–1950, 2021.

[126] Sofia McLeod, Gabriele Meoni, Dario Izzo, Anne Mergy, Daqi Liu, Yasir Latif, Ian Reid, and Tat-Jun Chin. Globally optimal event-based divergence estimation for ventral landing. In *2022 European Conference on Computer Vision (ECCV)*, 2022.

[127] Tat-Jun Chin, Samya Bagchi, Anders Eriksson, and Andre Van Schaik. Star tracking using an event camera. In *Proceedings of the IEEE/CVF Conference on Computer Vision and Pattern Recognition Workshops*, 2019.

[128] Mohsi Jawaid, Ethan Elms, Yasir Latif, and Tat-Jun Chin. Towards bridging the space domain gap for satellite pose estimation using event sensing. *arXiv preprint arXiv:2209.11945*, 2022.

[129] Nicholas Owen Ralph, Alexandre Marcireau, Saeed Afshar, Nicholas Tothill, André van Schaik, and Gregory Cohen. Astrometric

calibration and source characterisation of the latest generation neuromorphic event-based cameras for space imaging. *arXiv preprint arXiv:2211.09939*, 2022.

[130] Seth Roffe, Himanshu Akolkar, Alan D George, Bernabé Linares-Barranco, and Ryad B Benosman. Neutron-induced, single-event effects on neuromorphic event-based vision sensor: A first step and tools to space applications. *IEEE Access*, 9:85748–85763, 2021.

[131] Dario Izzo and Guido de Croon. Landing with time-to-contact and ventral optic flow estimates. *Journal of Guidance, Control, and Dynamics*, 35(4):1362–1367, 2012.

[132] Hann Woei Ho, Guido CHE de Croon, and Qiping Chu. Distance and velocity estimation using optical flow from a monocular camera. *International Journal of Micro Air Vehicles*, 9(3):198–208, 2017.

[133] Planet and asteroid natural scene generation utility product website. https://pangu.software/. Accessed: 2022/11/14.

[134] Guillermo Gallego, Henri Rebecq, and Davide Scaramuzza. A unifying contrast maximization framework for event cameras, with applications to motion, depth, and optical flow estimation. In *2018 IEEE/CVF Conference on Computer Vision and Pattern Recognition*, pages 3867–3876, 2018.

[135] Timo Stoffregen and Lindsay Kleeman. Event cameras, contrast maximization and reward functions: An analysis. In *2019 IEEE/CVF Conference on Computer Vision and Pattern Recognition (CVPR)*, pages 12292–12300, 2019.

[136] Naresh C Gupta and Laveen N Kanal. 3-d motion estimation from motion field. *Artificial Intelligence*, 78(1-2):45–86, 1995.

[137] Hugh Christopher Longuet-Higgins and Kvetoslav Prazdny. The interpretation of a moving retinal image. *Proceedings of the Royal Society of London. Series B. Biological Sciences*, 208(1173):385–397, 1980.

Artificial Intelligence for Spacecraft Location Estimation based on Craters

Keiki Takadama

Department of Informatics, The University of Electro-Communications, Tokyo, Japan

Fumito Uwano

Faculty of Environmental, Life, Natural Science and Technology, Okayama University, Okayama, Japan

Yuka Waragai

Department of Informatics, The University of Electro-Communications, Tokyo, Japan

Iko Nakari

Department of Informatics, The University of Electro-Communications, Tokyo, Japan

Hiroyuki Kamata

Department of Electronics and Bioinformatics, Meiji University, Kanagawa, Japan

Takayuki Ishida

Research and Development Division, Japan Aerospace Exploration Agency (JAXA), Kanagawa, Japan

Seisuke Fukuda

Institute of Space and Astronautical Science (ISAS), Japan Aerospace Exploration Agency (JAXA), Kanagawa, Japan

Shujiro Sawai

Institute of Space and Astronautical Science (ISAS), Japan Aerospace Exploration Agency (JAXA), Kanagawa, Japan

DOI: 10.1201/9781003366386-5

Shinichiro Sakai

Institute of Space and Astronautical Science (ISAS), Japan Aerospace Exploration Agency (JAXA), Kanagawa, Japan

CONTENTS

5.1	Introduction	162
5.2	SLIM mission	163
5.3	Related works on location estimation	164
	5.3.1 Category of location estimation methods	164
	5.3.2 Difficulty of crater matching	165
	5.3.3 Line segment matching (LSM)	166
5.4	Triangle similarity matching (TSM)	168
	5.4.1 Overview	168
	5.4.2 Algorithm of TSM	168
	5.4.3 Mechanisms	170
	5.4.4 Crater matching strategy	172
5.5	Improved TSM	174
	5.5.1 Appropriate crater selection	174
	5.5.2 Close craters elimination	175
5.6	Test case	176
	5.6.1 Cases in coasting phase	176
	5.6.2 Errors in images	177
5.7	Experiments	178
	5.7.1 Experimental design	178
	5.7.2 Evaluation criteria and parameter setting	179
	5.7.3 Experimental results	179
	5.7.4 Discussion	180
5.8	Conclusion	185
	Bibliography	188

THIS chapter focuses on our method, Triangle Similarity Matching (TSM), which estimates a spacecraft location by comparing the triangle (composed of the three detected craters) in the camera-shot image taken by the spacecraft with the triangle (composed of the three detected craters) in the crater map (stored in the spacecraft beforehand) as the information on the craters in the planetary surface. To improve an accuracy of a spacecraft location estimation, this chapter proposes the method for TSM which selects the "appropriate craters" and eliminates the "close craters" in the camera-shot image, and aims at investigating its effectiveness through the experiment. The proposed methods contribute to improving an accuracy of a spacecraft location estimation and preventing from wrongly estimating the spacecraft location when two or more craters are close to each other, where the wrong crater may be selected instead of the correct crater. In the experiment on the test-case employed in the Smart Lander for Investigating Moon (SLIM) mission of Japan Aerospace Exploration Agency (JAXA), the improved TSM which selects the "appropriate craters" and eliminates the "close craters" in the camera-shot image is compared with (the original) TSM and Line segment matching (LSM) proposed by JAXA. From the experiment the following implications have been revealed: (1) the improved TSM outperforms TSM and LSM in terms of an accuracy of spacecraft location estimation; (2) the improved TSM can prevent from the incorrect crater matching in comparison with TSM and LSM; and (3) the improved TSM is robust to both images with camera anomaly and those with large shadows due to a low sun elevation.

5.1 INTRODUCTION

In usual planetary landing, a spacecraft generally requires a large landing area without obstacles to land a safe area "where is *easy* to land." This means that the conventional approach is difficult make a spacecraft land at the area which is very close to an exploration target. In such a large landing area, even a rover is hard to surely reach at the exploration target because big rocks, craters, and caves may exist in a high possibility in the way to the target. This approach also requires a huge time to reach an exploration target which increases the mission cost and duration. To overcome this problem, Japan Aerospace Exploration Agency (JAXA) proposes the *pinpoint landing* on moon in the SLIM (Smart Lander for Investigating Moon) mission [1] which aims at establishing the technology of landing the pinpoint area "where is *desired* to land." Note that the pinpoint area is roughly the area within $100m$ from the target landing point. This approach contributes to future landing exploration by landing close to an investigating target area. To achieve this goal, it is indispensable for a spacecraft to estimate its current location by sensor data such as a camera-shot image.

One of approaches for a spacecraft location estimation is to match (a) the craters detected from the camera-shot image over the moon with (b) the craters in the crater map (database) which includes the locations (coordinates) of the craters on the moon obtained from "KAGUYA (SELENE)" satellite launched by JAXA [2]. To implement this approach, our previous research

Artificial Intelligence for Spacecraft Location Estimation based on Craters ■ 163

proposed the Triangle Similarity Matching (TSM) method [3], which estimates a spacecraft location by executing the following three procedures in turn: (1) the spacecraft takes the camera-shot image on the moon; (2) the craters are extracted from the camera-shot image, and (3) the extracted craters are compared with those in the crater map (stored in the spacecraft beforehand) as the information on the craters in the planetary surface. In detail, TSM compares the *triangle* (composed of the three detected craters) in the camera-shot image taken by the spacecraft with the *triangle* (composed of the three craters) in the crater map to estimate the spacecraft location. TSM focuses on the triangle matching because the triangle similarity does not depend on the height of a spacecraft.

What should be noted here is that an accuracy of a spacecraft location estimation by TSM cannot be guaranteed to be always sufficiently high because of the following reasons: (1) some craters may be detected even though there are no craters, while other craters may not be detected even though there are craters; and (2) it is difficult to determine which craters in the camera-shot image should be matched with those in the crater map especially in the case of many candidates of craters to be matched or the overlapped craters detected in the camera-shot image. To tackle this problem, this chapter improves TSM by selecting the "appropriate craters" as the candidates for matching and eliminating the "close craters" in the camera-shot image, and aims at investigating its effectiveness through the experiment.[1]

This paper is organized as follows. Section 5.2 explains the overview of the SLIM mission, and Section 5.3 introduces the related works. Section 5.4 describes TSM and Section 5.5 improves TSM in terms of selecting the "appropriate craters" and eliminating the "close craters" in the camera-shot image. Section 5.6 explains the experimental test case, and Section 5.7 conducts the experiment and shows the results. Finally, our conclusion is given in Section 9.4.

5.2   SLIM MISSION

As described in Section 10.1, the SLIM mission aims at establishing the technology of the *pinpoint landing* on the moon. To achieve the pinpoint landing, the following technologies should be established: (1) a surface topography matching based on the camera-shot image, (2) an automatically navigation of a spacecraft to a target area, and (3) a landing by avoiding obstacles. This chapter particularly tackles the issue (1).

---

[1]The preliminary version of these approaches were reported in [4, 5].

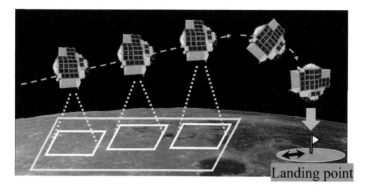

Figure 5.1   SLIM mission.

Figure 5.1 shows the landing sequence in the SLIM mission. The red broken line indicates the trajectory of the spacecraft. The yellow squares connected the orange dotted lines indicate the camera-shot range. The location with the flag is the goal and the circle around the flag indicates the range for the pinpoint landing. In the SLIM mission, a spacecraft modifies its course when taking the camera-shot image (*i.e.*, there are three coasting phases in Fig.5.1.)

5.3   RELATED WORKS ON LOCATION ESTIMATION

5.3.1   Category of location estimation methods

The location estimation approaches are roughly categorized as follows: (1) the location estimation for a rover and (2) the location estimation for a spacecraft/satellite. In detail, the major methods of the approach (1) include (1a) SLAM (Simultaneous Localization and Mapping) [6, 7] and (1b) the cross-view image matching [8], while the those of the approach (2) include (2a) the star matching [9, 10] and (2b) the crater matching [3, 11].

- **Location estimation for a rover**
  (1a) SLAM (Simultaneous Localization and Mapping) [6, 7] simultaneously creates the 2D or 3D map of a given environment and estimates a location of a mobile body (such as a rover). Concretely, the mobile body moves around in the environment to acquire the data around it by some sensors (such as a Lidar) which are needed for creating the map, and sequentially calculates its moving amount to estimate its location. As another method, (1b) the cross-view image matching method [8] estimates the location of a mobile body by matching the camera-shot

image taken by a mobile body (or a smartphone) as the street view image with the satellite image or the aerial photograph as the bird's eye view image. Since the locations of the landmarks in the satellite image or the aerial photograph are known/calculated beforehand, the location of the mobile body can be estimated from the correspondence of its location in the street view image to its location in the bird's eye view image.

- **Location estimation for a spacecraft/satellite**
  (2a) The star catalog matching method [9] [10] estimates the current location of a satellite according to the star pattern composed of the stars detected by the star sensor. As another method, (2b) the crater matching methods such as Line Segment Matching (LSM) [11] (described in subsection 5.3.3) and Triangle Similarity Matching (TSM) [3] (described in section 5.4) estimate the location of a spacecraft by matching the craters detected in the camera-shot image taken by the spacecraft with those in the crater map stored in the spacecraft beforehand. Like the cross-view image matching, the location of the spacecraft can be estimated from the correspondence of its location of the craters in the camera-shot image to its location in the crater map.

From the different features of the above approaches, SLAM and the cross-view image matching method are useful for a rover location estimation, but they are very difficult to be applied in to a spacecraft/satellite location estimation because it is hard to create the map in space due to a very small amount of landmarks and there is not enough time to create the map during the landing phase. The star catalog matching, on the other hand, is also hard to be applied in the SLIM mission because the accuracy of the star catalog matching is not enough high for the pinpoint landing which aims to land the area within $100m$ from the target landing point. From these difficulties, the crater matching is the only method for the SLIM mission.

5.3.2 Difficulty of crater matching

As described in the previous subsection, the crater matching has a potential of achieving the pinpoint landing, but it should overcome the problem that the craters in the crater map are not always correctly matched with those in the camera-shot image. To understand this difficulty, Figure 5.2 shows an example of the craters in both the camera-shot image and the crater map. The red and blue dots indicate the craters in the camera-shot image and the crater map, respectively. The red areas (a) include the only craters in the camera-shot

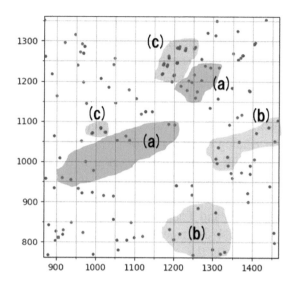

Figure 5.2 Craters (red craters in camera-shot image, blue craters in crater map).

image, while the blue areas (b) include the only craters in the crater map. The violet areas (c) include the craters matched in the camera-shot image and the crater map. From these crater placements, it is not easy to estimate the spacecraft location because (1) there are not enough craters matched in both the camera-shot image and the crater map image and (2) it is difficult to perfectly determine which the craters in the camera-shot image should be matched or not with those in the crater map especially in the case of many candidates of craters to be matched or the overlapped craters detected in the camera-shot image.

5.3.3 Line segment matching (LSM)

5.3.3.1 Overview

As one of the crater matching methods, JAXA proposed Line Segment Matching (LSM) [11] to estimate the spacecraft location by matching the line segments composed of two craters detected in the camera-shot image with those in the crater map, while satisfying the positional relation of the *external* craters (*i.e.*, the craters except for two craters in the line segment) in the camera-shot image and the crater map. To estimate the spacecraft location by LSM, the crater map of the planetary surface should be stored in the on-board computer of the spacecraft. More precisely, the line segment database (DB) created from the crater map is stored in the spacecraft in order to narrow down the target of the line segments among many other line segments created from all combination of craters, which contributes to reducing the matching time. This DB includes the coordinates of the two craters in the line segment, the coordinate of the center position of the line segment, and the length of the line segment.

5.3.3.2   Algorithm of LSM

As the algorithm of LSM, the following procedures are executed.

(1) Extracting craters
    After taking the camera-shot image by the spacecraft, the craters are extracted.

(2) Creating line segment
    The line segment composed of the two detected craters in the camera-shot image is created.

(3) Matching line segments
    The line segment in the camera-shot image is compared with those in the crater map (*i.e.*, the line segment DB) to find the matched line segments.

(4) Matching external craters
    After finding the matched line segments, all external craters in the camera-shot image are checked to match with those in the crater map. If the number of the matched external craters is larger than a certain threshold, then go to (5); otherwise return (3) by selecting the other line segment in the crater map. If all of the line segments in the crater map are not matched with the current line segment in the camera-shot image, then return (2) by selecting the next line segment in the camera-shot image.

(5) Estimating location
    The current spacecraft location is calculated from the matched craters of the line segment and the matched external craters by the point cloud matching method.

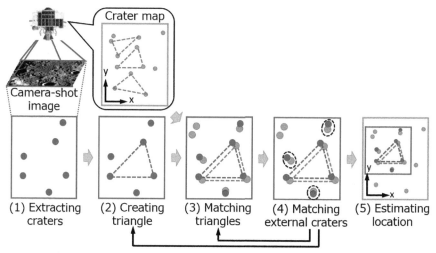

Figure 5.3   Algorithm of TSM.

5.4   TRIANGLE SIMILARITY MATCHING (TSM)

5.4.1   Overview

As the different approach of the crater matching methods, our previous research proposed Triangle similarity matching (TSM) [3] to estimate the spacecraft location by matching the triangle composed of the three craters detected in the camera-shot image with that in the crater map, while satisfying the positional relation of the *external* craters (*i.e.*, the craters except for three craters in the triangle) in the camera-shot image and the crater map. To estimate the spacecraft location by TSM, the crater map of the planetary surface should be stored in the onboard computer of the spacecraft like LSM. More precisely, the triangle database (DB) created from the crater map is stored in the spacecraft in order to narrow down the target of the triangles among many other triangles created from all combination of craters, which contributes to reducing the matching time. This DB includes the coordinates of the three craters in the triangle, the three angles of the triangle, the three lengths of the sides in the triangle in the ascending order, and the coordinates of the center gravity of the triangle.

5.4.2   Algorithm of TSM

As the algorithm of TSM, the following procedures are executed as shown in Fig. 5.3, which is similar to LSM.

(1) Extracting craters

After taking the camera-shot image by the spacecraft, the craters are extracted as shown in the red filled circles in Fig. 5.3 (1). (Note that any crater detection method can be employed. In this research, Takino's method [12] is employed).

(2) Creating triangle

The triangle composed of the three detected craters in the camera-shot image is created as shown in the red dashed triangle in Fig. 5.3 (2).

(3) Matching triangles

The triangle in the camera-shot image is compared with those in the crater map (*i.e.*, the triangle DB) as shown in the blue dashed triangle with the blue filled circles in Fig. 5.3 (3) to find the matched triangles.

(4) Matching external craters

After finding the matched triangles, all external craters in the camera-shot image are checked to match with those in the crater map as shown in the red and blue filled circles in Fig. 5.3 (4). If the number of the matched external craters as shown in the black dashed ovals in Fig. 5.3 (4) is larger than $TH_{num-matching}$ (see the detailed algorithm for matching strategy in Section 5.4.4), then go to (5); otherwise return (3) by selecting the other triangle in the crater map. If all of the triangles in the crater map are not matched with the current triangle in the camera-shot image, then return (2) by selecting the next triangle in the camera-shot image.

(5) Estimating location

The current spacecraft location is calculated from the matched craters of triangle and the matched external craters by the point cloud matching method as shown in Fig. 5.3 (5).

The essential difference between TSM and LSM is summarized as follows: (1) the triangle is a base of matching in TSM while the line segment is a base of matching in LSM as as shown in Fig. 5.4 where the red and blue triangles respectively indicate the triangles in the camera-shot image and the crater map while the orange and green lines respectively indicate the line segments in the camera-shot image and the crater map; (2) the correct triangle matching in TSM has one combination as shown in Fig. 5.4 (a) while the correct line segment matching in LSM has the multiple combinations as shown in Fig. 5.4 (b); and (3) since the number of combinations of matching triangles is

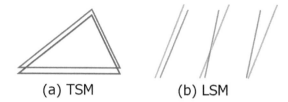

(a) TSM	(b) LSM

Figure 5.4   Different matching approaches between TSM and LSM.

smaller that that of line segments, TSM has a potential of finding the matched ones faster than LSM.

5.4.3   Mechanisms

In TSM, the following two mechanisms are important for estimating the spacecraft location: (i) Triangle similarity matching and (ii) External crater matching.

(i) Triangle similarity matching

The triangle in the camera-shot image is compared with that in the crater map from the viewpoint of the *length* of sides and the *angle* of triangles. Concretely, the two triangles are matched when satisfying Eqs. (5.1) and (5.2), meaning that the two triangles are matched when both the lengths of sides and the angles of triangles are mostly the same.

$$p_l \leq \frac{|\vec{d_i}|}{|\vec{d_i'}|} \leq p_u \tag{5.1}$$

$$\sum_{i=1}^{3}(|\cos\theta_i - \cos\theta_i'|) < TH_{triangle-matching} \tag{5.2}$$

In these equations, p_l and p_u are the lower and upper ratio of the i-th sides of the triangles in the camera-shot image $(\vec{d_i})$ and the crater map $(\vec{d_i'})$, where i is an arbitrary number from 1 to 3 which respectively corresponds to the long, middle, and short sides as shown in Fig. 5.5. θ_i and θ_i' are the i-th angles of the triangles in the camera-shot image and the crater map as shown in Fig. 5.5. $TH_{triangle-matching}$ is the maximum angle gap that can accept the triangles as similarity.

(ii) External crater matching

The external craters in the camera-shot image is compared with those in the crater map from the viewpoint of the *angle* from the vector of the

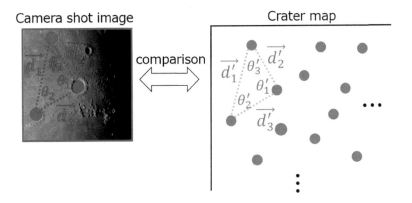

Figure 5.5 Triangle similarity matching.

long side of the triangle and the *distance* from the center of gravity of the triangle to the target external crater as shown in Fig. 5.6. Concretely, the two external craters are matched when satisfying Eq. (5.3).

$$\sqrt{I^2 + C^2} < TH_{crater-matching} \tag{5.3}$$

$$where \; I = \left| \overrightarrow{d_1} \cdot \overrightarrow{d_c} - \frac{\overrightarrow{d_1'} \cdot \overrightarrow{d_c'}}{\gamma^2} \right|, \tag{5.4}$$

$$C = \left| \overrightarrow{d_1} \times \overrightarrow{d_c} - \frac{\overrightarrow{d_1'} \times \overrightarrow{d_c'}}{\gamma^2} \right| \tag{5.5}$$

$$\gamma = \frac{|\overrightarrow{d_1'}|}{|\overrightarrow{d_1}|}, \tag{5.6}$$

This condition means that the two craters are matched when both the inner and cross products of the two vectors (*i.e.*, the vector of the long side of the triangle and the vector from the center of gravity of the triangle to the target external crater) in the camera-shot image and the crater map are close each other. Note that the inner and cross products of the two vectors roughly correspond to an angle and a length. In this equation, γ is the ratio of the long sides, $\overrightarrow{d_1}$ and $\overrightarrow{d_1'}$ (*i.e.*, the red and blue dotted arrows in Fig. 5.6), of the triangles in the camera-shot image and the crater map. $\overrightarrow{d_c}$ and $\overrightarrow{d_c'}$ (*i.e.*, the red and blue solid arrow in Fig. 5.6) are the vectors from the center of gravity of the triangles to the target external craters in the camera-shot image and the crater map. I and C

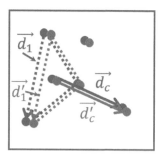

Figure 5.6   External crater matching.

respectively indicates the difference of the inner and cross products of two vectors (*i.e.*, $\overrightarrow{d_1}$, $\overrightarrow{d_c}$, and $\overrightarrow{d_1'}$, $\overrightarrow{d_c'}$). $TH_{crater-matching}$ is the maximum length and angle gaps that can accept to match the external craters.

5.4.4   Crater matching strategy

As the crater matching strategy of TSM, TSM *roughly* matches the triangles and external craters in the camera-shot image with those in the crater map. This is because (i) the spacecraft location cannot be estimated in the case of no matched triangles and (ii) its estimation accuracy becomes low or biased in the case of a small number of the matched external craters. From this reason, the hyper-parameters, $TH_{triangle-matching}$ (as the maximum angle gap that can accept the triangles as similarity) and $TH_{crater-matching}$ (as the maximum length and angle gaps that can accept to match the external craters), should not be set small for a rough matching. This means that these two hyper-parameters can be roughly set in comparison with the remaining hyper-parameter, $TH_{num-matching}$ (as the minimum number of the matched external craters for a location estimation) because an accuracy of a location estimation depends on the number of the matched craters. However, it is generally difficult to determine an appropriate threshold of $TH_{num-matching}$, *i.e.*, a location cannot be estimated with too large $TH_{num-matching}$ because the number of the matched craters hardly exceeds $TH_{num-matching}$ and its estimation accuracy becomes lows with too small $TH_{num-matching}$ because only small number the matched craters are used for the location estimation.

To tackle this problem, TSM differently estimates the spacecraft location according to the number of the matched craters as shown in Fig. 5.7. In detail, the following different procedures are executed according to the number of the matched, where the variable X is simplified to represent $TH_{num-matching}$

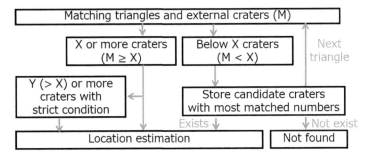

Figure 5.7   Crater selection according to the number of the matched craters.

for easy understanding and M indicates the number of the matched external craters.

- X or more number of the matched external craters $(M \geq X)$
  When $M \geq X$, the spacecraft location is differently estimated according to its number. Since an accuracy of the location estimation increases when employing the matched external craters that satisfy the *strict* condition (*i.e.*, the small $TH_{crater-matching}$) and when the number of the matched external craters increases, the location is *more accurately* estimated when more number of the external craters $Y(> X)$ are matched with the *strict* condition; otherwise it is estimated with the *normal* condition using the $(Y \geq)M(> X)$ number of the matched external craters. Note that $TH_{crater-matching}$ in the strict condition should be set as a smaller value than the normal condition.

- Below X number of the matched external craters $(M < X)$
  TSM stores the matched triangle and external craters even in a small number of the matched ones (*i.e.*, $M(< X)$) and continues to search the next triangle match. Note that the location is not estimated at this stage. In repeating this cycle, the location is estimated just after $M \geq X$ and completes the location estimation, while the stored triangle and external craters are updated when the number of the currently matched external craters is larger than the stored one. After searching all triangles in the camera-shot image, the location is estimated by the stored triangles and external craters if they are stored, while the proposed method outputs the result of "not found" if they are not stored.

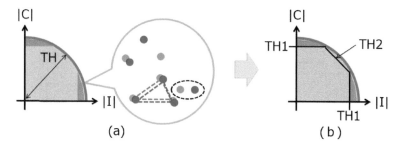

Figure 5.8   Appropriate crater selection.

5.5   IMPROVED TSM

5.5.1   Appropriate crater selection

In TSM, the external craters in the camera-shot image and the crater map are matched when satisfying $\sqrt{I^2 + C^2} < TH_{crater-matching}$ in Eq. (5.3), where I and C respectively indicates the difference of the inner and cross products of two vectors (*i.e.*, the vector of the long side of the triangle and the vector from the center of gravity of the triangle to the target external crater). This equation contributes to finding the matched external craters. However, the craters which are not close each other may be matched as shown in the black oval in Fig. 5.8 (a), where the horizontal and vertical axes respectively indicate the absolute values of the inner and cross products and the red and blue filled circles respectively indicate the external craters in the camera-shot and the crater map. Such an inappropriate matching occurs in the right or upper orange area in Fig. 5.8 (a) because the distance between the external craters in the camera-shot image and the crater map can be extended until the difference of the only inner (or cross) product due to $\sin \theta \approx 0$ (or $\cos \theta \approx 0$).

To overcome this problem, Eq. (5.7) is employed instead of Eq. (5.3) as shown in Fig. 5.8(b), where $TH1$ (corresponding $TH1_{crater-matching}$) and $TH2$ (corresponding $TH2_{crater-matching}$) are the new hyper-parameters instead of TH (corresponding $TH_{crater-matching}$). The area for the matching external craters is represented by the polygon within $TH1$ and $TH2$. This area contributes to selecting the appropriate external craters in the camera-shot image and the crater map which are close each other.

$$(I < TH1_{crater-matching}) \wedge (C < TH1_{crater-matching})$$
$$\wedge (I + C < TH2_{crater-matching}) \qquad (5.7)$$

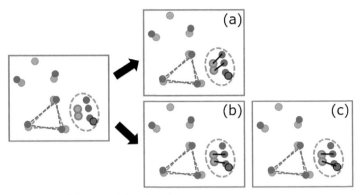

Figure 5.9   Possible matching in close craters.

5.5.2   Close craters elimination

In TSM, an accuracy of the spacecraft location estimation may decrease when the craters are dense or overlapped because it is difficult to find the correct matching craters in the camera-shot image and the crater map as shown in Fig. 5.9. In this figure, the red and blue filled circles respectively indicate the craters in the camera-shot image and the crater map, the red and blue dashed triangles respectively indicate the matched triangles in the camera-shot image and the crater map, and the gray dashed oval indicates the dense and overlapped area where craters are closely located. In this situation, the blue filled circles in the crater map can be matched with the upper and middle red filled circles in the camera-shot image (Fig. 5.9 (a)), the middle and lower left red filled circles in the camera-shot image (Fig. 5.9 (b)), or the middle and lower right red filled circles in the camera-shot image (Fig. 5.9 (c)). Even slight different crater matching affects the accuracy of the location estimation, which is very the serious problem for the pinpoint landing.

To overcome this problem, the close craters are eliminated in order not to incorrectly match the craters in the camera-shot image and the crater map. Note that the all close craters are eliminated because it is difficult to determine which crater is wrongly/correctly matched. To eliminate such doubtful craters, the following procedures are executed as shown in Fig. 5.10, where the red and blue filled circles and dashed triangles have the same meaning in Fig. 5.9.

(i) The triangle in the camera-shot image is matched with that in the crater map.

(ii) The averaged distance between craters in the camera-shot image, $AveDistance$ (*i.e.*, the averaged of all combination of the distances between craters (represented by the black arrow)), is calculated.

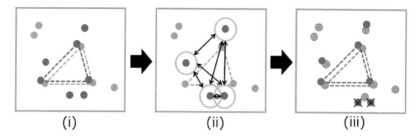

Figure 5.10 Eliminating craters.

(iii) The close craters (represented by the black cross mark) are eliminated when satisfying Eq. (5.8), where *Distance* is the distance between craters and α is the coefficient.

$$Distance < \alpha \cdot AveDistance \tag{5.8}$$

However, too much elimination of the close craters decreases the accuracy of the location estimation. To cope with this issue, the above elimination is not applied in the following cases, where the variables X and Y have the same meaning of Fig. 5.7.

- When the number of the matched external craters becomes less than Y from Y or more.

- When the number of the matched external craters becomes less than X from X or more.

5.6 TEST CASE

5.6.1 Cases in coasting phase

To investigate the effectiveness of the improved TSM, the coasting phase named VLD1 as the vertical descent phase in the SLIM mission is employed. In this phase, the following trajectories are planned as shown in Table 5.1.

- Nominal trajectory
  The following two cases are addressed: (i) no camera anomaly, which means that all images taken by the spacecraft include only the control and navigation errors (3σ) of its location and posture (No. 6710); and (ii) the camera anomaly, which means that all images include the camera-related errors (see Section 5.6.2) in addition to (i) (No. 6810).

TABLE 5.1 Test cases.

Trajectory	Case No.	Anomaly	Errors
Nominal (high sun elevation)	6710	None	Control and navigation errors of location and posture (3σ) ...(1)
	6810	Camera	(1)+all image errors
Backup (low sun elevation)	6760	None	Control and navigation errors of location and posture (3σ) ...(1)
	6860	Camera	(1)+all image errors

- Backup trajectory
  The following two cases are addressed like nominal trajectory: (i) no camera anomaly (No. 6760); and (ii) the camera anomaly (No. 6860).

The essential difference between the nominal and backup trajectories is a height of the sun elevation, *i.e.*, the sun elevation is high in the nominal trajectory while it is low in the backup trajectory. From this difference, the shape of the detected craters in the backup trajectory is not a circle but an oval and its shadow becomes large, which makes it difficult to accurately estimate the center of the craters, resulting in the bad accuracy of the spacecraft location estimation.

5.6.2 Errors in images

In the above trajectories, 10 kinds of errors are considered in images as shown in Table 5.2. In this table, the first and second columns indicate the target of the error and the kinds of errors, respectively.

In detail, the nominal pattern in the first row include the images modified by the control and navigation errors (3σ) of the spacecraft location and posture (which are employed in the nominal and backup trajectories with no camera anomaly (*i.e*, No. 6710 and 6760)). The camera anomaly categorized from the second to ninth rows include the images modified by the camera-related errors (which are employed in the nominal and backup trajectories with the camera anomaly (*i.e*, No. 6810 and 6860)). The camera-related errors are summarized as follows: brightness, contrast, blur, brightness fluctuation, radiation noise, distortion, camera shake, and limb darkening. The sun-related error in the tenth row includes the images modified by the error of sun elevation (which are employed in the nominal trajectory (*i.e*, No. 6710 and 6810) and backup trajectory (*i.e*, No. 6760 and 6860)). Note that the sun-related error is added in to camera-related errors.

TABLE 5.2 Error patterns.

Target	Errors
Nominal	Control and navigation errors of location and posture
Camera	Brightness
	Contrast
	Blur (out of focus)
	Brightness fluctuation
	Radiation noise
	Distortion
	Camera shake
	Limb darkening
Sun	Sun elevation

Figure 5.11   Examples of the images.

Figure 5.11 shows the example of the errors of the test pattern. The upper, middle, and lower images show the images of the nominal, brightness error, and contrast error, respectively.

5.7   EXPERIMENTS

5.7.1   Experimental design

To evaluate the improved TSM, the following four cases are tested as shown in Table 5.1: the nominal trajectory with no camera anomaly (No. 6710) and the camera anomaly (No. 6810) and the backup trajectory with no camera anomaly (No. 6760), and the camera anomaly (No. 6860). In all cases, the improved TSM is compared with TSM and LSM in 1000 camera shot images in each of four cases.

5.7.2 Evaluation criteria and parameter setting

As an evaluation, the following criteria are employed in this experiment.

- Number of location estimation

  - Great matching ($\Delta x, \Delta y < 3$): The difference between the estimated coordinate and the true coordinate is less than the three pixel (which roughly corresponds to $30m$) in terms of x and y-axes.
  - Good matching ($3 \leq \Delta x, \Delta y < 7$): The difference between the estimated coordinate and the true coordinate is the three pixels or more and less than the seven pixels in terms of x and y-axes.
  - Miss matching ($\Delta x, \Delta y \geq 7$): The difference between the estimated coordinate and the true coordinate is the seven pixels or more in terms of x and y-axes.
  - Not found: The spacecraft location cannot be estimated when the triangles and/or the external craters in the camera-shot image and the crater map are not matched in the improved TSM and TSM and when the line segments and/or the external craters in the camera-shot image and the crater map are not matched in LSM.

- Error of location estimation

  - The maximum error in the location estimation among all images
  - The averaged error in the location estimation among all images

As the parameter setting, Table 5.3 summarizes them in this experiment. Note that $\overrightarrow{d_1}$ indicates the long side of the triangle.

5.7.3 Experimental results

Table 5.4 shows the result of the improved TSM, TSM, and LSM. In this table, the columns indicates four cases (*i.e.*, No. 6710, No. 6810, No. 6760, and No. 6860), while the upper four rows indicate the number of location estimation (*i.e.*, "great", "good", and "miss" matching and "not found") and the lower two rows indicates the error of location estimation (*i.e.*, "maximum" and "averaged" errors). Note that the bold number is the best value among three methods.

TABLE 5.3 Parameter setting.

Parameter	Meaning	Value
X	The minimum number of the matched external craters (normal matching)	9
Y	The minimum number of the matched external craters (strict matching)	14
α	Coefficient in Eq. (5.8)	0.5
Both the normal and strict matching for X and Y		
$TH_{triangle-matching}$	The maximum angle gap of triangle similarity	93
The normal matching for X		
$TH_{crater-matching}$	The maximum length and angle gaps of the matched external craters (TSM)	$165 \cdot |\overrightarrow{d_1}|$
$TH1_{crater-matching}$	as above (improved TSM)	$150 \cdot |\overrightarrow{d_1}|$
$TH2_{crater-matching}$	as above (improved TSM)	$220 \cdot |\overrightarrow{d_1}|$
The strict matching for Y		
$TH_{crater-matching}$	The maximum length and angle gaps of the matched external craters (TSM)	$140 \cdot |\overrightarrow{d_1}|$
$TH1_{crater-matching}$	as above (improved TSM)	$100 \cdot |\overrightarrow{d_1}|$
$TH2_{crater-matching}$	as above (improved TSM)	$150 \cdot |\overrightarrow{d_1}|$

From the result, the improved TSM outperforms TSM which outperforms LSM. In detail, the number of the "great" crater matching in the improved TSM is larger than that of TSM and LSM, which derives the result that the number of the "good" crater matching in the improved TSM is smaller than that of TSM and LSM. The "maximum" error of the crater matching in the improved TSM is smaller than that in TSM except for No.6860 and is all smaller than that of LSM, while the "averaged" error of the crater matching in the improved TSM is equal or smaller than that of TSM and all is smaller than that of LSM.

5.7.4 Discussion

5.7.4.1 LSM vs TSM

To understand why the improved TSM outperforms LSM as shown in Table 5.4, Fig. 5.12 shows how the line segments, triangles, the external craters in the camera-shot image and the crater map are matched in LSM and the improved TSM. In Fig. 5.12 (a), the orange and green lines respectively indicate the matched line segments in the camera-shot image and the crater map, and

TABLE 5.4 Results of the improved TSM, TSM, and LSM.

(a) Improved TSM

	Nominal Trajectory (high sun elevation)		Backup Trajectory (low sun elevation)	
	None	Camera Anomaly	None	Camera Anomaly
	No. 6710	No. 6810	No. 6760	No. 6860
Great matching ($\Delta x, \Delta y < 3$)	**1000**	**997**	**1000**	**999**
Good matching ($3 \leq \Delta x, \Delta y < 7$)	**0**	**3**	**0**	**1**
Miss matching ($\Delta x, \Delta y \geq 7$)	0	0	0	0
Not found	0	0	0	0
Maximum error	**3.061**	**3.762**	**3.248**	3.535
Averaged error	**1.056**	**1.078**	**1.254**	**1.279**

(b) TSM

	No. 6710	No. 6810	No. 6760	No. 6860
Great matching ($\Delta x, \Delta y < 3$)	999	996	997	995
Good matching ($3 \leq \Delta x, \Delta y < 7$)	1	4	3	5
Miss matching ($\Delta x, \Delta y \geq 7$)	0	0	0	0
Not found	0	0	0	0
Maximum error	4.171	4.515	3.434	**3.377**
Averaged error	1.08	1.123	**1.254**	1.299

(*Continued on next page*)

TABLE 5.4 (Continued)

(c) LSM

	No. 6710	No. 6810	No. 6760	No. 6860
Great matching ($\Delta x, \Delta y < 3$)	998	990	992	974
Good matching ($3 \leq \Delta x, \Delta y < 7$)	2	10	8	26
Miss matching ($\Delta x, \Delta y \geq 7$)	0	0	0	0
Maximum error	3.534	7.15	3.893	4.874
Averaged error	1.581	1.609	1.631	1.697

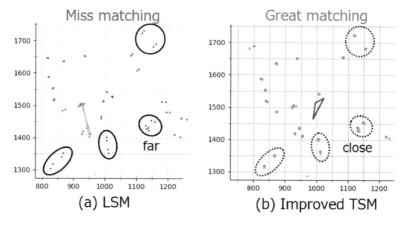

Figure 5.12   External crater matching in LSM and improved TSM.

the red and blue dots respectively indicate the matched external craters in the camera-shot image and the crater map in LSM. In Fig. 5.12 (b), on the other hand, the red and blue triangles respectively indicate the matched triangles in the camera-shot image and the crater map, and the orange and green dots respectively indicate the matched external craters in the camera-shot image and the crater map in the improved TSM.

The number of the matched external craters are large enough to estimate the spacecraft location in both LSM and the improved TSM. However, the matched external craters in the black solid circles are matched with a slightly far distance between craters in the camera-shot image and the crater map in LSM as shown in Fig. 5.12 (a), while those in the black dashed circles are

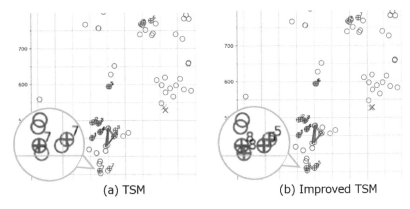

(a) TSM	(b) Improved TSM

Figure 5.13   External crater matching by appropriate crater selection.

matched with the close distance between craters in the camera-shot image and the crater map in the improved TSM as shown in Fig. 5.12 (b). This difference occurs because the crossed line segments in LSM as shown in Fig. 5.12 (a) are not correctly matched from the angle viewpoint in comparison with the parallel line segments as shown in Fig. 5.4. Compared with LSM, the triangles in the improved TSM (and TSM) tend to be correctly matched from the angle viewpoint because of one combination of the triangles matching, which increases the accuracy of the location estimation.

5.7.4.2   Effect of appropriate crater selection

To investigate an effect of the appropriate crater selection, Fig. 5.13 show how the external craters in the camera-shot image and the crater map are matched in TSM and the improved TSM. In this figure, the red and blue triangles respectively indicate the matched triangles in the camera-shot image and the crater map, and the red and blue circles respectively indicate the external craters in the camera-shot image and the crater map. In particular, the red and blue circles with the cross mark adding the number in the upper right are matched, while the simple circles without the cross mark are not matched. Note that the number indicates the identified number of the matched craters and the different number may be assigned in TSM and the improved TSM because the number of the matched external craters is different.

When focusing on the enlarged parts, the external craters of No. 7 are matched even in a slightly far distance between craters in the camera-shot image and the crater map in TSM as shown in Fig. 5.13 (a), while those of No. 5 and 8 are *separately* matched in the close distance between craters in the camera-shot image and the crater map in the improved TSM as shown in

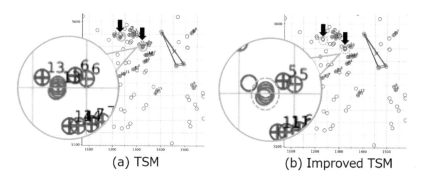

(a) TSM   (b) Improved TSM

Figure 5.14   External crater matching by close craters elimination.

Fig. 5.13 (b). This means that TSM incorrectly matches the craters of No. 7, while the improved TSM correctly matches the craters of No. 5 and 8 separately. Such incorrect matching can decrease by selecting appropriate craters which distance in the camera-shot image and the crater map is close each other.

5.7.4.3   Effect of close craters elimination

From the above analysis, the accuracy of the spacecraft location estimation increases by selecting the close craters in the camera-shot image and the crater map. However, such craters are not always good to be selected. To understand this meaning, Fig. 5.14 show how the external craters in the camera-shot image and the crater map are matched in TSM and the improved TSM. In this figure, the red and blue triangles, and the red and blue circles without/with the cross mark adding the number in the upper right have the same meaning of Section 5.7.4.2. When focusing on the craters indicated by the black arrows, the external craters are matched in TSM as shown in Fig. 5.14 (a), while those are not matched in the improved TSM as shown in Fig. 5.14 (b). By looking at the enlarged parts, in particular, the external craters of No. 13 are matched in addition to those of No. 6 in TSM, while those of the No. 5 are only matched in the improved TSM. Since the craters of No.6 in TSM and No. 5 in the improved TSM are the same, the close external craters of No. 13 are matched in TSM while those are not matched by removing them in the improved TSM. Due to such close craters, the external craters are incorrectly matched in TSM.

To investigate an effect of the close craters elimination, Fig. 5.15 shows the number of the improved and worsened matching in the improved TSM by contrast with TSM. In this figure, the vertical and horizontal axes

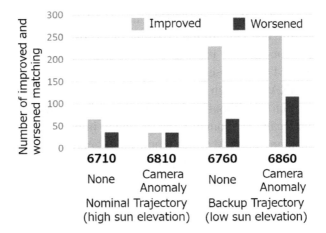

Figure 5.15   Number of the improved and worsened matching.

respectively indicate its number and four cases, and the blue and orange bars respectively indicate the number of the improved and worsened matching. This figure shows that the improved TSM works well by contrast with TSM in the backup trajectory (*i.e.*, No. 6760 and No. 6860) where the shape of the detected craters is mostly an oval and its shadow becomes large due to a low sun elevation. Even though it is difficult to accurately estimate the center of the craters in such a situation, the improved TSM cope with it by removing the craters which become close to other craters due to a low sun elevation.

5.8   CONCLUSION

This paper focused on our method, Triangle Similarity Matching (TSM), which estimated the spacecraft location by comparing the triangle (composed of the three detected craters) in the camera-shot image taken by the space-craft and the triangle (composed of the three detected craters) in the crater map (stored in the spacecraft beforehand) as the information on the craters in the planetary surface. To improve an accuracy of a spacecraft location estimation, this chapter improved TSM by selecting the "appropriate craters" and eliminating the "close craters" in the camera-shot image, and aims at investigating its effectiveness through the experiment. Concretely, the incorrect crater matching decreases by selecting the appropriate craters which distance in the camera-shot image and the crater map is close each other but by eliminating the close craters which are overlapped with other craters in the camera-shot image.

To investigate the effectiveness of the improved TSM, we conducted the experiment on the test case in the coasting phase of the Smart Lander for Investigating Moon (SLIM) mission of JAXA and compared the results of the improved TSM with those of TSM and Line Segment Matching (LSM) proposed by JAXA. Through the intensive experiment of four cases (*i.e.*, the nominal trajectory without/with camera anomaly and the backup trajectory without/with camera anomaly), the following implications have been revealed: (1) the improved TSM which employs the appropriate crater selection and the close crater elimination outperforms TSM and LSM. In detail, the number of "great" matching within three pixels in the improved TSM is larger than that in TSM and LSM, and the maximum and averaged errors of the crater matching in the improved TSM is smaller than those in TSM except for one case and all is smaller than those in LSM; (2) the improved TSM can prevent from the incorrect crater matching in comparison with TSM and LSM, which increases the accuracy of the spacecraft location estimation; and (3) the improved TSM is robust to both images with the camera anomaly and those with the large shadow of the detected craters due to a low sun elevation.

As future research, the following issues must be addressed in near the future: (1) since the spacecraft location cannot be estimated when the matched triangles are not found, new matching mechanism should be explored to overcome this problem; and (2) since the experiment conducted in PC, the experiment should conduct in Field Programmable Gate Array (FPGA) as the onboard computer in the spacecraft. The execution time and the required memory (or resource) should also be investigated in the onboard computer.

ACRONYM

FPGA Field Programmable Gate Array

JAXA Japan Aerospace Exploration Agency

LSM Line Segment Matching

SELENE Selenological and Engineering Explorer

SLAM Simultaneous Localization and Mapping

SLIM Smart Lander for Investigating Moon

TSM Triangle Similarity Matching

VLD Vertical Lunar Descent

GLOSSARY

Camera-shot image: The image taken by camera

Line Segment Matching (LSM): The location estimation method that compares the line (*e.g.*, composed of the two craters) in the camera-shot image with the line in the satellite image (map) or the aerial photograph

Triangle Similarity Matching (TSM): The location estimation method that compares the triangle (*e.g.*, composed of the three craters) in the camera-shot image with the triangle in the satellite image (map) or the aerial photograph

Point cloud matching: the registration or alignment of the two point clouds acquired from the different locations

Pinpoint landing: the landing at the area "where is *desired* to land.", which is roughly within $100m$ from the target landing point

FURTHER READING

J. Riedel, S. Bhaskaran, S. Desai, D. Hand, B. Kennedy, T. McElrath, and M. Ryne. (2000) "Autonomous optical navigation (Autonav) DS1 technology validation report", *Jet Propulsion Laboratory, California Institute of Technology, Tech. Rep.*.

A. Johnson, R. Willson, Y. Cheng, J. Goguen, C. Leger, M. Sanmartin, and L. Matthies. (2007) "Design through operation of an image-based velocity estimation system for Mars landing", *International Journal of Computer Vision*, Vol. 74, No. 3, pp. 319–341, 2007.

Y. Cheng, A. Johnson, and L. Matthies. (2005) "MER-DIMES: a planetary landing application of computer vision," in *IEEE Computer Society Conference on Computer Vision and Pattern Recognition (CVPR'05)*, Vol. 1, pp. 806–813.

H. Morita, K. Shirakawa, T. Kubota, T. Hashimoto, and J. Kawaguchi. (2006) "Hayabusa's real-time landmark tracking navigation for descents and touching-downs'," in *AIAA/AAS Astrodynamics Specialist Conference and Exhibit*.

T. Hashimoto, T. Kubota, S. Sawai, J. Kawaguchi, and M. Uo. (2006) "Final autonomous descent based on target marker tracking'," in *AIAA/AAS Astrodynamics Specialist Conference and Exhibit*.

A. E. Johnson, Y. Cheng, J. Montgomery, N. Trawny, B. E. Tweddle, and J. Zheng. (2016) "Design and analysis of map relative localization for access to hazardous landing sites on mars'," in *AIAA Guidance, Navigation, and Control Conference*.

D. A. Lorenz, R. Olds, A. May, C. Mario, M. E. Perry, E. E. Palmer, and M. Daly. (2017) "Lessons learned from OSIRIS-REx autonomous navigation using natural feature tracking", in *IEEE Aerospace Conference.*

G. Jonniaux, K. Regnier, and D. Gherardi. (2016) "Maturing vision-based navigation technologies for JUICE", in *International Planetary Probe Workshop.*

A. Bidaux-Sokolowski, J. Lisowski, P. Kicman, O. Dubois-Matra, and T. Voirin. (2017) "GNC design for pinpoint landing on phobos", in *10th International ESA Conference on Guidance, Navigation & Control Systems.*

BIBLIOGRAPHY

[1] Shujiro Sawai, Seisuke Fukuda, Shinichiro Sakai, Kenichi Kushiki, Tetsuhito Arakawa, Eiichi Sato, Atsushi Tomiki, Keisuke Michigami, Taro Kawano, Shun Okazaki, Akio Kukita, Yu Miyazawa, Satoshi Ueda, Hirobumi Tobe, Yusuke Maru, Haruhiko Shimoji, Yasuhiro Shimizu, Yusuke Shibasaki, Sadanori Shimada, Takahiro Yokoi, Takeshi Yabushita, Kenichiro Sato, Kazuyuki Nakamura, Takahiro Kuhara, Tsuyoshi Takami, Nobuhiko Tanaka, and Katsumi Furukawa. Preliminary system design of small lunar landing demonstrator SLIM. *Areaspace Technology Japan, Japan Society for Aeronautical and Space Sciences*, 17:35–43, 2018 (In Japanese).

[2] Makoto Hareyama, Yoshiaki Ishihara, Hirohide Demura, Naru Hirata, Chikatoshi Honda, Shunichi Kamata, Yuzuru Karouji, Jun Kimura, Tomokatsu Morota, Hiroshi Nagaoka, Ryousuke Nakamura, Satoru Yamamoto, Yasuhiro Yokota, and Makiko Ohtake. Global classification of lunar reflectance spectra obtained by kaguya (SELENE): Implication for hidden basaltic materials. *Icarus*, 321:407–425, 2019.

[3] Haruyuki Ishii, Yuta Umenai, Kazuma Matsumoto, Fumito Uwano, Takato Tatsumi, Keiki Takadama, Hiroyuki Kamata, Takayuki Ishida, Seisuke Fukuda, Shujiro Sawai, and Shinichiro Sakai. How to detect essential craters in camera shot image to increase the number of spacecraft location estimation while improving its accuracy? In *the 14th International Symposium on Artificial Intelligence, Robotics and Automation in Space (i-SAIRAS 2018)*, 2018.

[4] Fumito Uwano, Takato Tatsumi, Akinori Murata, Keiki Takadama, Hiroyuki Kamata, Takayuki Ishida, Seisuke Fukuda, Shujiro Sawai, and Shinichiro Sakai. How to select appropriate craters to estimate location

accurately in comprehensive situations for slim project. In *the 32nd International Symposium on Space Technology and Science (ISTS 2019) & 9th Nano-Satellite Symposium (NSAT)*, 2019.

[5] Yuka Waragai, Iko Nakari, Yousuke Hayamizu, Keiki Takadama, Hiroyuki Kamata, Takayuki Ishida, Seisuke Fukuda, Shujiro Sawai, and Shinichiro Sakai. Towards accurate spacecraft self-location estimation by eliminating close craters in camera-shot image. In *the 15th International Symposium on Artificial Intelligence, Robotics and Automation in Space (i-SAIRAS 2020)*, 2020.

[6] Hugh Durrant-Whyte and Tim Bailey. Simultaneous localization and mapping: part I. *IEEE Robotics & Automation Magazine*, 13(2):99–110, 2006.

[7] Tim Bailey and Hugh Durrant-Whyte. Simultaneous localization and mapping (SLAM): part II. *IEEE Robotics & Automation Magazine*, 13(3):108–117, 2006.

[8] Yicong Tian, Chen Chen, and Mubarak Shah. Cross-view image matching for geo-localization in urban environments. In *2017 IEEE Conference on Computer Vision and Pattern Recognition (CVPR)*, pages 22–25, 2017.

[9] James Richard Wertz. *Spacecraft Attitude Determination and Control*, volume 73. Springer Dordrecht, 1980.

[10] Toshiro Sasaki and Michitaka Kosaka. A star identification method for satellite attitude determination using star sensors. In *the 15th International Symposium on Space Technology and Science (ISTS 1986)*, pages 1125–1130, 1986.

[11] Kazuki Kariya, Takayuki Ishida, Shujiro Sawai, Tomoo Kinoshita, Kunihiro Kajihara, Osamu Iwasa, and Seisuke Fukuda. Position estimation using crater-based linear features for pinpoint lunar landing. *Aerospace Technology Japan, The Japan Society for Aeronautical and Space Sciences*, 17:79–87, 2018 (In Japanese).

[12] Tatsuya Takino, Izuru Nomura, Misako Moribe, Hiroyuki Kamata, Keiki Takadama, Seisuke Fukuda, Shujiro Sawai, and Shinichiro Sakai. Crater detection method using principal component analysis and its evaluation. *Transaction of The Japan Society for Aeronautical and Space Sciences, Aerospace Technology Japan*, 14:7–14, 2016.

CHAPTER 6

Artificial Intelligence for Space Weather Forecasting

Enrico Camporeale

CIRES, University of Colorado & NOAA Space Weather Prediction Center, Boulder, CO, USA

CONTENTS

6.1	Introduction	191
6.2	Why Space Weather Matters	192
6.3	Conventional Forecasting Methods and their limitations	193
6.4	Machine Learning Space Weather Forecasting	196
	6.4.1 Speed of execution	197
	6.4.2 Accuracy	199
	6.4.3 Actionability	200
	6.4.4 Geomagnetic indices	200
	6.4.5 Solar flares	201
	6.4.6 Coronal Mass Ejection arrival time	203
	6.4.7 Forecast of solar wind speed	204
6.5	Conclusions	205
Bibliography	....	210

S PACE Weather is broadly defined as the study of the space conditions surrounding Earth and their sudden variations due to solar events. space weather 'events' can be caused by solar flares, coronal mass ejections, or fast solar wind streams, that in turn cause variations of the plasma an electromagnetic field conditions around Earth. Those events can have catastrophic

DOI: 10.1201/9781003366386-6

effects on space-borne and ground-based infrastructures, such as satellites and electric power grid. The ability of forecasting space weather events become therefore necessary in order for our society to be resilient against adverse space weather effects, which can potentially result in huge economic losses. The field of space weather forecasting is relatively new when compared, for instance, to terrestrial weather. However, it is rapidly advancing, particularly by adopting new artificial intelligence technologies. In this chapter we discuss the state of the art of several space weather application that have been aided by artificial intelligence, and the interplay between traditional physics-based approaches and modern data-driven methods.

6.1 INTRODUCTION

Space weather refers to the dynamic conditions in the near-Earth environment that result from the interactions between the solar wind, the Earth's magnetic field, and the Earth's atmosphere. This can include events such as solar flares, coronal mass ejections (CMEs), and geomagnetic storms, which can impact satellite and communication systems, navigation systems, and power grids. A continual stream of charged particles, primarily protons and electrons, is discharged from the Sun's outermost atmosphere, known as the corona, forming a phenomenon called the solar wind. This wind generally flows at rates of 300 to 1000 km/s. An occurrence known as a CME is from time to time produced from the Sun due to a shift in the magnetic topology, which leads to the liberation of magnetic energy from the corona. These CMEs can hurl up to 10 billion tons of plasma and magnetic fields into the environment, travelling at a rapid velocity of 3 million miles per hour [1]. CMEs can take between 1 and 4 days to get to Earth and can produce significant changes in the magnetic field of our planet. These ejections can be of varying sizes, shapes, and directions and can take place on their own or combined with other solar events, like solar flares.

Solar flares are intense bursts of radiation that are released during magnetic storms on the Sun's surface. These events can be classified according to their strength using a system that ranges from A to X-class flares, with X-class flares being the most powerful. X-rays travel at the speed of lights and therefore reach Earth in about 8 minutes.

Each of these effects can cause a geomagnetic storm, which is a disturbance in the Earth's magnetosphere or atmosphere [2]. Geomagnetic storms can cause a range of effects on the Earth, including auroras, power outages, and disruptions to communication and navigation systems. Geomagnetic storms can also cause an increase in radiation levels that can be harmful to astronauts and passengers on high-altitude flights.

6.2   WHY SPACE WEATHER MATTERS

The economic impact of space weather is estimated to be substantial, with some estimates suggesting it could be in the billions of dollars for industries such as telecommunications and electric power [3–5]. It is therefore important to understand and forecast space weather to minimize its impact on various industries, including telecommunications, navigation, and insurance. Acknowledging this, studies have been conducted to assess the cost of a severe space storm. According to a report from the National Research Council of the National Academies in 2008 [6], a storm such as the one which occurred in September 1859 (the so-called Carrington event) would cost an estimated one trillion USD in the present day and take 4-10 years to recover from - an order of magnitude greater than the damage caused by Hurricane Katrina. More recently, Eastwood et al. (2017) [3] estimated that the total economic loss could range between 0.5 and 2.7 trillion USD, based on disruption to the global supply chain. With a different method, they found a total loss of 140-613 billion USD. According to the National Research Council [6], power outages can have far-reaching consequences across multiple sectors, particularly in banking, finance, government services, and emergency response. Schulte et al. (2014) [7] and Riley et al. (2018) [8] have also emphasized that all economic sectors would be affected by such an event. In addition to sectoral vulnerability due to dependence on electricity, Schulte et al. (2014 [7] has further noted that the global economic production system is increasingly susceptible due to just-in-time production, reduced inventories, and long-distance supply chains.

Oughton et al. (2017) [9] examined the economic consequences, both direct and indirect, of extreme space weather scenarios in the mainland United States by focusing on supply chain impacts upstream and downstream. In particular, the study investigated the potential costs of four different storm scenarios, characterized by their varying geomagnetic latitude and corresponding levels of power loss, with losses ranging from 15% to 100% of the daily U.S. GDP. The corresponding economic impact on the U.S. population and economy ranges from $6.2 billion to $41.5 billion per day.

The Helios Solar Storm Scenario introduced in Oughton et al. (2017) [10] was the first space weather stress test designed for the global insurance industry. The Helios Solar Storm Scenario aimed to provide a tool for analyzing the impact of different restoration periods on the economic losses and exposure of global insurance companies. According to that study, a severe power outage resulting from a space weather event in the US could have significant global economic consequences. The authors estimated that the associated supply chain disruptions could result in economic losses ranging from 0.5-2.7 trillion USD. They used an integrated economic model to account for the post-event dynamic responses of global trade, and projected a potential decrease in global GDP of up to 1.1 trillion USD over a five-year period.

In the case of a strong space weather event, the US manufacturing industry is expected to take a major hit, with estimated losses of USD 350 billion. Roughly half of these losses would be indirect, and divided equally between upstream and downstream disruptions to the supply chain. The insurance industry in the US would also face significant losses, with estimates of up to USD 334 billion. Property insurance policies would be particularly affected, accounting for 90% of the losses caused by service interruption.

These findings highlight the potential economic impact of a space weather event on the US and the world. A detailed analysis of the global scale impact of historical space weather events can be found, e.g. in Sokolova et al. (2021) [11]. Those studies underscore the need for better preparedness and resilience strategies to minimize the damage and speed up the recovery process in the event of such an occurrence.

Finally, although the public response to such an event is challenging to model, it is critical to consider. Hapgood et al. (2021) [12] have proposed that a power grid disruption caused by a severe space weather event could lead to panic buying and hoarding of essential goods, including petrol, bottled water, non-perishable food, and toilet paper. The long-term economic implications of such behaviors have yet to be determined. It is imperative to recognize the potential societal responses to a prolonged power outage caused by a space weather event to design effective preparedness and response strategies.

6.3 CONVENTIONAL FORECASTING METHODS AND THEIR LIMITATIONS

Space weather is continuously monitored and forecast by multiple dedicated centers globally. These centers utilize various technologies and models to analyze and predict space weather events. They collect data from various sources, such as ground-based instruments, satellites, and telescopes, and use it to generate space weather forecasts. Additionally, these centers collaborate with international organizations to share data and observations to enhance their forecasting capabilities.

One of the primary centers for space weather monitoring and forecasting is the Space Weather Prediction Center (SWPC) in the United States, which is part of the National Oceanic and Atmospheric Administration (NOAA) National Weather Service (NWS). SWPC is responsible for monitoring and forecasting space weather conditions, including solar flares, geomagnetic storms, and radiation storms, and disseminating information to stakeholders in various sectors, such as aviation, navigation, and power grids. Recently, a survey on customer needs and requirements for space weather products was commissioned by SWPC to respond to the National Space Weather Strategy (NSWS) and Space Weather Action Plan (SWAP) released in 2015 by the White House Office of Science and Technology Policy (OSTP). The report is available at `https://repository.library.noaa.gov/view/noaa/29107/noaa_29107_DS1.pdf`. Moreover, SWPC real-time forecasts and warnings are avaible at `https://www.swpc.noaa.gov/`.

Similarly, the European Space Agency (ESA) operates the Space Situational Awareness (SSA) program, which includes the Space Weather Coordination Centre (SSCC). The SSCC provides space weather forecasts and warnings for ESA missions and the European space industry, including satellite operators and spacecraft designers. See `https://swe.ssa.esa.int/current-space-weather`.

In Asia, the Japan Meteorological Agency (JMA) operates the Space Weather Forecast Centre, which provides space weather forecasts and warnings for Japan and the Asia-Pacific region (`https://swc.nict.go.jp/en/`). The China Meteorological Administration (CMA) also operates a space weather forecasting center, which provides space weather forecasts and warnings for China and the surrounding areas.

Other countries and international organizations also operate dedicated centers for space weather forecasting, including Canada's Space Weather Prediction Centre, Russia's Institute of Terrestrial Magnetism, Ionosphere, and Radio Wave Propagation, and the International Space Environment Service (ISES), which is a collaboration of space weather centers from multiple countries.

Overall, these dedicated centers play a critical role in monitoring and forecasting space weather events, which have significant impacts on various technological assets and human activities. Their work enables stakeholders to take necessary precautions and mitigate risks associated with space weather events.

Physics-based models are often employed in traditional space weather forecasting methods. One of the most commonly used physics-based models is the magnetohydrodynamic (MHD) simulation. MHD simulations utilize mathematical equations to model the behavior of a plasma, which is a highly conductive gas that makes up much of space.

In MHD simulations, the plasma is treated as a fluid with magnetic properties, and the equations used to model its behavior include the laws of conservation of mass, momentum, and energy. The equations are solved numerically on a computer, and the resulting model provides a highly detailed representation of the plasma's behavior. One significant advantage of MHD simulations is their ability to model the three-dimensional structure of the plasma, which is essential for predicting the behavior of space weather events accurately. The resulting simulations are used to predict the behavior of space weather events, particularly regarding CME propagation, and the dynamics of the magnetosphere.

However, MHD simulations also have limitations. They come with a high computational cost, making them time-consuming to develop and operate. For example, the Space Weather Modeling Framework developed by the University of Michigan [13], which is run operationally at the NOAA Space Weather Prediction Center, uses more than 400,000 CPU-hours per year, with an estimated cost of over $15,000/year, to provide a forecast with a lead time of 20-60 minutes (see `https://www.youtube.com/watch?v=rHBeM2JPRTO`).

In contrast, machine learning methods can provide faster and more cost-effective solutions for space weather forecasting. For instance, neural networks can provide forecasts with a lead time of 1-6 hours ahead of a geomagnetic index at a total cost of less than $10/year (and approximately 250 CPU-hours) (see `https://swx-trec.com/dst/`). Thus, machine learning methods can provide highly accurate forecasts at a fraction of the cost of conventional methods.

Due to the intricate and multifaceted nature of the physical phenomena driving space weather, conventional forecasting methods based on physics-based models frequently rely on a series of simplifying assumptions, which can significantly reduce the accuracy of their forecasts. These assumptions can result in significant uncertainties in the predictions, and may also lead to a lack of robustness, making the models sensitive to small variations in their input parameters.

For instance, Vršnak et al. (2014) [14] has compared the accuracy of the operational MHD model ENLIL (`http://ccmc.gsfc.nasa.gov/requests/SH/ENLIL-1cl/`)with the solution provided by an analytical drag-based model (DBM). The analysis presented in that study shows that the numerical model ENLIL and the analytical model DBM provide similar results in calculating Sun-Earth travel times (TTs) for certain drag-parameter values and solar wind speeds. The difference between the two models is generally less than 10%, and the average difference in absolute value is usually below 8 hours. However, during periods of higher solar activity, the difference between the two models becomes more significant, with standard deviations ranging from approximately 6 to 9 hours for low-activity periods and increasing to approximately 10-11 hours for high-activity periods with complex background solar winds. This is due to ENLIL's ability to consider a more realistic background solar-wind structure, which is not currently included in DBM.

The development of a first-principle approach for forecasting space weather events is unfeasible due to the large scale of space and time, the short time lag between cause and effect, and the significant computational cost of physics-based models. The space weather community has a good understanding of the missing physics in approximations, and which parts of the prediction chain could benefit from coupling with a data-driven approach. This makes space weather an ideal choice for a gray-box approach. The gray-box paradigm exists between two opposing approaches. Black-box methods are completely data-driven, looking for correlations between variables, and do not employ physical information; Machine Learning also falls within this category. White-box models, on the other hand, are based on assumptions and equations, not necessarily related to data; although, it is important to note that physical laws are in fact based on data validation.

6.4 MACHINE LEARNING SPACE WEATHER FORECASTING

Machine learning can help overcome some of the limitations of standard forecasting techniques (particularly the ones based on physics simulations) by using advanced algorithms to analyze large amounts of data and identify patterns that can be used to make more accurate forecasts. Moreover, the availability of large and diverse data sets collected over decades of space missions make space weather a promising application for machine learning. This is due to the availability of large and diverse data sets collected over decades of space missions. The Advanced Composition Explorer (ACE), Wind, and the Deep Space Climate Observatory (DSCOVR) provide in situ plasma data near the first Lagrangian point (L1), while the Solar and Heliospheric Observatory (SOHO), the Solar Terrestrial Relations Observatory (STEREO), and the Solar Dynamics Observatory (SDO) offer Sun images at different wavelengths, magnetograms, and coronographs, covering a 20-year period. In addition, data from the Van Allen Probes, Geostationary Operational Environmental Satellites (GOES), Global Positioning System (GPS), Polar Operational Environmental Satellites (POES), the Defense Meteorological Satellite Program (DMSP), and ground-based magnetometers are available,

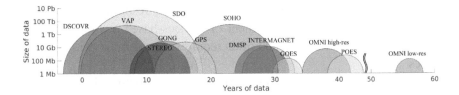

Figure 6.1 Size and volume of some space weather data. Each mission/instrument is represented by a semicircle, where its radius (vertical axis)denotes the order of magnitude of the amount of data available thus far, and the center (horizontal axis) denotes how many years of data are available for each dataset.

offering a wealth of information for space weather predictions. The integration of machine learning algorithms with this data could enable the identification of patterns and relationships that may not be apparent to human experts. A snapshot of the amount and volume of several space weather data sets is shown in Figure 6.1

In a review by Camporeale (2019) [15], the challenges and opportunities of machine learning in space weather were discussed (see also [16]). Here, we center our focus on three applications that have been extensively studied: forecasting of geomagnetic indices, relativistic electrons at geosynchronous orbits, and solar eruptions (flares and CMEs). This review is not intended to be comprehensive. Our goal is to provide an overview of the diverse range of techniques that have been explored over the years, rather than providing in-depth commentary on individual studies.

Our perspective is that machine learning will play an increasingly important role in all aspects of space weather in the near future. Furthermore, machine learning has already shown to be more effective than traditional and standard methods, such as those based on physics-based simulations. Specifically, we focus on three critical aspects of space weather forecasting: speed of execution and lead-time, accuracy, and actionability. We contend that machine learning based methods outperform traditional approaches in all of these areas, making them the preferred tool for space weather forecasting.

6.4.1 Speed of execution

The task of forecasting an event is highly dependent on the ability of computational methods to perform faster than real-time using real-time observations. However, this presents a significant challenge to physics-based models that demand high-resolution discretizations, resulting in millions of degrees of freedom. The development of physics-based models with such high resolution is challenging and often requires considerable computing resources.

While recent advances in High Performance Computing (HPC) have allowed for large-scale simulations of the Sun-Earth environment that can run faster than real-time, these simulations are often limited to simple MHD models, which provide only approximate results. Despite these limitations, MHD models have demonstrated their utility in forecasting and predicting space weather events. However, as the models are simplified, they may not be sufficient for more complex events or scenarios.

Another significant challenge of physics-based models is the long computing times that can make it difficult to quickly prototype new features or debug new codes in operational environments. With the increasing reliance on space weather predictions, there is a growing demand for faster and more sophisticated models that can take into account the complex interplay between various physical processes. Thus, there is a need to develop more efficient and accurate computational methods to better forecast space weather events. Recent advances in Artificial Intelligence (AI) based forecasting tools hold promise in addressing some of these challenges. These tools are capable of processing large volumes of data more efficiently, and can provide faster and more accurate predictions. Additionally, these models can incorporate data from a variety of sources, such as ground-based observations and space-based instruments, making them more comprehensive and robust.

In recent years, significant progress has been made in the fields of computational fluid dynamics and numerical weather predictions. One notable development is the creation of machine learning emulators, which aim to simulate physics-based simulations with comparable accuracy but much faster execution times. These emulators can be either entirely data-driven or physics-guided and physics-informed, with the latter designed to follow physical laws and constraints.

According to Kochkov et al. (2021) [17], promising speed-ups of the order of 103-104 have been achieved for emulators of turbulent three-dimensional Navier-Stokes equations. A comprehensive overview of machine learning techniques for computational fluid dynamics, including approaches to combine machine learning and physics-based models, can be found in Vinuesa and Bruntor's (2022) review [18]. This work also discusses the gray-box approach, where a machine learning model is guided by physics-based constraints, such as those presented in Camporeale et al. (2020) [19], which applied a gray-box approach to space weather forecasting.

Training a machine learning model can be a time-consuming process that requires a heuristic approach, but executing a machine learning model involves running a large number of pointwise operations with millions of parameters, making even complex and deep neural networks executable in seconds. Another advantage of machine learning models is their ability to extend lead times beyond those possible with physics-based models. The lead time of a prediction in a physics-based model is determined by the availability of real-time data and the prediction window. In contrast, a machine learning model can forecast events several hours in advance by training models with longer lead times or incorporating additional sources of input, such as solar images, which can achieve a lead time of 1-2 days, as demonstrated in Hu et al. (2022) [20].

6.4.2 Accuracy

Machine learning methods seek to find mathematical relationships between inputs and outputs that can accurately predict future outcomes based on patterns in historical data. However, one of the key challenges in machine learning is to avoid overfitting or underfitting, which can lead to poor generalization performance on new data. Overfitting occurs when the model is too complex and starts to fit the noise in the training data, while underfitting occurs when the model is too simple and cannot capture the underlying patterns in the data.

In space weather forecasting, the goal of machine learning models is to minimize the discrepancy between predictions and observations, which is typically done by optimizing a cost function. By doing so, the machine learning model seeks to maximize accuracy for a given output of interest, such as solar flares or CMEs. However, unlike physics-based models, which depend on numerous factors such as initial and boundary conditions and modeling assumptions, machine learning models aim to find general relationships that can be applied to new data.

Despite the challenges of overfitting and underfitting, many machine learning based space weather forecasting models have been developed and claimed to outperform traditional physics-based models. This is because machine learning models can learn from large amounts of data and identify patterns that may be difficult to discern using traditional physics-based methods. However, it is important to note that the accuracy of machine learning models depends on the quality and quantity of data used to train them, as well as the specific input features and model architecture chosen. Therefore, careful validation and testing of machine learning based space weather forecasting models are necessary to ensure their reliability and robustness.

6.4.3 Actionability

To ensure that a space weather forecasting model is actionable, two critical properties must be met. Firstly, the model must provide sufficient lead time to allow users to take appropriate mitigating actions. The required lead time will depend on the specific application, and it is essential that users and stakeholders communicate their needs to the research community and forecast providers. Secondly, the model must be capable of accurately quantifying and expressing uncertainties associated with its predictions as probabilities. Accurate uncertainties are crucial to facilitate a risk/cost analysis of the mitigating actions. Machine learning tools are becoming increasingly popular for achieving these two properties in space weather forecasting.

To produce a day or more lead time, remote observations of the solar disk are required as the magnetosphere has a limited memory of about 10 hours. Ground-based magnetograms provided by the Global Oscillation Network Group (GONG) and Air Force Data Assimilative Photospheric Flux Transport (ADAPT), EUV (extreme ultraviolet) images from the SDO, SOHO, GOES/SUVI, and coronagraphs are some of the information sources used for this purpose. Recent proof-of-principle results have shown that machine learning techniques can extract useful information from these images to forecast geomagnetic indices and solar wind velocity a few days in advance.

Uncertainty quantification has seen significant growth in recent years, and machine learning applications have contributed significantly to this area. Several studies have applied uncertainty quantification techniques to space weather forecasting, specifically to quantify uncertainties associated with geomagnetic indices and solar wind predictions.

Furthermore, machine learning can be used as a post-processing tool to calibrate physics-based simulations. This approach has been successfully demonstrated in the context of numerical weather predictions and has recently been applied to space weather forecasting as well.

6.4.4 Geomagnetic indices

Geomagnetic indices are used to condense a large amount of information about the state of the magnetosphere into a single number. While there are many different indices available, some of the most commonly used include Kp and distributed storm time (Dst), as well as a variety of others like auroral electrojet (AE, AL, AO, AU), ap, am, Inter-Hour Variability index (IHV), Ap, Cp, C9, SYMH, and ASYH, each designed to capture a different aspect of geomagnetic activity. One recent attempt to create a more comprehensive index is described in the work by Borovsky and Denton (2018) [21].

In Hu et al. (2022) [20] a new model is presented for predicting Dst with a lead time of 1 to 6 hours. The model uses a Gated Recurrent Unit network trained with solar wind parameters, and its uncertainty is estimated using the ACCRUE (ACCurate and Reliable Uncertainty Estimate) method. A multi-fidelity boosting method is also developed to improve the model's accuracy and reduce its uncertainty. The model can predict Dst 6 hours ahead with a root mean square error (RMSE) of 13.54 nT, which outperforms a persistence model and a simple gated recurrent units (GRU) model.

Tasistro-Hart et al. (2021) [22] proposes a neural network architecture that combines data from the L1 point and solar disk to improve forecast reliability. The architecture also estimates uncertainty in multiple-hour ahead forecasts, improving the accuracy of the predictions. The focus is on forecasting the external component of geomagnetic storms, Est, instead of the conventional Dst index.

Gruet et al. (2018) [23] presents a method that combines a Long Short-Term Memory (LSTM) recurrent neural network with a Gaussian process (GP) model to provide up to 6 hours ahead probabilistic forecasts of the Dst geomagnetic index. The model is trained using the hourly OMNI and Global Positioning System (GPS) databases, both of which are publicly available. The LSTM network is first developed to get a single-point prediction of Dst. The model shows great accuracy in forecasting the Dst index from 1 to 6 hours ahead, with a correlation coefficient always higher than 0.873 and a root-mean-square error lower than 9.86. However, the model is poor in predicting intense storms (Dst < −250 nT) 6 hour in advance. To improve the model and to obtain probabilistic forecasts, the LSTM model is combined with a GP. The hybrid predictor is evaluated using the receiver operating characteristic curve and the reliability diagram, and the study concludes that this methodology provides improvements in the forecast of geomagnetic storms, from 1 to 6 hour ahead.

6.4.5 Solar flares

Solar flares, characterized by sudden and intense releases of magnetic energy that result in large fluxes of X-rays, pose a significant threat to space weather and associated technological systems. As a result, solar flare prediction has become a major area of research in space weather. Most studies that use solar images focus on predicting solar flares, which are categorized based on their X-ray peak flux measured by the GOES. Flares are categorized into A-, B-, C-, M-, or X-classes, depending on the peak X-ray flux. The consequences of solar flares include radio communication blackouts and an increase in satellite drag, which can lead to a range of issues in technological systems that rely on satellite communication. Given these consequences, the prediction of solar flares has become a critical area of study in space weather research.

Florios et al. (2018) [24] describes a new method for predicting >M1 and >C1 solar flares using classic and modern machine learning techniques such as multi-layer perceptrons, support vector machines, and random forests. The predictor variables used were based on the SDO/HMI SHARP data product, which was available from 2012. The sample used in the study included all calendar days within a five-year period of Solar Cycle 24 (2012-2016), with predictor variables recorded every 3 hours.

The study found that the random forest (RF) method was the most effective for predicting both >M1 and >C1 flares, with a probability threshold of 15% for >M1 flares resulting in mean RF True Skill Statistics (TSS) of 0.74 ± 0.02 and Heidke Skill Score (HSS) of 0.49 ± 0.01, and a probability threshold of 35% for >C1 flares resulting in mean TSS=0.60 ± 0.01 and HSS=0.59 ± 0.01. The respective accuracy values were ACC=0.93 and ACC=0.84. In terms of probabilistic skill scores, the random forest method was ranked the highest, followed by multi-layer perceptrons and support vector machines for both >M1 and >C1 flares.

Nishizuka et al. (2018) [25] describes the development of a deep neural network (DNN) model called Deep Flare Net (DeFN) for predicting solar flares. The model calculates the probability of flares occurring in the next 24 hours in each active region and determines the most likely maximum classes of flares via binary classification. The model uses 79 features for each region and was trained on a dataset from 2010-2014 and tested on a dataset from 2015. The model was able to predict M-class flares with a TSS of 0.80 and C-class flares with a TSS of 0.63. The authors note that the DeFN model differs from usual DNN models in that the features are manually selected and can be analyzed for their effectiveness in prediction.

Sun et al. (2022) [26] proposes a deep learning approach to predict M- and X-class solar flares using line-of-sight magnetograms and active region parameters. Two deep learning algorithms, convolutional neural network (CNN) and LSTM, and their stacking ensembles are trained and evaluated. The study finds that LSTM trained on data from two solar cycles achieves higher true skill scores than LSTM trained on data from a single solar cycle. A stacking ensemble combining predictions from LSTM and CNN using the TSS criterion achieves a higher TSS than the "select-best" strategy. The study also uses a visual attribution method called "integrated gradients" to attribute the CNN's predictions of flares to the emerging magnetic flux in the active region and reveals a limitation of CNNs in treating the polarity artifact of line-of-sight magnetograms as positive evidence of flares.

Kaneda et al. (20220 [27] presents the Flare Transformer (FT), a method for predicting the maximum solar flare class that will occur in the next 24 hours. The study proposes several contributions, including the use of the FT method to handle line-of-sight magnetograms and physical features, the implementation of a transformer attention mechanism to model temporal relationships between input features, the introduction of Gandin–Murphy–Gerrity score (GMGS) and Brier skill score (BSS) losses to balance major metrics in solar flare prediction, and demonstration that the model's predictions outperform those of human experts in terms of GMGS and TSS.

6.4.6 Coronal Mass Ejection arrival time

CMEs are explosive outbursts of magnetized plasma that originate from the surface of the Sun and can travel at speeds of up to 1,000 km/s. One of the major obstacles of space weather forecasting is predicting the behavior of CMEs as they propagate away from the Sun and towards Earth [28]. The speed, magnetic field strength, and orientation of the plasma are intimately connected to the onset of geomagnetic storms [29], which can have severe consequences for Earth's technological infrastructure. To model CME propagation, space weather forecasters generally rely on MHD, a mathematical framework that characterizes the behavior of magnetized fluids. By employing numerical simulations of the MHD equations with proper boundary and initial conditions, forecasters can predict CME trajectories. Nonetheless, estimating the uncertainties linked to numerical simulations remains a major challenge since they are impacted by uncertain initial and boundary conditions [30]. Machine learning techniques, particularly in the "gray-box" paradigm, can help address this challenge by refining numerical simulations through the integration of observational data.

Liu et al. (2018) [31] proposes a new approach called CAT-PUMA for predicting the arrival time of partial-/full halo CMEs using machine learning algorithms. The prediction engine uses previously observed geo-effective CMEs and algorithms of the Support Vector Machine. The proposed approach is accurate and fast with a mean absolute prediction error of ∼5.9 hour within the CME arrival time. Comparisons with other models reveal that CAT-PUMA has a more accurate prediction for 77% of the events investigated and can be carried out very quickly. The paper provides a practical guide containing the CAT-PUMA engine and the source code of two examples.

Wang et al. (2019) [32] has utilized a CNN regression model to analyze transit times of geoeffective CME events over the past 30 years. Unlike previous studies, the CNN regression model does not require manual feature selection, time-consuming feature collection, or expert knowledge. The model solely utilizes white-light observations of CMEs as input. The mean absolute error of the constructed CNN regression model is 12.4 hours, comparable to the average performance of previous studies. As more CME data become available, the authors expect the CNN regression model to reveal better results.

Alobaid et al. (2022) [33] presents an ensemble framework (CMETNet) for predicting the arrival time of CME events using five machine learning models. The experimental results showed that CMETNet outperforms individual machine learning models, as well as two previously published machine learning methods and three physics-based methods. However, accurately predicting CME arrival time remains challenging, and further improvement can be achieved by combining machine learning with physics-based methods.

6.4.7 Forecast of solar wind speed

Chandorkar et al. (2020) [34] introduces a new regression problem, Dynamic Time-Lag Regression (DTLR), which aims to predict near-Earth solar wind speed based on estimates of the Sun's coronal magnetic field. DTLR differs from mainstream regression and sequence-to-sequence learning as it lacks ground truth and contains irrelevant information. The paper proposes a Bayesian approach with theoretical justifications based on linear stability analysis, presents synthetic problem proof of concept, and improves on the state of the art in solar wind forecasting through empirical results.

Sun et al. (2022) [35] introduces the Graph-Temporal-AR model (GTA) for solar wind speed prediction. GTA uses a graph attention module to learn the complex dependencies among features, dilated causal convolution to extend the receptive field, and an autoregressive model to solve the scale insensitive problem of neural networks. The model combines OMNI data and extreme ultraviolet images to predict the solar wind speed. Compared to baseline models, GTA shows significant performance improvements and can visualize the relationships between multiple variables without prior domain knowledge. The data and code are publicly available.

Upendran et al. (2020) [36] developed a model to predict solar wind speed at L1 using extreme ultraviolet images and OMNIWEB data from NASA. Their Graph Convolutional Network model outperformed autoregressive and naive models with a best fit correlation of 0.55 ± 0.03. The model's visualization revealed that it was able to learn associations between coronal and solar wind structure, indicating a potential to discover unknown relationships in heliophysics data sets.

6.5 CONCLUSIONS

Machine learning has the potential to transform space weather forecasting and give stakeholders the information they need to make informed decisions and reduce their losses. However, there are certain challenges and considerations that must be taken into account when implementing machine learning in space weather forecasting, such as data quality and reliability, the potential for bias in the models, and the need for ongoing monitoring and improvement of the models.

Data quality and reliability are of utmost importance for machine learning applications. Data must be complete, accurate, and up-to-date. Furthermore, the data itself may have inherent bias and thus the models could be inadvertently biased. Therefore, the data and the models must be continually monitored and improved.

The "black-box" nature of machine learning models can also present a challenge. In conventional physical models, it is often straightforward to explain why a particular model produces a certain result. In machine learning, it is much more difficult to explain why a model makes certain predictions or decisions. This lack of transparency can make it difficult to assess the accuracy and reliability of the models and to identify and address potential problems.

In addition, developing and deploying machine learning models in space weather forecasting requires interdisciplinary collaborations between space scientists and data scientists. Understanding the language and tools of machine learning is critical for space physicists to be able to effectively use machine learning in their research.

Finally, machine learning models must be continually updated and improved to ensure accuracy and reliability. As the data used in machine learning models change over time, the models need to be regularly updated to ensure that they remain accurate and reliable.

In addition to the aforementioned challenges and considerations, machine learning applications for space weather forecasting are expected to provide a cheaper alternative to traditional physics-based simulation forecasts. This is because machine learning models can be developed and trained on large datasets of observational data, which can be readily available and cheaper to acquire than creating physics-based simulations. Additionally, once trained, machine learning models can be deployed and maintained at a lower cost compared to traditional numerical simulations.

The commercial applications of space weather forecasting are also expected to increase in the future due to the growing demand for accurate space weather predictions across several industries. For example, in the aviation industry, space weather events can disrupt communication and navigation systems, leading to delays, cancellations, and reroutes. Telecommunications and energy sectors can also experience similar disruptions due to space weather events, leading to significant economic losses. Accurate space weather predictions can enable these industries to take proactive measures to mitigate the impact of space weather events on their operations.

Furthermore, with the increasing interest in space exploration and the potential for long-duration human missions to the Moon and Mars, the need for reliable space weather forecasting is paramount. Astronauts and their equipment will be exposed to the harsh space environment, including solar flares and cosmic radiation, and accurate forecasting will be crucial for ensuring their safety and the success of their missions.

ACRONYM

ACE Advanced Composition Explorer

ACCRUE ACCurate and Reliable Uncertainty Estimate

ADAPT Air Force Data Assimilative Photospheric Flux Transport

AI Artificial Intelligence

BSS Brier skill score

CMA China Meteorological Administration

CME Coronal mass ejection

CNN Convolutional neural network

CPU Central processing unit

DBM Drag-based model

DeFN Deep Flare Net

DNN Deep neural network

DMSP Defense Meteorological Satellite Program

DSCOVR Deep Space Climate Observatory

Dst Disturbance Solar Storm index

DTLR Dynamic Time-Lag Regression

ESA European Space Agency

EUV Extreme ultraviolet

FT Flare Transformer

HPC High performance computing

HSS Heidke skill score

IADC Inter-Agency Space Debris Coordination Committee

GDP Gross domestic product

GMGS Gandin–Murphy–Gerrity score

GOES Geostationary Operational Environmental Satellite

GONG Global Oscillation Network Group

GP Gaussian process

GPS Global Positioning System

GRU Gated recurrent units

GTA Graph-Temporal-AR model

JMA Japan Meteorological Agency

LSTM Long Short-Term Memory

MHD Magnetohydrodynamic

ML Machine learning

NOAA National Oceanic and Atmospheric Administration

NSWS National Space Weather Strategy

NWS National Weather Service

POES Polar Operational Environmental Satellites

RMSE Root mean square error

RF Random forest

SDO Solar Dynamics Observatory

SOHO Solar and Heliospheric Observatory

SSA Space Situational Awareness

STEREO Solar Terrestrial Relations Observatory

SWAP Space Weather Action Plan

SWCC Space Weather Coordination Centre

SWPC Space Weather Prediction Center

TSS True skill statistics

TT Travel time

GLOSSARY

ACCRUE: Cost function that is used to estimate the input-dependent variance, given a black-box "oracle" mean function, by solving a two-objective optimization problem.

Coronal mass ejection: Ejections of high-mass clouds of solar material in the Sun's corona into the heliosphere. They are mainly associated with disappearing solar filaments, erupting prominences, and solar flares.

Dst index: The disturbance storm index is a measure of the horizontal magnetic field variations due to the presence of the enhanced equatorial ring current.

ENLIL: A time-dependent 3D MHD model of the heliosphere. It solves equations for plasma mass, momentum and energy density, and magnetic field, using a Flux-Corrected-Transport (FCT) algorithm

Geomagnetic storm: A worldwide disturbance of the Earth's magnetic field, distinct from regular diurnal variations. A storm is precisely defined as occurring when Dst index becomes less than -50 nT.

Lagrange point: The Lagrangian points are the five positions in an orbital configuration where a small object affected only by gravity can theoretically be part of a constant-shape pattern with two larger objects (such as a satellite with respect to the Sun and Earth).

Solar flare: A sudden release of energy in the solar atmosphere lasting minutes to hours, from which electromagnetic radiation and energetic charged particles are emitted.

Solar wind: Particle radiation emitted constantly by the Sun in all directions with the velocity ranging from less than 10 km/s to approximately 400-500 km/s.

Space weather: Phenomenon determined by the most varied interactions between the Sun, interplanetary space, and the Earth.

FURTHER READING

Chen, Y., Maloney, S., Camporeale, E., Huang, X., & Zhou, Z. (2023) "Machine Learning and Statistical Methods for Solar Flare Predictions". *Frontiers in Astronomy and Space Sciences*, 10, 74.

McGranaghan, R. M., Thompson, B., Camporeale, E., Bortnik, J., Bobra, M., Lapenta, G., ... & Delouille, V. (2022) "Heliophysics Discovery Tools for the 21st Century: Data Science and Machine Learning Structures and Recommendations for 2020-2050". *arXiv preprint arXiv:2212.13325.*

McGranaghan, R. M., Camporeale, E., Georgoulis, M., & Anastasiadis, A. (2021) "Space Weather research in the Digital Age and across the full data lifecycle: Introduction to the Topical Issue". *Journal of Space Weather and Space Climate,* 11, 50.

Bortnik, J., E. Camporeale (2021) "Ten ways to apply machine learning in Earth and space sciences". *EOS,* 102, https://doi.org/10.1029/2021EO160257. Published on 29 June 2021.

Nita, G., Georgoulis, M., Kitiashvili, I., Sadykov, V., Camporeale, E., Kosovichev, A., ... & Yu, S. (2020) "Machine learning in heliophysics and space weather forecasting: a white paper of findings and recommendations". *arXiv preprint arXiv:2006.12224.*

BIBLIOGRAPHY

[1] Karl Schindler. *Physics of space plasma activity*. Cambridge University Press, 2006.

[2] DN Baker. Effects of the sun on the earth's environment. *Journal of Atmospheric and Solar-Terrestrial Physics*, 62(17-18):1669–1681, 2000.

[3] JP Eastwood, E Biffis, MA Hapgood, L Green, MM Bisi, RD Bentley, Robert Wicks, L-A McKinnell, M Gibbs, and C Burnett. The economic impact of space weather: Where do we stand? *Risk analysis*, 37(2):206–218, 2017.

[4] Edward J Oughton, Mike Hapgood, Gemma S Richardson, Ciarán D Beggan, Alan WP Thomson, Mark Gibbs, Catherine Burnett, C Trevor Gaunt, Markos Trichas, Rabia Dada, et al. A risk assessment framework for the socioeconomic impacts of electricity transmission infrastructure failure due to space weather: An application to the united kingdom. *Risk Analysis*, 39(5):1022–1043, 2019.

[5] Ilan Noy and Tomáš Uher. Four new horsemen of an apocalypse? solar flares, super-volcanoes, pandemics, and artificial intelligence. *Economics of Disasters and Climate Change*, 6(2):393–416, 2022.

[6] Space Studies Board, National Research Council, et al. *Severe space weather events: Understanding societal and economic impacts: A workshop report*. National Academies Press, 2009.

[7] H Schulte in den Bäumen, D Moran, M Lenzen, I Cairns, and A Steenge. How severe space weather can disrupt global supply chains. *Natural Hazards and Earth System Sciences*, 14(10):2749–2759, 2014.

[8] Pete Riley, Dan Baker, Ying D Liu, Pekka Verronen, Howard Singer, and Manuel Güdel. Extreme space weather events: From cradle to grave. *Space Science Reviews*, 214:1–24, 2018.

[9] Edward J Oughton, Andrew Skelton, Richard B Horne, Alan WP Thomson, and Charles T Gaunt. Quantifying the daily economic impact of extreme space weather due to failure in electricity transmission infrastructure. *Space Weather*, 15(1):65–83, 2017.

[10] E Oughton, J Copic, A Skelton, V Kesaite, ZY Yeo, SJ Ruffle, and D Ralph. Helios solar storm scenario: Cambridge risk framework series, centre for risk studies. *Cambridge, UK: University of Cambridge*, 2016.

[11] Olga Sokolova, Nikolay Korovkin, and Masashi Hayakawa. *Geomagnetic Disturbances Impacts on Power Systems: Risk Analysis and Mitigation Strategies.* CRC Press, 2021.

[12] Mike Hapgood, Matthew J Angling, Gemma Attrill, Mario Bisi, Paul S Cannon, Clive Dyer, Jonathan P Eastwood, Sean Elvidge, Mark Gibbs, Richard A Harrison, et al. Development of space weather reasonable worst-case scenarios for the uk national risk assessment, 2021.

[13] Tamas I Gombosi, Yuxi Chen, Alex Glocer, Zhenguang Huang, Xianzhe Jia, Michael W Liemohn, Ward B Manchester, Tuija Pulkkinen, Nishtha Sachdeva, Qusai Al Shidi, et al. What sustained multi-disciplinary research can achieve: The space weather modeling framework. *Journal of Space Weather and Space Climate*, 11:42, 2021.

[14] Bojan Vršnak, M Temmer, T Žic, A Taktakishvili, M Dumbović, C Möstl, AM Veronig, ML Mays, and D Odstrčil. Heliospheric propagation of coronal mass ejections: comparison of numerical wsa-enlil+cone model and analytical drag-based model. *The Astrophysical Journal Supplement Series*, 213(2):21, 2014.

[15] Enrico Camporeale. The challenge of machine learning in space weather: Nowcasting and forecasting. *Space Weather*, 17(8):1166–1207, 2019.

[16] Enrico Camporeale, Simon Wing, and Jay Johnson. *Machine learning techniques for space weather.* Elsevier, 2018.

[17] Dmitrii Kochkov, Jamie A Smith, Ayya Alieva, Qing Wang, Michael P Brenner, and Stephan Hoyer. Machine learning–accelerated computational fluid dynamics. *Proceedings of the National Academy of Sciences*, 118(21):e2101784118, 2021.

[18] Ricardo Vinuesa and Steven L Brunton. Enhancing computational fluid dynamics with machine learning. *Nature Computational Science*, 2(6):358–366, 2022.

[19] Enrico Camporeale, Michele D Cash, Howard J Singer, Christopher C Balch, Z Huang, and Gabor Toth. A gray-box model for a probabilistic estimate of regional ground magnetic perturbations: Enhancing the noaa operational geospace model with machine learning. *Journal of Geophysical Research: Space Physics*, 125(11):e2019JA027684, 2020.

[20] A Hu, E Camporeale, and B Swiger. Multi-hour ahead dst index prediction using multi-fidelity boosted neural networks. *arXiv preprint arXiv:2209.12571*, 2022.

[21] Joseph E Borovsky and Michael H Denton. Exploration of a composite index to describe magnetospheric activity: Reduction of the magnetospheric state vector to a single scalar. *Journal of Geophysical Research: Space Physics*, 123(9):7384–7412, 2018.

[22] Adrian Tasistro-Hart, Alexander Grayver, and Alexey Kuvshinov. Probabilistic geomagnetic storm forecasting via deep learning. *Journal of Geophysical Research: Space Physics*, 126(1):e2020JA028228, 2021.

[23] Marina A Gruet, M Chandorkar, Angélica Sicard, and Enrico Camporeale. Multiple-hour-ahead forecast of the dst index using a combination of long short-term memory neural network and gaussian process. *Space Weather*, 16(11):1882–1896, 2018.

[24] Kostas Florios, Ioannis Kontogiannis, Sung-Hong Park, Jordan A Guerra, Federico Benvenuto, D Shaun Bloomfield, and Manolis K Georgoulis. Forecasting solar flares using magnetogram-based predictors and machine learning. *Solar Physics*, 293(2):28, 2018.

[25] Naoto Nishizuka, Komei Sugiura, Yuki Kubo, Mitsue Den, and Mamoru Ishii. Deep flare net (defn) model for solar flare prediction. *The Astrophysical Journal*, 858(2):113, 2018.

[26] Zeyu Sun, Monica G Bobra, Xiantong Wang, Yu Wang, Hu Sun, Tamas Gombosi, Yang Chen, and Alfred Hero. Predicting solar flares using cnn and lstm on two solar cycles of active region data. *The Astrophysical Journal*, 931(2):163, 2022.

[27] Kanta Kaneda, Yuiga Wada, Tsumugi Iida, Naoto Nishizuka, Yûki Kubo, and Komei Sugiura. Flare transformer: Solar flare prediction using magnetograms and sunspot physical features. In *Proceedings of the Asian Conference on Computer Vision*, pages 1488–1503, 2022.

[28] Emilia KJ Kilpua, Noe Lugaz, M Leila Mays, and M Temmer. Forecasting the structure and orientation of earthbound coronal mass ejections. *Space Weather*, 17(4):498–526, 2019.

[29] John T Gosling. The solar flare myth. *Journal of Geophysical Research: Space Physics*, 98(A11):18937–18949, 1993.

[30] C Kay and Natchimuthuk Gopalswamy. The effects of uncertainty in initial cme input parameters on deflection, rotation, b z, and arrival time predictions. *Journal of Geophysical Research: Space Physics*, 123(9):7220–7240, 2018.

[31] Jiajia Liu, Yudong Ye, Chenglong Shen, Yuming Wang, and Robert Erdélyi. A new tool for cme arrival time prediction using machine learning algorithms: Cat-puma. *The Astrophysical Journal*, 855(2):109, 2018.

[32] Yimin Wang, Jiajia Liu, Ye Jiang, and Robert Erdélyi. Cme arrival time prediction using convolutional neural network. *The Astrophysical Journal*, 881(1):15, 2019.

[33] Khalid Alobaid, Yasser Abduallah, Jason Wang, Haimin Wang, Haodi Jiang, Yan Xu, Vasyl Yurchyshyn, Hongyang Zhang, HÜSEYİN ÇAVUŞ, and Ju Jing. Predicting cme arrival time through data integration and ensemble learning. *Frontiers in Astronomy and Space Sciences*, 9, 2022.

[34] Mandar Chandorkar, Cyril Furtlehner, Bala Poduval, Enrico Camporeale, and Michèle Sebag. Dynamic time lag regression: Predicting what and when. In *ICLR 2020-8th International Conference on Learning Representations*, 2020.

[35] Yanru Sun, Zongxia Xie, Haocheng Wang, Xin Huang, and Qinghua Hu. Solar wind speed prediction via graph attention network. *Space Weather*, 20(7):e2022SW003128, 2022.

[36] Vishal Upendran, Mark CM Cheung, Shravan Hanasoge, and Ganapathy Krishnamurthi. Solar wind prediction using deep learning. *Space Weather*, 18(9):e2020SW002478, 2020.

CHAPTER 7

Using Unsupervised Machine Learning to make new discoveries in space data

Giovanni Lapenta

Center for mathematical Plasma Astrophysics, Department of Mathematics, Katholieke Universiteit Leuven (KU Leuven), Belgium

Francesco Califano

Dipartimento di Fisica "Enrico Fermi", University of Pisa, Italy

Romain Dupuis

Naval Group Belgium, Belgium

Maria Elena Innocenti

Institut für Theoretische Physik, Ruhr-Universität Bochum, Germany

Giorgio Pedrazzi

Dipartimento High Performance Computing (HPC), Cineca, Italy

CONTENTS

7.1 The AIDA project 216
7.2 Data discovery in magnetospheric simulations 219
    7.2.1 The unsupervised clustering technique 221
    7.2.2 Using SOM products for physical investigation 226
    7.2.3 Perspectives on the use of unsupervised clustering methods in the analysis of simulated datasets 229
7.3 Finding Reconnection using unsupervised machine learning ... 230
    7.3.1 Machine Learning pipeline 233

DOI: 10.1201/9781003366386-7

7.3.2   Perspectives on the use of unsupervised machine
learning for the analysis of very large databases 238
7.4   Characterizing turbulence with unsupervised machine learning   239
   7.4.1   DBSCAN Identification of current layers and their
properties 241
   7.4.2   Link between current layers and reconnection 243
   7.4.3   Perspectives on the DBSCAN method for the study of
turbulence 247
7.5   Characterizing velocity distribution functions with
unsupervised machine learning 247
   7.5.1   Magnetic reconnection signatures and agyrotropy 248
   7.5.2   Fitting particle velocity distributions 249
   7.5.3   Experiments on 2.5D PIC simulation 251
   7.5.4   Analysis of the particle velocity distribution fit 254
   7.5.5   Perspectives on the Gaussian Mixture Model method
for the study of reconnection 256
7.6   Summary and key takeaway messages 257
Bibliography 261

MODERN research in space science is accumulating an exponentially growing amount of data. The size of the datasets produced by simulations has grown in time with the speed of the computers, following its own Moore's law. At the same time, also data from space mission has grown, with more images, more time series and more complex datasets such as spectrograms and velocity distribution function. Mining information and making discoveries out of this wealth of data has far transcended the ability of the human mind. Artificial intelligence (AI), instead, has taken off in many areas of research and applications. Space science can benefit from this new trend. We report here on the recent developments on the application of AI to the analysis of space data. Without limiting ourselves to it, we focus especially on the work done in the context of the AIDA Horizon 2020 project (Artificial Intelligence for Data Analysis, www.aida-space.eu). We show how AI developments are transforming the way we analyze space data. In particular, we find promise in unsupervised Machine Learning (ML) to give optimism for fundamental new discoveries. Supervised ML can expand the range of known methods of analysis, learning from human training and treat vast amount of data. But unsupervised ML, a method where the learning is not guided by previous knowledge, can truly find unexpected new discoveries. Unsupervised ML provides techniques to treat large data sets and discover within them features, correlations and physical laws that would escape traditional approaches. In this chapter we review some of the applications of unsupervised ML developed by AIDA.

7.1 THE AIDA PROJECT

The project AIDA[1] is a project initiated with funding from Horizon 2020 by the European Commission. AIDA is Coordinated by Giovanni Lapenta at the KULeuven in Belgium and it involved Università della Calabria , Università di Pisa and Cineca in Italy the CWI (Centrum Wiskunde & Informatica) in the Netherlands, the CNRS in France and Space Consulting International in the United States.

AIDA uses some of the most advanced data analysis techniques to treat the data available from space missions and from space simulations. AIDA started from acknowledging three major trends in the space field of research and in the wider society.

First, there has been a tremendous accumulation of data from space studies. Every new space mission accumulates new data. But every new mission also increases the amount of data, its quality and resolution compared with previous missions, leading to an ever increasing quantity of data. Similarly, the computer simulations of space science are becoming ever more complex and their results are more detailed and rich leading to a rapidly increasing mass of data available.

All this data is becoming progressively more accessible. The data from space missions is now regularly available to the public and even the results of simulations are started to be shared in public databases. General databases of space data are also being created. AIDA promotes this progress by supplementing these tools to make space data more readily accessible to the public, creating common standards for storing data when such standards do not yet exist while using community standards whenever possible. This advancement is achieved by federating existing tools and creating new ones into AIDApy[2], the main outcome of the AIDA project: a software tool to gather and analyze space-relevant data from observations and simulations.

Second, the vast amount of space data accumulated by past missions and past simulation studies and the new data generated every day by ongoing missions and by new simulations is becoming so large that a single human brain cannot grasp its entire complexity. Luckily, a trend in sweeping society to

---

[1] See www.aida-space.eu.
[2] See https://gitlab.com/aidaspace/aidapy.

address this very same problem arising in every aspect of human activities: machine learning and artificial intelligence (AI). AIDA brings the latest developments in AI to the space arena (AI in AIDA is indeed Artificial Intelligence for Data Analysis, AIDA). The competences and passions needed to be a space scientist are quite different from those of an expert in artificial intelligence. AIDA brings together expertise in computer science, AI and space science to create a new tool AIDApy that will make the use of AI simpler for the space scientist and even the general public.

Third, there is an ongoing trend in all big data analysis and in AI in particular to move towards a free open source language for computers called python. Python is a widely used versatile language, the de-facto standard in AI programming. AIDA develops its main tool, AIDApy, in python. The space community is traditionally linked primarily with MATLAB and IDL. These are commercial tools that come at a very high cost, although non profit organizations and students can obtain better deals. To truly democratize science, including the so-called trend in citizen science, to promote its spread in developing countries and to simplify access to the data and its analysis to all, python provides a free alternative. Python potentially has even more features than those commercial alternatives and many science communities are currently making the transitions to it. AIDA is fostering this same transition to python also for the space community, collecting in AIDApy (where py stands for python) tools previously based on non-free languages. And developing new tools based on the most advanced tools in the state of the art.

As this brief summary clarifies, the aim of AIDA is as much scientific as it is directed at transforming the way society can access the results of space missions and space simulations, making the data accessible to all and giving to all, including non expert citizens, access to the most modern tools to analyze the data. Just as amateur astronomers have in the past discovered moons, comets and astronomical processes, AIDApy is empowering citizens to use the most advanced tools in data analysis and AI to explore space science.

Space is an important growing area in our economy, it is the final frontier. Companies are moving to mine space and extend economic growth to space assets. But space is a hazardous environment and understanding and predicting what happens in the Earth space environment and beyond is critical for this sector of our economy. But it is also critical in other activities, even on the ground, because space can have dangerous impact on the ground, for example damaging power lines and large high tension transformers. This domain called space weather has become one of the focuses of societal impact of science. The tools of AIDA will help the experts make their predictions and understand the physics behind it. But AIDA, thanks to its reliance on a freely

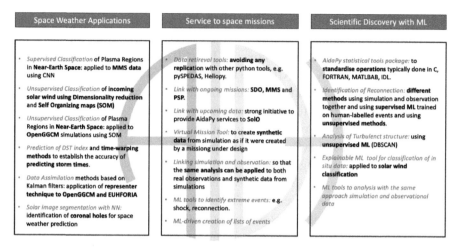

Space Weather Applications	Service to space missions	Scientific Discovery with ML
• *Supervised Classification* of Plasma Regions in **Near-Earth Space: applied to MMS data using CNN**	• *Data retireval tools:* **avoiding any replication with other python tools, e.g.** pySPEDAS, Heliopy.	• *AidaPy statistical tools package:* **to standardise operations typically done in C,** FORTRAN, MATLBAB, IDL.
• *Unsupervised Classification* of **incoming solar wind using Dimensionality reduction and Self Organizing maps (SOM)**	• *Link with ongoing missions:* **SDO, MMS and PSP.**	• *Identification of Reconnection:* **different methods using simulation and observation together and using supervised ML trained on human-labelled events and using unsupervised methods.**
• *Unsupervised Classification* of Plasma Regions in **Near-Earth Space: applied to OpenGGCM simulations using SOM**	• *Link with upcoming data:* **strong initiative to provide AidaPy services to SolO**	• *Analysis of Turbulent structure:* **using unsupervised ML (DBSCAN)**
• *Prediction of DST index* and **time-warping methods to establish the accuracy of predicting storm times.**	• *Virtual Mission Tool:* **to create synthetic data from simulation as if it were created by a missiong under design**	• *Explainable ML tool for classification of in situ data:* **applied to solar wind classification**
• *Data Assimilation* **methods based on Kalman filters: application of representer technique to OpenGGCM and EUHFORIA**	• *Linking simulation and observation:* **so that the same analysis can be applied to both real observations and synthetic data from simulations**	• *ML tools to analysis with the same approach simulation and observational data*
• *Solar image segmentation with NN:* **identification of coronal holes for space weather prediction**	• *ML tools to identify extreme events:* **e.g. shock, reconnection.**	
	• *ML-driven creation of lists of events*	

Figure 7.1   Summery of the main components of AIDApy.

available and easy to use language, python, aims at democratizing this area of science. It will make space data and the tools to analyze it available freely to all. To scientists in underprivileged conditions who cannot afford costly software. But also to normal amateur citizens interested in testing some tools and running comparisons and obtaining statistical trends. AIDA will achieve in this manner also a pedagogical and social dimension.

The main objective of AIDA to achieve this goal was the creation of AIDApy, a software tool written in the free language python that gives the user access to data of space science and to the methods needed to analyze it, chief among them the tools of AI now revolutionizing every aspect of life. The goal is to provide the users and even common citizens with examples to learn how to access and study space data. To reach this end, two schools were organized with teaching material and exercises all designed in the free language python that anybody can freely use and obtain freely from the WWW.AIDA-SPACE.EU website.

AIDApy includes packages to retrieve data from remote sources, such as space missions and space simulations, packages to make statistical analysis of space data, packages to synthesize virtual satellite observations within computer simulations, packages to deploy AI techniques to the analysis of space data.

Figure 7.1 summarizes visually all tools present in AIDApy, including:

1. Link AIDApy with some of the most important space databases (e.g. OMNIweb) and space missions (MMS mission of NASA) to retrieve data from remote sources;

2. Interfaces with computer codes for space simulations such as OpenG-GCM for modeling the Earth environment and iPic3D for modeling energy releases in space;

3. Tools to deploy advanced statistical tools to space data;

4. Tools to synthesize observations from computer simulations that have the same format and can be directly compared with space missions;

5. AI techniques to analyze space data and to make forecast of space weather events.

6. Help improve space weather predictions with methods to include data in space weather physics-based simulations and use AI to make space weather predictions

7.2 DATA DISCOVERY IN MAGNETOSPHERIC SIMULATIONS

In this section, we review and summarize the results of two works conducted as part of the AIDA project, and aimed at assessing the effectiveness of unsupervised clustering procedures in the processing of simulated data. The interested reader is encouraged to read the full contributions at [1] and [2].

In both cases, we aimed at answering the following questions: can clustering methods meaningfully identify different simulated regions, where we know, from previous knowledge, that different physical processes are bound to take place? Can these methods assist in the physical interpretation of simulation results? How robust and reliable are these methods?

We tested the unsupervised clustering procedure first introduce in [3] on 13 years of ACE data on two, rather different simulations of different aspects of magnetospheric physics. We focus on the magnetosphere due to two concurrent sets of reasons: first, the wealth of available in-situ magnetospheric data, produced by decades of magnetospheric missions such as Cluster [4], Time History of Events and Macroscale Interactions during Substorms (THEMIS, [5]) and Magnetospheric Multiscale Mission (MMS, [6]). Second, the ever-growing amount of works rooted in Machine Learning and investigating magnetospheric dynamics. Among them, [7–12] dwell in particular in the classification of large-scale magnetospheric regions and detection of boundary crossing, chiefly bow-shock and magnetopause crossings.

The first work we present is the clustering of data points from a global magnetospheric simulation run with the OpenGGCM-CTIM-RCM code, a

MHD-based (MagnetoHydroDynamics – MHD) model that simulates the interaction of the solar wind with the magnetosphere–ionosphere–thermosphere system [13–15]. OpenGGCM-CTIM-RCM was the code selected in the AIDA project for the simulation of global magnetospheric processes and was also employed to generate the simulations used for the Domain of Influence analysis presented in another work associated with the AIDA project, [16]. The simulation was run using a stretched Cartesian grid [13] with 325x150x150 cells and extending from $-3000\ R_E$ to 18 R_E in the Earth-Sun direction, from $-36\ R_E$ to $+36\ R_E$ in the y and z direction. R_E is the Earth's mean radius, the Geocentric Solar Equatorial (GSE) coordinate system was used in this study. The simulation was initialized with solar wind conditions observed starting from May 8th, 2004, 09:00 UTC, denoted as t_0. After a transient, the magnetosphere is formed by the interaction between the solar wind and the terrestrial magnetic field. We clustered data points from time $t_0 + 210$ minutes.

The second work we present is the clustering of data points from a fully kinetic simulation of tearing instability, an instability active in both the collisional and collisionless regime (albeit with fundamental differences) which swiftly breaks up preferentially long and thin current sheets into multiple magnetic islands later subject to non-linear evolution [17, 18]. The tearing instability is fundamentally related to magnetic reconnection and to particle heating and acceleration, both in heliospheric (and hence: magnetospheric) and astrophysical environments [19–22]. The simulation was run with the semi-implicit, energy-conserving code ECSim [23–25] from an initial equilibrium composed of the oppositely directed, force-free current sheets. We set $v_A = 0.2\,c$, $\omega_{pe}/\Omega_{ce} = 1$, $\beta_e = \beta_i = 0.02$ and $B_g = 0.03\,B_{0,x}$, with $v_A = B_{0,x}/(4\pi n_0 m_i)$ the Alfvén velocity, $\omega_{pe} = \sqrt{4\pi e^2 n_0/m_e}$ the electron plasma frequency, $\Omega_{ce} = eB_{0,x}/(m_e c)$ the electron cyclotron frequency, $\beta_{e,i} = 8\pi n_0 T_0/B_{0,x}^2$ the ratio between electron/ion pressure and magnetic pressure and c the speed of light in vacuum. The simulation box was $L_x = L_y = 200\,d_i$, we use 2128×2128 cells and $dt = 0.16\,\Omega_{ci}^{-1}$ with $\Omega_{c,i}$ the ion cyclotron frequency. We clustered data from time $\Omega_{ci}t = 320$.

We intend to analyze the results of our clustering on these two, rather different simulations to assess the effectiveness of the method across scales and simulations approaches. The first simulation, in fact, models the plasma at global magnetospheric scales, using a modeling approach, MagnetoHydroDynamics, that represents the plasma as a single fluid characterized by large-scale, low-frequency processes and by equilibrium velocity distribution functions. The second simulation, instead, is a fully kinetic simulation of non equilibrium plasma dynamics run with a Particle In Cell code. There, computational particles contribute self-consistently to shape the electric and magnetic field in which they are immersed through the moments (density and current) that they deposit on a grid. A clustering method which produces meaningful results on numerical approaches so different must work, in principle, for everything in between.

Furthermore, in the second case, the simulation products overlap so well, in terms of spatial and temporal resolutions, with spacecraft observations, to advocate for the future extension of the methods to Cluster, THEMIS, MMS observations as well.

However, at this stage, we work on simulations, however different, with the objective of overcoming one of the main difficulties in unsupervised clustering methods: result validation. In the absence of a "ground truth" it is in fact sometimes challenging to validate results of unsupervised clustering. This issue is mitigated here, where we have full knowledge of all ongoing processes in the domain and we can simply print and visualize clustering results, and compare them with our pre-existing knowledge of the physical processes at work.

This contribution is organized as follows. First, in Sect. 7.2.1, we describe the unsupervised clustering technique we use, with minimal differences, in [1] and [2]. Then, in Sect. 7.2.2, we describe our clustering results and the clustering products we can use to illuminate the physics ongoing in the different clusters we identify. In Sect. 7.2.3, we speculate on perspectives on the use of unsupervised clustering methods in the analysis of simulated datasets.

7.2.1 The unsupervised clustering technique

The unsupervised clustering technique we use for both simulations is the following:

1. preliminary data inspection and data pre-processing

2. training of a Self-Organizing Map (SOM, [26–28])

3. k-means clustering [29] of the trained SOM nodes

4. clustering of the simulated data points, based on the k-means cluster of their Best Matching Unit (BMU)

5. analysis of clustering results and of the SOM products

We will further comment here on these points, in the spirit of giving hopefully helpful advice to colleagues who may want to apply similar analysis techniques to problems of their interest.

Preliminary data inspection, that may include e.g. violin and correlation plots as in [1], is a useful tool to predict the outcome of clustering techniques and to select the features to use for the clustering. Violin plots highlighting bi- or multi-peaked distributions of specific features obviously bode better for clustering than distributions exhibiting no peaks. Features that are highly-correlated to others may be dropped from the list of features used for the clustering, since they will not add remarkably new information to the task. This is demonstrated in [1], section 4.3. There we show that features such as the Alfvén speed, the Mach number and the plasma beta, i.e. *derived* features, do not significantly change clustering results when added to a feature list that already includes the base features (e.g., density, pressure, magnetic field) that one needs to calculate them. Another important aspect of preliminary data inspection is understanding the range of variation of the different features. Features that span several orders of magnitude may benefit from logarithmic representation, as it is the case for density, pressure and temperature in [1]: compare the clustering results in [1], Figure 4 (logarithmic values are used) and Figure 8, panel a (logarithmic values are *not* used), for a rapid visual assessment of the benefits of the former choice.

A further pre-processing activity to evaluate carefully is the choice of the *scaler*. Scaling is needed whenever clustering activities rely on distance metrics, as it is the case for both SOMs and k-means: without scaling, the largest-amplitude features will dominate the clustering. In [2], we show how much the choice of scaler influences clustering results. The *MinMax*, *Standard* and *Robust* scalers, all from the scikit-learn library [30], have been used for the pre-processing of the data whose clustering results are depicted in Figure 4, panel a, b and c respectively. The *MinMax* scaler scales all features to the same interval, here [0, 1]. The *Standard* scaler rescales the distribution to zero mean and unit variance. The *Robust* scaler removes the median and scales the data according to a quantile range, here the interquartile range between the 1st and 3rd quartile. This third scaler is presented by the scikit-learn library as the robust choice in the case of outlier-rich datasets. We verify that this is the case with our dataset, composed of both outlier-poor and outlier-rich features. In Figure 2 in [2] we observe that the amplitude of the distribution of an outlier-poor feature such as the B_y magnetic field component does not change significantly with the three scalers. Instead, the amplitude of an outlier-rich feature, such as the z component of the electron current, $J_{z,e}$, changes significantly in the three cases. One can see how, especially in the case of the *MinMax* scaler, the majority of the $J_{z,e}$ points are compressed towards the zero, making them of poor impact in the clustering. This has a significant impact in the clustering results, since $J_{z,e}$ is often used as a proxy for magnetic reconnection, and helps understanding why the clustering depicted in Figure 4, panel a (*MinMax* scaler), is less physically informative than the one depicted in Figure 4, panel c (*Robust* scaler).

A further data preprocessing activity that may be employed is the reduction of the feature number, to be achieved through tools such as Principal Component Analysis ([31], used in [1]), or more advanced techniques to the same effect that do not rely on linear correlation between the features. We have to remark, however, that this activity is of poor utility, since the training time of a SOM depends mainly on the number of data points and on the number of iterations, not significantly on the number of features associated with each data point.

The core of our unsupervised clustering technique are Self-Organizing Maps [26–28], that can be viewed both as a clustering and as a dimensionality reduction procedure. With SOMs we can represent a large set of high-dimensional data as an *ordered* lattice made of $q = y \times x$ nodes, with x and y the number of rows and columns respectively. Each node i is characterized by a position in the (usually 2D) lattice and by a weight $\mathbf{w}_i \in \mathcal{R}^n$, where n is the number of features associated with each data point, and hence with each weight. At the end of the training, the weights of each node have been modified so that they represent local averages of the input data associated to that node (as it is the case, e.g., of the centroids in k-means) and they are topographically ordered with respect to each other according to similarity relation and the concept of distance. This last is the defining characteristics of SOMs that differentiate them with respect to other clustering techniques: nearby nodes are more "similar" among each other than further away nodes, and one can move to "similar" or "different" data points simply by navigating the map. This characteristic is used in the scientific exploitation of SOM products. The SOM is trained by presenting the input data to the map a certain number of times (epochs). Each time a new input is presented, weights are modified as follows:

1. The most similar node to the input point, the Best Matching Unit (BMU), is identified as the one that minimizes the distance between the input and the weights of all nodes. This is the *competition* stage of the training

2. The weight of the BMU is modified to be closer to the input point. Since the aim is to obtain an ordered representation of the data, not only the weights of the BMU, but also those of a certain number of neighbors is modified. The affected neighbors are selected through a neighborhood

function parametrized by a lattice neighborhood width, σ, and the magnitude of the correction is parametrized by a learning rate ϵ: the closer the node to the BMU, the more its weights are modified. Both σ and ϵ depend on the iteration number τ, and are reduced as the training progresses. The rationale is that the map should change significantly, both in terms of the number of nodes affected by each weight update and of the relative magnitude of the update itself, during the initial phases of the training. At later stages, only minor modifications are required. This is the *collaboration* stage of the training. We refer the interested reader to [2] for a more detailed description of the training procedure, and to [1] for a visual representation of how the SOM node change to adapt to the distribution of the data in feature space (Figure A1 and related movie).

Fundamental aspects in SOM training are both the choice of the SOM hyper-parameters (chiefly, the number of SOM nodes, of rows and columns, the initial neighborhood width σ_0 and initial learning rate ϵ_0) and of the features to use for the training. [28] offers rules of thumb e.g. for the choice of σ_0 and of the total number of nodes q. The latter should be of the order of $q \sim 5 \times \sqrt{m}$, with m the total number of samples, and the former should be of the order of 20 % of the longer side of the map. Independently of these choice, tests should be carefully conducted to investigate how much the hyper-parameter choice affects clustering results, as done in [1] and [2], Appendix A. Similarly, one should investigate how the choice of features affects clustering results. We conducted this analysis with particular care in [1]. The result of the clustering was evaluated by visual inspection of the clustered simulated points, plotted in 2D space. We obtained surprisingly "common sense" results: once the feature are appropriately pre-processed (for example, one has to take the logarithm of features spanning several orders of magnitude), different features give different types of clustering results. While all results are somehow meaningful, some map better to our understanding of magnetospheric dynamics, and are therefore considered preferable. The hyper-parameters and features that give our so-defined "best" results are the following, in [1] and [2] respectively: 1) $\sigma_0 = 1$, $\epsilon_0 = 0.25$, $q = 10 \times 12$, features: the three components of the magnetic field, the three components of the fluid velocity, the logarithm of the density, pressure, temperature and 2) $\sigma_0 = 19$, $\epsilon_0 = 0.5$, $q = 93 \times 71$, features: the three components of the electric and magnetic field, all moments for electrons and ions from the density up to the six unique component of the pressure tensor. For the MHD simulation, we cluster 1 % of the 5557500 data points at $x/R_E > --41$ and $t = t0 + 210$ min, using the serial SOM

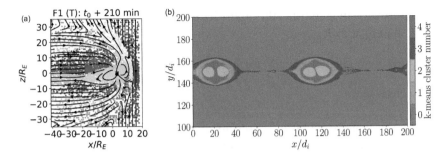

Figure 7.2　Clustering of (a) a global magnetospheric simulation, modified from Innocenti, M. E. et al., 2021, and (b) a PIC tearing instability simulation, modified from Koehne, S. et al., 2023.

implementation found in [32]. For the PIC simulations, we cluster all simulated data point in the upper half of the simulation at time $\Omega_{ce}t = 320$, with Ω_{ce} the electron cyclotron frequency, using the parallel CUDA implementation found in [33]. The serial and parallel SOM implementation has been carefully benchmarked against each other.

The trained SOM nodes are then clustered with k-means in a lower number of cluster chosen with the Satopaa method [34]. We obtain $k = 7$ in [1] and $k = 5$ in [2]. We use this second round of clustering to reduce the number of SOM clusters (i.e., the SOM nodes) to one that can be meaningfully represented in the simulated domain. By visually inspecting the simulated data points, colored according to the k-means cluster their BMU falls in, we can evaluate the results of the clustering. In Fig. 7.2, panel (a) and (b), we depict the clustering obtained in [1] and [2] respectively.

We compare the result of the clustering with our pre-existing knowledge of the two types of processes, the magnetospheric system and the evolution of tearing instability simulations. We map the clusters in Fig. 7.2, panel a, to the following macroscale magnetospheric regions:

0. (purple) pristine solar wind

1. (blue) magnetosheath, immediately downstream the bow shock, mainly at $z/R_E < 0$

2. (cyan) boundary layers: plasma separating the magnetosheath and the lobes, and northern and southern lobe

3. (green) inner magnetosphere

4. (brown) magnetosheath, immediately downstream the bow shock mainly at $z/R_E > 0$

5. (orange) magnetosheath, further away from the bow shock

6. (red) lobes

We see that these clustering results, obtained without guiding the unsupervised procedure in any way, maps well to our knowledge of the magnetospheric system, build in decades of observations and theoretical investigation.

Similarly, we can map the clusters in Fig. 7.2, panel (b), to regions developing in tearing instability simulations:

0. (blue) inflow region, where the plasma is minimally affected by plasmoid development

1. (orange) plasmoid: intermediate region between the outer plasmoid region and the merging regions

2. (green) plasmoid: inner region

3. (red) plasmoid: outer region, separatrices

4. (purple) plasmoid: merging region, where the signatures of plasmoid merging are stronger

Also in this case, the clustering procedure produces results that map well to ongoing processes, and points grouped together map to regions where we expect similar processes to take place, as highlighted e.g. in [2], Figure 9.

At this stage, the clustering procedure that we described in this section seems to offer little more than, e.g., k-means. We have verified that the clustering we obtain is meaningful, but we have not exploited at all SOM products. Again encouraging the reader to read carefully the contributions in [1] and [2], we illustrate, in the next section, how SOM products can be used to interpret clustering results and try to glimpse the physics at work in each cluster.

7.2.2 Using SOM products for physical investigation

In this section, we will briefly illustrate how SOM products can be used to infer the physics at work in the different clusters. We will focus in particular on *feature maps* and on the *Unified Distance Matrix*.

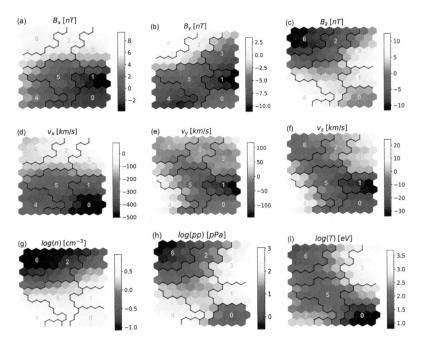

Figure 7.3 Feature maps for the clustering depicted in Fig. 7.2, panel a: variation of the feature values across the trained SOM nodes. Cluster boundaries and numbers are superimposed to the maps (from Innocenti, M. E. et al., 2021).

Feature maps, see Fig. 7.3, from [1], show how the different features change across the SOM nodes. In Fig. 7.3, we draw the k-means cluster boundaries, with corresponding labels, across the SOM node. While examining the Fig. 7.3, we compare the feature values in clusters 1 and 4, the two clusters in the magnetosheath immediately downstream the bow shock. It is somehow perplexing that this plasma should be divided into separate clusters, since we expect substantially similar processes to occur throughout the entire downstream of the bow shock. We see from Fig. 7.3 that the two features that vary most across the two cluster are the fluid velocity in the y and, critically, z direction. The other features are remarkably similar between the two clusters, as we expect. We ascribe the difference in v_z to the structure of the magnetospheric circulation system in the north and south lobe, and we understand why the boundary between cluster 1 and 4 tend to be somehow unstable in different classification experiments at different times, see [1], Figure 6: the plasma in those two clusters is in fact quite similar, apart from the two features v_y and v_z. Significantly more interesting is to explore on the feature map the transition

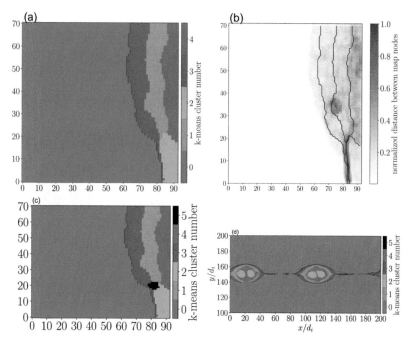

Figure 7.4   An example to illustrate how the Unified Distance Matrix, UDM, can be used to highlight smaller-scale regions of interest. (a) SOM nodes, colored according to their k-means clusters; (b) UDM: the darker nodes are quite dissimilar to their nearest neighbors. In black, points in the simulated domain – panel (d) – corresponding to nodes depicted in black in the SOM of panel (c). Modified from Koehne, S. et al., 2023.

between cluster 0, bottom right corner, pristine solar wind, and the above-mentioned cluster 1 and 4. We know, from previous magnetospheric knowledge, that this corresponds to moving from unshocked solar wind to shocked magnetosheath plasmas. Shocked plasma is slower, denser, hotter and higher pressure. We can see a jump in these directions moving from cluster 0 to 1 and 4 in Fig. 7.3 panel d, g, h and i. This is a powerful example of how we can get a feeling for the physical processes occurring in a simulation by examining the feature maps, at the macro-scale, at this stage.

In Fig. 7.4 we provide an example from [2] to illustrate how the Unified Distance Matrix, UDM, can be used to highlight smaller-scale regions of interest. The UDM depicts the SOM nodes in shades of grey: the darker a node, the more "distant" it is from its nearest neighbors. We remind the reader that the nearest neighbors are, as expected in SOMs, the most similar nodes to each node in the entire map: dissimilar nearby nodes may codify information of interest.

We see in Fig. 7.4, panel (b), the UDM associated with the clustering procedure depicted in Fig. 7.2, panel (b), and described in [2]. We see dark regions (i.e., very dissimilar nodes) at the boundary between cluster 2 and 3, the inner plasmoid region, green, and the separatrix region, red. Indeed, analysis of the feature map shows that the inner plasmoid region is the most dissimilar to the other. In Fig. 7.4, panel (c), we highlight in black a subset of nodes at the boundary between clusters 1,2, and 3. We depict in panel (d) the points in the 2D simulated space associated with those nodes: we see that they map to very interesting regions, namely the small scale plasmoids dotting the current sheet between larger scale plasmoids and, somehow surprisingly, some points at the boundary between the inner plasmoid region (green) and the cluster associated with plasmoid merging (purple).

We have then shown that nodes highlighted in the UDM map to regions of interest in the simulated domain. In [2] we show other similar examples. Hence, when confronted with new, unknown data points, we can quickly identify potential regions of interest by checking if their BMU corresponds to one of the "interesting" nodes in the SOM. This task is facilitated by the significant robustness of the trained SOM to temporal variations in the simulation, as checked both in [1] and [2]: we can train a SOM on data from a certain simulated time (or from several simulated times, mixed), and use it for analysis even at a rather distant time, with meaningful results.

7.2.3 Perspectives on the use of unsupervised clustering methods in the analysis of simulated datasets

In this section, we have presented an unsupervised clustering method based on Self Organizing Maps and developed as part of the AIDA project. We demonstrated this method on two rather diverse simulations, a global magnetospheric simulation, described in [1], and a fully kinetic simulation of tearing instability, described in [2]. We have demonstrated that this procedure produces clustering results which, a posteriori, appears meaningful in cases, as the ones described, where we are already familiar with the physical processes at work in the simulations. While presenting our work, we describe simulation analysis methods which appear promising investigation tools in cases where we approach unknown processes. We advocate for the inclusion of similar investigation tools in the pipeline adopted when analyzing simulation results and, in perspective, observed datasets.

7.3 FINDING RECONNECTION USING UNSUPERVISED MACHINE LEARNING

Magnetic reconnection is a fundamental process in space plasmas, by far, the most studied in plasma physics. Magnetic reconnection is ubiquitous in space, from the Earth's magnetosphere to the solar corona, and is constantly invoked in astrophysics, particularly for its role in energetic processes, most notably particle acceleration. Considered for many decades as the driver of solar flares and coronal mass ejections, just to name two well-known examples. It is also studied in the laboratory, particularly in fusion devices. Magnetic reconnection occurs when the magnetic field, which constrains the plasma connections at large fluid scales (the so-called *frozen-in condition* of the ideal MHD regime), locally builds up very large gradients due to the formation of a singularity. As a result, non ideal effects come into play, violating the *ideal Ohm's law* allowing for the filed lines to break and to rejoin eventually producing a new large-scale configuration where now previously unconnected plasma regions do exist.

The process of line reconnection is very localized but has a global impact by changing the global magnetic topology and by releasing a very large amount of energy. This process develops spontaneously to allow the system to relax to low energy magnetic states, a transition formally forbidden by the ideal large-scale MHD dynamics.

The initial magnetically sheared region, where the reconnection process starts and develops, corresponds to the presence of a current layer or current sheet (CS), although not all CSs are regions, where reconnection develops. Typically a CS is an elongated structure very thin in the transverse direction, of the order d_i and/or ϱ_i, where the current varies very rapidly. By defining L and δ as the typical length and thickness of the CS, their aspect ratio (AR) is a characteristic parameter identifying the structure where reconnection occurs.

The main feature that characterizes this multi-scale, multi-physics process is the formation of two thin layers. In the first (larger) one, namely the ion-diffusion layer (IDR), the dynamics develops at the ion-kinetic scale. The second one, namely the electron-diffusion layer (EDR), is embedded into the first one and is characterized by a dynamics at the electron-kinetic scale. In the IDR the ion demagnetize but it is in the EDR that the magnetic field can slip, break and reconnect with respect to the plasma, because only here the electrons demagnetize.

At low frequencies this behavior is summarized by the so-called Ohm's law, Eq. (7.1), which takes into account the fluid response at the ion and electron scale, blue and red terms, respectively, i.e. at scales smaller than the ideal

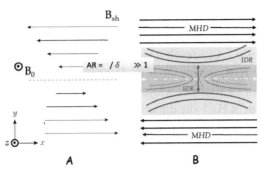

A   B

Figure 7.5   Magnetic reconnection: A initial configuration. B reconnected layer where the IDR and EDR micro-layers are strongly enlarged for the sake of clarity.

MHD scales (green terms).

$$E + U \times B = \frac{1}{n}j \times B - \frac{1}{n}\nabla P_e + \frac{d_e^2}{n}\left[\nabla \cdot \left(uj + ju - \frac{jj}{n}\right) - \nabla^2 E\right] \qquad (7.1)$$

The Ohm's law is particularly well suited to the study of reconnection because of its clear physical meaning and because it is a fluid equation. It is worth noting that the non-ideal terms that violate the ideal MHD law, the RHS terms of Eq. (7.1), become important on scales of the order of even smaller than the kinetic terms. However, these small-scale fluid terms included in the Ohm's law are the only ones that allow reconnection to occur in the context of a global low frequency dynamics of the system, thus motivating the Ohm's law model. In Fig. 7.5, we give a simple but representative picture of 2D reconnection. In the left frame (A), we plot the initial sheared magnetic field, which is characterized by an inversion neutral line (i.e. the region where the magnetic field reverse its direction) representing the typical model for the development of the reconnection instability. The (reconnecting) in-plane magnetic field $\mathbf{B}_{sh}$ is directed along the x axis and varies in the transverse y direction. The so-called "guide field" $\mathbf{B}_0$ is instead homogeneous and directed in the perpendicular z-direction. The right frame (B) shows the reconnection instability at play in the EDR (red color) embedded in the IDR (blu color). These two layers are of microscopic size with respect to the global system, which is dominated by the low frequency MHD dynamics. The Ohm's law is a simple but very effective marker of the regime. The *ideal*-MHD regime corresponds to the LHS terms (green ones) which dominate and can be assumed equal to zero: $\mathbf{E}'_{MHD} = \mathbf{E} + \mathbf{U} \times \mathbf{B} \simeq 0$. Under these conditions the so-called MHD *connection theorem* holds and the magnetic field is *frozen-in* into the plasma (and vice versa). Ions demagnetize when the system, because of the presence of the singularity, reaches a scale comparable to the ion skin depth d_i and/or the ion Larmor radius ϱ_i. As a result a layer of thickness of the order d_i (ϱ_i) is created, namely the IDR, where the ions demagnetize. In this region, $\mathbf{E}'_{MHD}$ is no longer zero (or negligible) while the electron "decoupling" electric field, $E'_{dec,e} = |(\mathbf{E} + \mathbf{u}_e \times \mathbf{B})_z|$ (in dimensionless units), still remains very small because the electrons are still *frozen-in* (where $\mathbf{u}_e$ is the electron velocity). Embedded in the IDR then forms the EDR of thickness $\sim d_e$ (ϱ_e), where now $E'_{dec,e}$ is no longer "small", let's say zero, because the electron inertia terms (highlighted in red in the Ohm's law) come into play and allow the electrons to demagnetize as well. It is in this last sublayer that magnetic lines break and reconnect changing the global magnetic topology and converting magnetic energy into flow, particle acceleration, heating etc. The configuration is now characterized by a so-called x-point at the center where the perpendicular electric field is a measure of the ongoing reconnection process. Mathematically this corresponds to the formation of a boundary layer structure with the outer solution obeying ideal MHD (i.e. smooth gradients) and an inner solution where the ideal Ohm's law is violated on a very short scale length. For the sake of mathematical simplicity, reconnection was studied in the distant past (but not completely anadoned even today) by adding a resistive term in the RHS of the Ohm's law (especially in space, largely negligible) imposing a much large artificial value of the resistivity coefficient. In this case, but not considered here, both ions and electrons now demagnetize together in one layer, namely the resistive layer. Going back to our sketch image of the IDR-EDR reconnection layer together with $E'_{dec,e}$, there exist a number of "physical markers" that are the direct consequence of the very strong gradients forming into the microscopic kinetic layers. Each of these is a signature of the reconnection event, but is by no means sufficient on its own to say that reconnection has occurred. An approach such as machine learning based on the global signatures of the ensemble of the markers, therefore, represents a very promising approach especially in 3D where the reliable definition of a reconnection layer is still a matter of debate. Here, we list the main reconnection markers typically used in numerical simulations but also available experimentally from satellite measurements:

a) the current density magnitude $J = |\mathbf{J}|$,

b) the magnitude of the in-plane electron fluid velocity $|\mathbf{u}_{e,\text{in-plane}}|$,

c) the magnitude of the in-plane magnetic field $|\mathbf{B}_{\text{in-plane}}|$,

d) the magnitude of the electron vorticity $\Omega_e = |\mathbf{\Omega}_e| = |\nabla \times \mathbf{u}_e|$, *e*) the electron decoupling term $E_{\text{dec},e} = |(\mathbf{E} + \mathbf{u}_e \times \mathbf{B})_z|$,

f) energy conversion term $\Upsilon \equiv \mathbf{J} \cdot (\mathbf{E} + \mathbf{u}_e \times \mathbf{B})$.

Identifying magnetic reconnection in a numerical simulation or in space is not an easy task, because the physics underlying the reconnection process cannot be identified simply and accurately by one of the above markers. As a result, a physicist, much like a doctor, has to interpret the different markers together to make a diagnosis. In addition, the main difficulty lies in the flood of data produced by numerical simulations or by high-resolution satellite instruments to be analyzed in the hunt for reconnection events, which, as discussed above, occur in very thin regions of negligible size in relation to the whole system. This is the optimal territory for running machine learning algorithms.

Magnetic reconnection in a dynamic system, especially in a turbulent system, evolves randomly in space and time, but can be detected with confidence in the limit of 2D geometries, while it becomes much more complex in 3D. Since we want to demonstrate the point of principle of the efficiency of machine learning in automatic reconnection detection, we limit our simulation dataset to 2D geometries, leaving the 3D case for future work.

Machine learning algorithms can be divided into two main categories, supervised and unsupervised. In the first category, a labelled dataset must be provided for training, with the labelling being done by the researchers. This approach has been used, for example, in [35], in particular with a convolutional neural network algorithm, and has shown great promise even in ambiguous cases where even expert researchers do not have a global consensus. Nevertheless, the process of building the labelled dataset is a relatively lengthy one, requiring the analysis of very large datasets and possibly requiring the synergy of a number of experts in the field. The unsupervised approach, on the other hand, aims to identify patterns and structures, such as a reconnection layer, without the need to fish in an ad-hoc dataset.

7.3.1 Machine Learning pipeline

To identify CSs and reconnecting structures, we have developed a pipeline (referred as "AML" or "Unsupervised Machine-Learning" in the subsequent discussion) based on unsupervised clustering algorithms, (see e.g. [36]). We have compared the results obtained from "AML" with those generated by two alternative algorithms that do not employ machine learning, namely "A1" and "A2", respectively. The following quantities previously introduced,

J, $u_{e,\text{in-plane}}$, $B_{\text{in-plane}}$, Ω_e, $E_{\text{dec},e}$, Υ are used as the marker variables to detect magnetic reconnection by the AML algorithm.

The current density, the in-plane electron velocity and magnetic field are directly linked to the geometry of the CS, while the electron decoupling term is responsible for the de-magnetization of the magnetic lines at the electron scale. The energy conversion term is a proxy accounting for energy dissipation nearby a reconnection site [37]. The alternative methods that a researcher typically may use to detect reconnection are characterized by a use of less number of variables. In particular, we take J for A1 and J, Ω_e and $E_{dec,e}$ for A2.

A crucial physics parameter that we take into account related to the development of the magnetic reconnection process, is the aspect ratio (AR) of the reconnecting current sheet (see Fig. 7.5). In particular, all our algorithms include a common step, which involves setting a threshold for the AR of the selected CSs. The AR is defined for each selected structure as the ratio of its characteristic length L (in 2D parallel to the magnetic field lines nearby the neutral line) to its characteristic thickness δ (perpendicular to the field lines before reconnection starts):

$$AR = L/\delta \tag{7.2}$$

This ratio is typically very large, i.e. $L \gg \delta$. For any given CS, we have calculated its width and length using the methodology discussed in [38] and references therein. It is worth noticing that for methods A1 and A2, the reconnecting current sheets (CSs) are defined as regions where the current density J exceeds a given thresholds defined a priori, as discussed in [39]. On the other hand, when applying the AML method, the CSs are defined in a different way and we have employed distinct techniques to identify the CSs in physical space, which will be explained in detail in the subsequent sections. Finally for the set of points belonging to the CS, we calculate the Hessian matrix H using the current density values close to the current peak.

Our pipeline combines two machine learning algorithms: K-means and DBSCAN. Both algorithms are clustering techniques designed to identify patterns and groupings within a given dataset. First, similar to its use in image segmentation, we have employed K-means clustering in order to group data points into K clusters, highlighting their similarity in the variables space. Then, to identify different structures based on their physical spatial location, we have used DBSCAN, a popular clustering algorithm. DBSCAN groups closely packed data points together, while also identifying outliers. It works by defining a neighborhood around each data point and labeling dense regions as clusters, while points that do not belong to any cluster are considered noise.

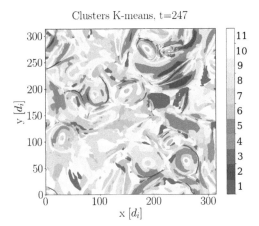

Figure 7.6   The shaded iso-contours of the eleven clusters calculated by K-means (in the variables space). The regions that comprise all 11 clusters in the variable space are depicted where cluster "1" is highlighted in red, and the remaining clusters are represented using different shades of blue. Figure from Sisti et al., The Astrophysical Journal, **908**,107, 2021: *"Detecting reconnection events in kinetic vlasov hybrid simulations using clustering techniques"*, DOI 10.3847/1538-4357/abd24b (©AAS. "Reproduced with permission").

For implementing the pipeline, a systematic approach has been undertaken, consisting of the following steps:

i) To tune the "K" parameter in the K-means algorithm we have followed the cross-validation-like approach described in [40] with the Davis-Bouldin index as the internal cluster metric applied to the predicted clusters. Then the tuning procedure has been applied by setting the "K" value at about one eddy-turnover time ($t \simeq 250 \sim 1/\Omega_{ci}$). This choice turns out to be well suited since during this phase the well formed CSs still do not interact one with each other and can be therefore considered as "isolated CS structures". We found $K = 11$ as the best value for the "K" parameter since the corresponding Davis-Bouldin index is minimized.

ii) We have applied the K-means algorithm to the variables normalized between zero and one. As a result, we got eleven clusters in the variable space. To identify potential reconnection sites, we focused on the cluster with the highest mean value of current density, denoted as cluster 1. The presence of a peak in the current density value is a necessary condition (but no sufficient) for reconnection to occur. Cluster 1 involves several structures in the physical space of our box.

In Fig. 7.6, we draw the simulation (x, y) space domain. In particular, we show the shaded iso-contours of the eleven clusters calculated by K-means (in

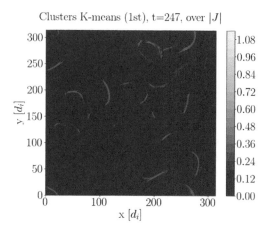

Figure 7.7 The shaded iso-contours of cluster 1 regions (red) superimposed on the contour shaded plots of the current density $|\mathbf{J}|$. Cluster "1" regions are displayed in red. These regions are superimposed onto the contour plot of the current density, providing a visual indication of their correlation. Figure from Sisti et al., The Astrophysical Journal, **908**,107, 2021: *"Detecting reconnection events in kinetic vlasov hybrid simulations using clustering techniques"*, DOI 10.3847/1538-4357/abd24b (©AAS. "Reproduced with permission").

the variables space). Cluster 1 is depicted by the red color, while the other ones are represented by different blue tones. In Fig. 7.7, we illustrate the shaded iso-contours of cluster 1 regions (red) superimposed on the contour shaded plots of the current density $|\mathbf{J}|$. This indicates that cluster 1 approximately overlaps to the collective set of CSs.

iii) By analyzing Fig. 7.7, we observe that cluster 1 within the variable space corresponds to multiple structures in the physical space. Consequently, an algorithm capable of discerning distinct structures based on their physical locations becomes necessary. In this context, DBSCAN turns out to be a suitable algorithm to be exploited for this purpose. In particular, the x and y-coordinates of the points belonging to cluster 1 serve as input for its use. In this approach, the DBSCAN algorithm utilizes an exploration radius of $\epsilon = 50$ grid points, approximately equivalent to $5d_i$, and a minimum number of points, $Min_pts = 100$, as basic parameters.

Figure 7.8 showcases the outcomes achieved through the implementation of the DBSCAN algorithm. The figure exhibits different colors to depict the various structures that have been identified.

It is important to mention here that in this particular context DBSCAN is not employed for image processing purposes. Instead, its application

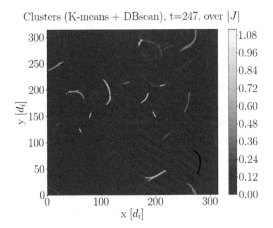

Figure 7.8 The contour plot of the current density, denoted as $|\mathbf{J}|$, is illustrated. The various structures identified by the DBSCAN algorithm applied on cluster "1" points are highlighted in different colors, allowing for clear differentiation and visualization of these distinct structures. Figure from Sisti et al., The Astrophysical Journal, **908**,107, 2021: *"Detecting reconnection events in kinetic vlasov hybrid simulations using clustering techniques"*, DOI 10.3847/1538-4357/abd24b (©AAS. "Reproduced with permission").

revolves around partitioning cluster 1 into separate subsets that align with distinct structures in the physical space. To enhance the understanding of the relationship between these subsets and potential reconnection sites within the CSs, we present Figures 1.4 and 1.5 providing a visual representation of the correlation between the identified subsets and the CSs in the physical space.

iv) A threshold is applied to the AR of the identified structures. The AR for each structure is computed as discussed previously, see Eq. (7.2). To optimize performance, various threshold values for the AR have been considered. Specifically, the following thresholds have been tested: 10, 12.5, 20, 30, 50, and 70.

The performance of the different methods (AML, A1, A2) is analyzed at different time instants during the simulation representing different phases. Two quality parameters, precision and nMR-precision, were introduced to quantitatively evaluate the algorithm performance. Precision measures the capability of the method to accurately select good reconnection sites, while nMR-precision provides a measure about the exclusion of potentially valuable sites. The results show that the unsupervised machine-learning algorithm (AML) performs competitively with the threshold-based methods. However, the precision and non-magnetic reconnection precision are affected by the threshold values and by the complex non-linear dynamics going on during the turbulent phase because of the interactions between the different CSs. The turbulence complicates the identification of reconnection sites due to advection, shrinkage, and deformation of the current sheets (CSs). Despite some degradation in precision during the turbulent phase, all methods perform well in the initial and central phases of the simulation. An optimal threshold value is determined for AML and A1, balancing precision and the exclusion of reconnection sites. Based on the results, the AML method emerges as the most successful approach. It has a better performance in detecting reconnecting CSs, suggesting the effectiveness of unsupervised machine learning techniques for this task. These findings contribute to the field of turbulence simulation analysis and provide a promising avenue for automating the detection of reconnection phenomena in future studies.

7.3.2 Perspectives on the use of unsupervised machine learning for the analysis of very large databases

Unsupervised machine learning now offers scientists a great opportunity to help them to deal with very large data sets such as in the context of space plasma physics. In this context, a single satellite or a fully kinetic numerical simulation integrating the Vlasov-Maxwell system of equations can generate can generate an impressive amount of data that would take a researcher years to fully analyze the physics of interest. However, as for instance happens for a chess move, the large majority of data (movements in a chess game) do not need to be analyzed at all, as a good chess player knows well, drastically reducing the amount of data to be examined.

One is left with "small" sub-data sets of data, where the relevant physics is at play, thus strongly optimizing the possibility of playing with only the interesting processes to be studied. In other words, machine learning offers the possibility of automatically selecting the data relating to a given structure or, more generally, of a physical process of interest. Here, we have limited ourselves to the detection of magnetic reconnection in 2D-3V kinetic numerical simulations of a turbulent plasma with parameters explicitly tuned to the space plasma regime. However, it is undeniable that this type of approach has a wide range of possible applications and will be increasingly exploited in the coming years.

In conclusion, unsupervised learning algorithms applied to satellite mission data or to large-scale numerical simulations prove to be a very promising approach for future studies and research providing a very powerful tool for the analysis of very large databases.

7.4 CHARACTERIZING TURBULENCE WITH UNSUPERVISED MACHINE LEARNING

Turbulence properly defined is a process characterized by chaotic changes that when properly investigated reveal the presence of scale-invariant mathematical structures presenting self-similarity at different scales [41]. Turbulence properly defined is a process characterized by chaotic changes that when properly investigated reveal the presence of scale-invariant mathematical structures presenting self-similarity at different scales [41].

Here we consider the specific case of turbulence induced by high speed flows generated by magnetic reconnection [42]. Specifically, we consider the 3D full kinetic simulation of reconnection outflows conducted with the iPic3D PIC code [43] reported in [44, 45]. Reconnection is initiated in a modified Harris equilibrium [46]:

$$\mathbf{B}(y) = B_0 \tanh(y/L)\hat{\mathbf{x}} + B_g\hat{\mathbf{z}} \tag{7.3}$$

$$n(y) = n_0 \operatorname{sech}^2(y/L) + n_b \tag{7.4}$$

where the thickness $L/d_i = 0.5$ and the following parameters $m_i/m_e = 256$, $v_{the}/c = 0.045$, $T_i/T_e = 5$ $B_g/B_0 = 0.1$, $n_b/n_0 = 0.1$ are assumed. Here B_0 is the asymptotic in-plane field and n_0 is the peak Harris density. Reconnection is initialized with a uniform perturbation along z. The evolution is described in our previous works [42, 44, 47, 48]: the initial perturbation produces a primary reconnection region that results in a strong outflow that leads to turbulence [42, 48–50] and secondary reconnection sites [44, 51], recently confirmed experimentally by observations of the magnetospheric multiscale (MMS) mission [52].

Figure 7.9 shows the state at the end of the simulation, when turbulence is fully developed. The outflow of reconnection transitions from laminar in close proximity of the central electron diffusion region is progressively more chaotic as the jet evolves outward. Studies of correlations and structure functions show the development of Kolmogorov scaling at large scales and the presence of other power law indices at the ion and the electron dissipation scales [42].

Traditional methods of investigation rely on statistical correlations, Fourier analysis or structure functions of different orders to capture the scales and nature of turbulence [53].

a) J_e volume rendering

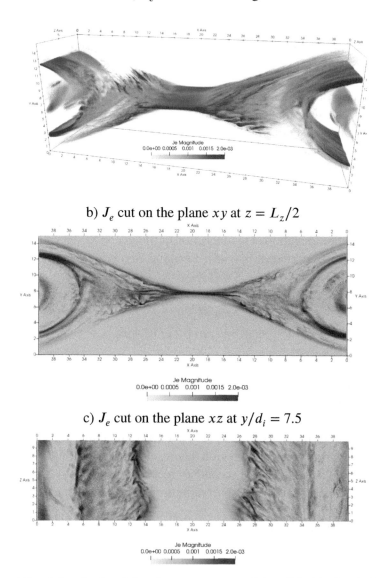

b) J_e cut on the plane xy at $z = L_z/2$

c) J_e cut on the plane xz at $y/d_i = 7.5$

Figure 7.9   Magnitude of the electron current, J_e in a iPic3D simulation for a box of size $L_x = 40d_i$, $L_y = 15d_i$ and $L_z = 10d_i$. Panel a shows a 3D volume rendering. Panel b and c show two cross sections in the central xy plane (b) and a xz plane passing just below the electron diffusion region.

The project AIDA [45] showed how machine learning can provide an alternative way to look at turbulence: it can collect individual information about all features at all scales. We consider here the electron current amplitude, but other positive definite features could be treated with the same method. ML can collect all the current layers formed from the smallest to the largest and identify their properties to investigate their deeper physical meaning.

In 3D large scale simulations, identifying each current layer is humanly impossible, but it becomes feasible for artificial intelligence. For this task, we use the same method used in the previous Sect. 7.3 to analyze reconnection [36]: the Density-Based Spatial Clustering of Applications with Noise, DBSCAN [54].

DBSCAN collects data points and aggregates them in tight groups that form aggregates of near-neighbors. Outliers in low density regions are not spuriously aggregated with others. DBSCAN is automatic but requires a definition of a distance function, a neighborhood radius ϵ to construct the aggregates and a minimum number of points n_{min} to distinguish a cluster from isolated outliers. Reference [45] used the Cartesian norm with $\epsilon = 5$ and $n_{min} = 5$, a natural choice for the specific simulation method iPic3D that uses a particle smoothing function with support of 5 cells.

The first step is to define what a current layer is. The current amplitude ($J = \sqrt{J_x^2 + J_y^2 + J_z^2}$) varies from zero to infinity and defining current layers requires to set a threshold above which currents are considered small and above which a simulation point is considered part of a current layer. Figure 7.9 shows cuts of the magnitude of the electron current J_e. If we use a too low value, all points will be considered current carrying. If we set a too high value, we miss important current regions. In the results below, we choose as threshold 1/10 of the maximum value. All cells of the computational box where the current exceeds 1/10 of the maximum are considered current-carrying. The domain then is divided in a binary choice between non-current carrying (0) and current-carrying (1) pixels. DBSCAN is used to cluster all pixels with binary value 1 to identify all current layers of any shape and size.

7.4.1 DBSCAN Identification of current layers and their properties

With the use of DBSCAN, we no longer look for statistical correlations but we identify every current layer. Once each layer is identified, its physical properties can be measured. It is literally like using machine learning for searching a haystack for each needle in it and grabbing it. A relentlessly boring task for a human but perfect for AI.

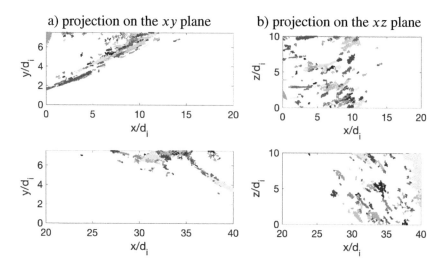

Figure 7.10   Four orthogonal views of the current layers identified by the DBSCAN clustering algorithm. Each cluster is shown with a different color chosen not for visual beauty but for ease of disticntion between nearby cludsters. The big central laminar current sheet at $x = L_x/2$ is not shown to avoid overcrowding the figure. Each pixel shown is one cell of the computational domain flagged as carrying a significant electron current. The left ($x < L_x/2$) is shown on top and the right ($x > L_x/2$) below. The xy plane is shown on the left panels and xz in the left.

Figure 7.10 reports four views of the 3D clustering provided by DBSCAN. The area shown is only half of the system ($y < L_y/2$) by virtue of the cross symmetry of reconnection in presence of guide fields (left-top symmetric with right-bottom and vice-versa). The DBSCAN implementation described above finds 290 current layers (including the big central laminar one, not shown in Fig. 7.10).

It is clear from preliminary inspection that some current structures are on the ion scale, but others are smaller, on the electron scale. DBSCAN excludes isolated pixels as outliers which may actually be small current layers on the electron shell, but may also be noise. The DBSCAN setup used excludes clusters of less than 5 cells.

Having now the precise definition of all current layers, we can study the properties of each one by averaging physical quantities over each cluster. Instead of dealing with statistical correlations computed generically over all 1200×450/2×300 (i.e. 81 million) cells, the problem is now reformulated as analysis of 290 entities. The DBSCAN approach reduces dramatically the number of the individual entities we need to analyze and visualize.

7.4.2 Link between current layers and reconnection

To illustrate the possible application of the DBSCAN method we consider the much discussed question of the association between current structures created by turbulence and reconnection. We want to investigate what is the occurrence of reconnection in current layers generated by turbulence. DBSCAN has identified all current layers in the system, we can then look in each cluster identified by DBSCAN for signatures of reconnection. This identification will then allow us to study what is the link between the size of the current layers and the likelihood of it reconnecting or the intensity of energy conversion ongoing.

Figure 7.11 reports a scatter plot of all identified currents. Each current layer is measured for its size defined by fitting each layer with an ellipsoid identified by the three primary orthogonal axis of each current structure: σ_1, σ_2, σ_3 in decreasing order. In Fig. 7.11, each current structure is represented by a dot of size proportional to he number of pixel in the corresponding cluster identified by DBSCAN. A scatter plot of all current layers is presented in the plane formed by its main size (current length) and aspect ratio (between minimum and maximum axis of the ellipsoid).

In each panel the dots are colored based on three indicators of reconnection [55] computed as average over each cluster: the parallel electric field, $E_{\parallel}$ (panel a), the electron agyrotropy (panel b) [56] and the dissipation in the electron frame, $\mathbf{J} \cdot \mathbf{E}'$ where $\mathbf{E}'$ is the electric field in the electron frame, $\mathbf{E} + \mathbf{v}_e \times \mathbf{B}$ (panel c) [57].

The DBSCAN analysis allows to observe directly the parallel electric field and agyrotropy are large primarily at small scales, i.e. at the electron skin depth scale or less. The parallel electric field (panel a) and the agyrotropy have their highest values for the smallest current layers, both indicating stronger reconnection on electron scale current layers.

Energy dissipation, $\mathbf{J} \cdot \mathbf{E}'$, is also largest (positive and negative) for the same smaller current layers at the electron scale.

This analysis suggests then that turbulence produces a cascade of smaller and smaller electron scale, i.e. at the electron skin depth scale and smaller. These smallest layers are the ones where reconnection plays the largest role and energy dissipation is most intense.

While we have reasonable confidence in the identification of reconnection via the parallel electric field and the electron agyrotropy, recently, an alternative way has been proposed for detecting regions of reconnection: the

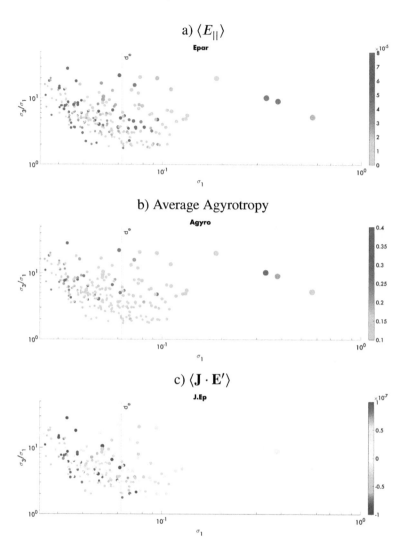

Figure 7.11 Scatter plot of current layers identified by DBSCAN. Each cluster of current carrying pixels is characterized with an ellipsoid with 3 axis: σ_1, σ_2, σ_3, ordered in decreasing size. The plot shows in hte x axis the major axis (length of the current layer) and on hte y axis the aspect ratio of the smallest to the largest axis of each current cluster. From top to bottom: a) parallel electric field, b) agyrotropy, c) dissipation in the electron co-moving frame, all averaged over the cells comprising the current layer.

Lorentz indicator [51]. At a reconnection point, the speed of the local Lorentz transformation in the plane orthogonal to the local electric field that eliminates the in-plane magnetic field is zero because reconnection by definition

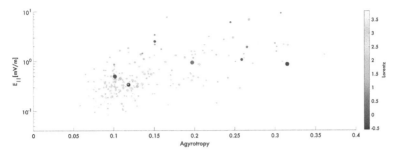

Figure 7.12   Scatter plot of the identified current layers based on their average agyrotropy and average parallel electric field. The dot size is proportional to the number of cells forming the current and their color is proportional to the measured Lorentz reconnection indicator, in logarithmic scale, $\log_{10}(v_L/c)$ (from Laptenta, G., 2021).

eliminates the in-plane component of the magnetic field. In the vicinity, the value increases from zero and becomd very large as the distance increases, in fact, surpassing the speed of light (i.e. the transformation becomes impossible). The Lorentz indicator that is simply the speed of the local Lorentz transformation orthogonal to the laboratory electric field that eliminates the component of the magnetic field normal to the electric field:

$$\mathbb{L} = \frac{v_L}{c} = c \left| \frac{\mathbf{E} \times \mathbf{B}}{E^2} \right| \tag{7.5}$$

When the Lorentz indicator $\mathbb{L}$ becomes small, a reconnection site is identified in the form of an x-point or a o-point in 2D or one in a class of six different types of points in 3D [51].

To test the results of the DBSCAN analysis, we have computed for each cluster the lowest value of the Lorentz indicator within the cluster and in a radius of 5 cells around it in each direction to account for the fact that reconnection tends to break current layers and therefore we expect reconnection to be happening *near the current* but *not in the current*.

Figure 7.12 shows the Lorentz indicator for all identified current layers. The diagram reports a scatter plot of all clusters in the plane formed by their average agyrotropy and magnitude of the parallel electric field. The dots are colored according to the Lorentz indicator.

The distribution is clearly bimodal. At low agyrotropy and low parallel electric field (lower-left part), the Lorentz indicator shows a very high speed for the transformation signaling that here is no reconnection. At high

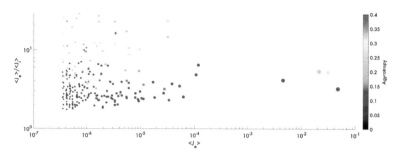

Figure 7.13 Scatter plot of the current layers found by DBSCAN in the plane formed by their average electron current and by the ratio between their electron and ion current. The size of each dot is proportional to the volume of the current layer and their color represents the average agyrotropy.

agyrotropy and high parallel electric field (upper-right part) the current layers show a small velocity for the Lorentz indicator signaling reconnection.

The results from the Lorentz indicator confirm the conclusions reported regarding the concurrency of reconnection and small scale current layers.

A further investigation can then answer the questions as to whether the reconnection happening at the electron scale is the so-called electron-only reconnection or it involves also the ions.

Electron-only reconnection has been identified in observational data [58, 59] and in computer simulations [60–63] and refers to a reconnection regime, where only the electrons show substantial acceleration, while the ions remain extraneous to the process.

In a recent study based on the use of the DBSCAN technique, [45] the 209 clusters have been analyzed to ascertain whether reconnection is electron-only or not. When the reconnection process accelerates only the electrons, the expectation is that the ion speed (and therefore the ion current) remains small, while in the case where both ions and electrons are involved in reconnection, the ion and electron speed equilibrate in the outflow and the two currents, ion and electron, should also equilibrate at a distance from the reconnection site.

Figure 7.13 shows for all current clusters the ratio of the electron to ion current, averaged over each cluster identified by DBSCAN. The dots are colored according to the average electron agyrotropy within the cluster. The size of the dots is again proportional to the size of the cluster. Clearly, all cases where the agyrotropy is high show a very dominant electron current suggesting electron-only reconnection.

Following the analysis typical for observational data [58], Ref. [45] investigated a few selected clusters to make a visual human confirmation of electron-only reconnection. The results of the human analysis confirm the outcome of the DBSCAN automatic analysis.

7.4.3 Perspectives on the DBSCAN method for the study of turbulence

In the context of the AIDA project (www.aida-space.eu) a new method based on DBSCAN was introduced to analyze turbulent space flows using ML [45].

The method is based on DBSCAN clustering and identifies each current structure from the biggest to the smallest as a cluster of current-carrying pixels contiguous to each other and separated (by a prescribed radius of 5 cells in the present example) from the others. In the example shown, the new method identifies 290 clusters, a number that can be easily handled by computer analysis and visualization.

The advantage of the approach is that it provides the user with information that is classified in individual clusters. The initial information made of 81 million individual points is digested by ML into a structured set of 291 clusters. With this information we can correlate the size and physical properties of the current layers to make in depth physical analysis of the ongoing processes.

As an example, the information provided by the method was used to correlate the size of the clusters with the processes of reconnection. The smallest current clusters at the electron scale are the most active in terms of reconnection and energy dissipation. Within these clusters the process of electron-only reconnection is predominant. In this regime, the outflow kinetic energy from reconnection is carried by the electrons, while the ions remain unaffected.

This is only one example of the possibilities. More applications and variations are possible. DBSCAN has much potential also in other applications of data mining from large scale simulations. Besides the applications highlighted in this chapter, DBSCAN has been used previously also for clustering large numbers of particles in extreme scale simulations [64].

7.5 CHARACTERIZING VELOCITY DISTRIBUTION FUNCTIONS WITH UNSUPERVISED MACHINE LEARNING

Magnetic reconnection was already described in Sect. 7.3. Here, we describe a specific method for characterizing magnetic reconnection using particle distributions in Particle-In-Cell (PIC) simulations. When kinetic models are used to describe plasmas, the data set that researchers must analyze becomes six-dimensional. PIC methods provide information about the distribution of hundreds or thousands of particles per cell, resulting in a data set that can contain trillions of particles. Although analysis typically focuses on the electromagnetic fields and the moments of the particle distribution, which are three-dimensional and easier to manage. This approach overlooks valuable information contained within the particles. We propose using a Gaussian mixture model to automatically extract information from the particle distribution to analyze magnetic reconnection without the need for human intervention.

In the context of magnetic reconnection, recent research has shown renewed interest in particle-scale kinetic physics, particularly at the electron-scale. Electron distributions have been found to be effective indicators of magnetic reconnection with various shapes observed in different regions around the reconnection site. For instance, crescent-shaped distributions can be detected near the electron stagnation point for asymmetric reconnection [65].

However, a unique specific distribution cannot be used as a signature for reconnection as it does not reflect the phenomenon for all possible external conditions. Therefore, developing a machine learning-based detection algorithm capable of detecting non-Maxwellian features is very interesting. Such methods can analyze complex shapes in the electron velocity distributions and be combined with other classical detection methods based on field quantities such as agyrotropy [66].

7.5.1 Magnetic reconnection signatures and agyrotropy

The literature has highlighted a variety of signatures for reconnection using different sources of data. Readers are referred to a recent review of all proposed signatures [55].

Moments of plasma species, such as the relative drift between plasma and field lines or energy dissipation on the electron frame, can be used to identify reconnection. We focus on the agyrotropy, a measure of the complexity of the particle velocity distribution, which is linked to electron pressure non-gyrotropies. Direct in situ measurements have recently demonstrated that velocity distributions that are not gyrotropic are important during the magnetic reconnection process [65].

The measure of agyrotropy called Q is defined by the following equation: [67]:

$$Q = \frac{P_{12}^2 + P_{13}^2 + P_{23}^2}{P_{\parallel} + 2P_{\perp}}, \tag{7.6}$$

where $P_{\parallel}$ and $P_{\perp}$ are the diagonal terms of the tensor and P_{12}, P_{13}, and P_{23} are the sub and upper diagonal terms of the symmetric tensor. Regions with complex velocity distributions, such as crescents, have high agyrotropy values.

Computing the agyrotropy will help to assess the method developed in this chapter.

7.5.2 Fitting particle velocity distributions

Fitting distributions to a sample of data involves selecting a probability distribution and estimating its parameters based on mathematical and physical arguments. Commonly used models for plasma particle velocity distributions include the Maxwellian and the bi-Maxwellian, as well as the Kappa distribution, which can describe both low-energy Maxwellian cores and suprathermal tails. Previous studies have used various methods, such as one-dimensional cuts [68] or two-dimensional distributions [69]. However, these approaches require a strong understanding of the physical phenomena and are not always applicable. Density estimation techniques, such as parametric and nonparametric methods, have become popular for automatically fitting complex distribution functions. The Gaussian Mixture Model (GMM) is a popular model fitting the data with a sum of Gaussian distributions. Specific shapes can be approximated by the sum of Gaussian distributions such as beams or non-Maxwellian features, that can be related to reconnection sites. There may be other distributions that have not yet been discovered, so an optimal algorithm for detecting reconnection should not be dependent on distributions.

7.5.2.1 Gaussian Mixture Models

Details on the GMM framework can be found in [70]. The GMM method approximates the particle distribution with a weighted sum of K Gaussian distributions. Each k-th Gaussian distribution is described by its mean μ_k, its covariance matrix Σ_k, and its weight w_k. The number of Gaussians distribution K is an input to the GMM algorithm. There are methods to estimate this value, such as cross-validation, elbow method, and information criteria, which provide a trade-off between the goodness of fit and the complexity of the model. The two main information criteria are the Akaike Information Criterion (AIC) and Bayesian Information Criterion (BIC). BIC is preferred for simulations due to noisy data and significant numbers of particles. It is important to note that the physical meaning of K and the parameters associated with each Gaussian should be analyzed carefully and not necessarily interpreted as specific beams or electron populations.

7.5.2.2 Link between the sum of Gaussians and thermal energy

The thermal energy $E_{thermal}$ for a single velocity distribution is given by its variance:

$$E_{thermal} = \frac{1}{N_p} \sum_{i=1}^{3} \left[\sum_p \left(V_p - \langle V_p \rangle \right)^2 \right]_i, \text{ with } \langle V_p \rangle = \sum_p \frac{V_p}{N_p}. \quad (7.7)$$

The variance $(\sigma^2)^{(K)}$ for K multiple Maxwellians is given by:

$$(\sigma^2)^{(K)} = \sum_{i=1}^{3} \left[\sum_{k=1}^{K} w_k^2 \left(\sigma_k \right)^2 + \sum_{k=1}^{K} w_k \left(\mu_k \right)^2 - \left(\sum_{k=1}^{K} w_k(\mu_k) \right)^2 \right]_i. \quad (7.8)$$

The first term can be interpreted as the mixture of the variances and is related to the thermal energy per unit mass of the mixture:

$$E_{thermal}^{(K)} = \frac{1}{2} \sum_{i=1}^{3} \sum_{k=1}^{K} w_k^2 \left[\sigma_k^2 \right]_i. \quad (7.9)$$

To differentiate between particle acceleration into beams and heating, a calculation called the thermal energy ratio E_{drop} is used:

$$E_{drop} = \frac{E_{thermal}^{(K)}}{E_{thermal}}. \quad (7.10)$$

This ratio determines the decrease in the thermal speed of particles and is computed by comparing the mixture of the variance and the variance of the velocity distribution.

This metric is always below 1. If the ratio of thermal energy for a mixture of particles to the thermal energy computed directly from the definition is low, it suggests that the overall distribution's second order moment is not a good indicator of the present conditions. This situation can occur when two cold beams, each having zero thermal spread and only relative mean velocity, are combined to form a broad thermal spread. The measure discussed here identifies such distributions that consist of interpenetrating beams.

The final two components in Eq. (7.8) can be interpreted as the deviation of each mean compared to the overall mixture mean:

$$E_{dev}^{(K)} = \sum_{i=1}^{3} \left[\sum_{k=1}^{K} w_k \left(\mu_k \right)^2 - \left(\sum_{k=1}^{K} w_k(\mu_k) \right)^2 \right]_i. \quad (7.11)$$

This deviation is always positive as it corresponds to a weighted variance. This is the thermal energy of the center of all beams, measuring the distance between them. A second metric E_{dev}, called thermal velocity deviation, defines the ratio between the velocity deviation for the mixture and the classical thermal velocity of the distribution:

$$E_{dev} = \frac{E_{dev}^{(K)}}{E_{thermal}}. \tag{7.12}$$

The quantity mentioned is always positive and provides a way to understand different mixtures. A large value indicates that the components are significantly far apart from each other, which suggests that they may have different origins and identities. On the other hand, a small value indicates that the components are closely packed together, which suggests that they may not have a meaningful separation. This has been discussed in [71].

7.5.3 Experiments on 2.5D PIC simulation

7.5.3.1 Double Harris sheet

A scenario with a double Harris sheet [46] is studied, where two distinct reconnection sites are easily identifiable and analyzed. The double Harris sheet is considered a paradigm in reconnection. This problem is also known as the GEM Challenge [72], and it is chosen to evaluate the performance of the diagnostic. Here it is defined by:

$$B_x(y) = B_0(-1 + \tanh(y - y_1) - \tanh(y - y_2)), \tag{7.13}$$

with the location of the two current layers at $y_1 = L_y/4$ and $y_1 = 3L_y/4$ [73]. Pressure balance is kept by a uniform temperature but a non uniform density:

$$n_s(y) = n_0(-1 + \operatorname{sech}(y - y_1)^2 + \operatorname{sech}(y - y_2)^2) + n_b, \tag{7.14}$$

where a background density equal to $n_b = n_0/10$ is added.

We conducted the simulation in 2.5D, which means that all vectors are three-dimensional, but their spatial variation is restricted to a two-dimensional plane that does not depend on the dawn-dusk (Z) direction. We tested two different guide field values (0.1 and 1.0), and all the presented quantities are normalized. The simulations are performed with the fully kinetic massively parallel implicit moment method Particle-in-Cell code iPic3D [74, 75].

The equilibrium is defined by the thickness $L/d_i = 0.5$ and with the parameters $m_i/m_e = 256$, $v_{the}/c = 0.045$, $T_i/T_e = 5$. With these choices, the asymptotic in plane field B_0 is set by the ratio $\omega_{ci}/\omega_{pi} = 0.0097$ and the peak Harris density $n_0 = 1$ is imposed by the normalisation used that results in the ion plasma frequency and ion inertial length to be unitary. The coordinates are chosen with the initial Harris magnetic field along x with size $L_x = 30d_i$, the initial gradients along y with $L_y = 40d_i$. The third dimension, where the initial current and guide field are directed, is invariant. Periodicity is assumed in all directions.

The Cartesian mesh has a size of 769×1025 and about $196.600.000$ particles with varying weights are injected in the computational domain, representing approximately 250 particles by cell. The particle distributions are analyzed in a frame of reference driven by the local magnetic field, in addition to the Cartesian system, as suggested in [55].

7.5.3.2 GMM identification of magnetic reconnection

The GMM algorithm for detecting magnetic reconnection and regions of interest is applied to the particle distributions from the simulation with a guide field value of 0.1. The decomposition into several Gaussians by the GMM algorithm is compared with measures of agyrotropy.

In Fig. 7.14, the left column shows the number of components identified by the detection algorithm, while the right column displays the measure of agyrotropy at various time steps. The goal is to illustrate how these two quantities behave as reconnection grows. The detection algorithm is capable of identifying not only the electron diffusion region (EDR) but also other structures such as inflows, ion and electron diffusion regions, outflow, and separatrix boundaries. It can even detect regions where the influence of reconnection is weak. However, the algorithm works best for distributions with one component. At the first time step of $t = 8,000$, a background with two components surrounds the EDR, which is mainly composed of mixtures with 4 and 5 components. The GMM analysis is expected to be related to Swisdak's agyrotropy, as both methods focus on the complexity and non-Maxwellianity of distributions. Downstream from the EDR in the outflow, there is a C-shaped structure characterized by distributions with 4 and 5 components connecting the EDR with the separatrix region. The separatrix region is mainly composed of distributions with 2 and 3 components.

For the time steps $t = 12,000$, $t = 16,000$, and $t = 20,000$, similar structures and behaviors are observed. The size of the EDR slightly increases in the x-direction over time, while the inflow region remains constant in size. The outflow region is well-identified and its location remains stable. The reconnection creates a magnetic island on the right-hand side of the figure. The

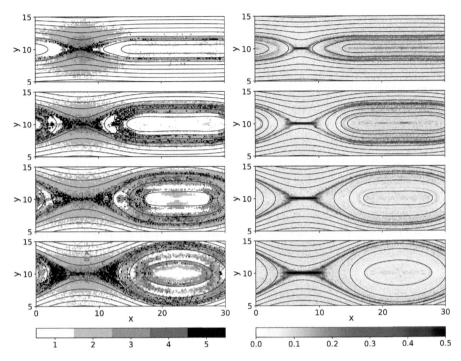

Figure 7.14 Magnetic reconnection detection for the Double Harris sheet case with a guide field value of 0.1 at four different time steps, from top to bottom: $t = 8,000$, $t = 12,000$, $t = 16,000$, and $t = 20,000$. The left-hand column presents the number of components provided by the BIC optimization and the right-hand column shows the measure of agyrotropy $\sqrt{Q}$. The red rectangles indicate the location where specific distributions are observed. Figure from Dupuis et al., The Astrophysical Journal, **889**.1 , 2020: "*Characterizing magnetic reconnection regions using Gaussian mixture models on particle velocity distributions*", DOI 10.3847/1538-4357/ab5524 (©AAS. "Reproduced with permission")

thickness of the region around the O point increases significantly in the y-direction as the reconnection grows. There are various types of distributions, resulting in a noisy mix with a background having 2 components and some having 3 and 4 components. Furthermore, secondary structures appear near the O point, linking the bottom and top layers of the island. At $t = 20,000$, two concentric ellipses made up of 3 components for the outer ellipse and 2 components for the inner ellipse can be seen. It is important to note that the detection algorithm does not impose any spatial constraints or correlations, so all structures identified by the BIC minimization may exist in the distributions.

In Fig. 7.14, the results obtained by the detection algorithm are compared to the values of agyrotropy in the right-hand column for the same time steps. Some similarities are observed, such as the clear identification of the EDR with peak agyrotropy values above 0.5. The mapping of the topological boundaries of the reconnection almost coincides with the boundaries of the GMM algorithm but different behaviors are also exhibited when compared to the detection algorithm. For example, the agyrotropy measure does not diagnose the region surrounding the EDR, as well as the outflow and inner structures around the O point. Small artifacts are present within the topological boundaries, but the background noise hinders their clear identification. Unlike the detection algorithm, the measure of gyrotropy is not exactly zero for regions far away from the reconnection, and a background noise around 0.1 is present. The detection algorithm, however, clearly identifies single distributions.

7.5.4 Analysis of the particle velocity distribution fit

Figure 7.15 presents the particle distributions corresponding to the five red rectangles shown in Fig. 7.14 and the associated GMM fit with several observations:

- **Inflow region (box 'A')**: The velocity distributions in this region exhibit a strong anisotropy, indicating heating along the parallel direction. The GMM algorithm does not describe the data with a single anisotropic Gaussian distribution but instead uses two components to capture the broad mode with a short tail. This suggests that the electron distribution deviates from a Maxwellian within the inflow region. The presence of the short tail, approximated by two components, leads to high values of energy drop as each mixture represents a relatively small portion compared to the overall second moment. The distributions perpendicular to the magnetic field show a Gaussian shape.

- **Electron distribution region (box 'B')**: The marginal distribution in the $v_{\perp 1} - v_{\perp 2}$ plane exhibits a crescent shape, which explains the higher number of components identified by the algorithm. All the mixtures are spread across the distribution with a narrow width compared to the overall second moment, making it challenging to properly fit such a complex distribution. The $v_{\parallel} - v_{\perp 1}$ and $v_{\parallel} - v_{\perp 2}$ projections display triangular shapes.

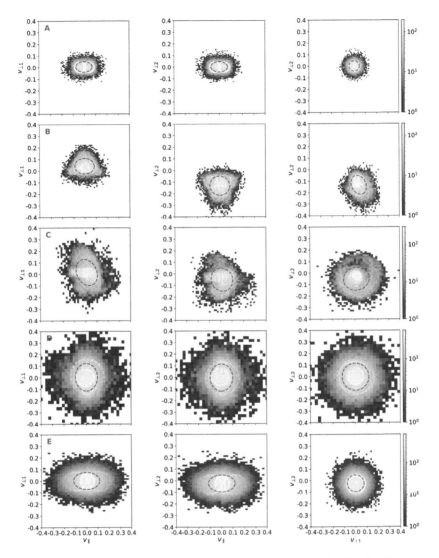

Figure 7.15　Electron velocity distribution for the Double Harris sheet case at $t = 20,000$. Each row corresponds to one of the five red rectangles depicted in Fig. 7.14. Three two dimensional marginal distributions are presented: $v_\parallel - v_{\perp 1}$, $v_\parallel - v_{\perp 2}$, and $v_{\perp 1} - v_{\perp 2}$. The white ellipses illustrate the different Gaussians of the mixtures in each distribution. The transparency is determined by the weight of each Gaussian: no transparency for a weight of 1 and full transparency for a zero weight. The red ellipses are drawn from the mean and variance of the data. Figure from Dupuis et al., The Astrophysical Journal, **889**.1, 2020: "*Characterizing magnetic reconnection regions using Gaussian mixture models on particle velocity distributions*", DOI 10.3847/1538-4357/ab5524 (©AAS. "Reproduced with permission")

- **Outflow region (box 'C')**: Similar to the EDR, a crescent shape is observed in the outflow region, justifying the similar number of components identified by the algorithm. In the $v_{\perp 1} - v_{\perp 2}$ projection, a narrow Gaussian with a high weight captures the zero-centered mode, while scattered Gaussians with low weights represent the crescent shape. The other projections are complex due to the crescent shape, making it challenging for the algorithm to approximate. Therefore, five components are required: one for the core and four for the crescent.

- **Intermediate region (box 'D')**: The crescent shape disappears in this region, and the distribution only slightly deviates from a Maxwellian. Thin core distributions with strong weights are accompanied by wider distributions with smaller weights. This combination of core and tail distributions improves the overall fitting, especially for the long tail.

- **Outer ellipse near the O point (box 'E')**: The distribution in this region exhibits strong anisotropy along the parallel direction, similar to the inflow region. However, the distribution has a significant mode around zero coupled with a very broad tail. This explains the use of three components by the algorithm: one approximates the mode, while the other two fit the tail on each side along the parallel direction.

7.5.5 Perspectives on the Gaussian Mixture Model method for the study of reconnection

The Gaussian Mixture Model (GMM) is an automated way to detect magnetic reconnection from particle velocity distributions. It has proven to be successful in identifying various regions around reconnection sites in a PIC simulation with a guide field of 0.1. This approach allows for a physical interpretation of the heating and distribution of particles, beyond just the statistical properties of the GMM. However, the algorithm is not a unique solution and requires calibration to properly set the resolution. The algorithm has only been tested on 2.5D simulations with weighted particles, but it could potentially be applied to other simulation types. Testing the algorithm on three-dimensional simulations is a necessary next step. Additionally, the algorithm could be applied to other fields of study, such as turbulence analysis. Further work will explore applying the algorithm to local observational data of particle distribution function, which is limited in comparison to simulations that have access to complete plasma descriptions.

7.6 SUMMARY AND KEY TAKEAWAY MESSAGES

Artificial intelligence encompasses many activities, machine learning being the one that has received most attention in recent years. Machine learning can include both supervised and unsupervised methods.

In supervised methods, the algorithm is trained using previous examples – usually generated by human labelling and processing of data. This approach is very successful in replacing human activities with ML tools, saving effort but with the potential risk of affecting the job market.

We focus instead here on unsupervised ML. In this case, algorithms are designed to operate without previous human directions. In this case new methods can emerge and ways to analyse data can emerge that humans had not conceived before.

One example, on which we focused in this chapter, is clustering. Clustering can be described in a simple example. We can cluster images of pets into cats and dogs. A human labels pictures of cat and dogs and supervised neural networks learn to mean the distinction. It is several years now that ML is more accurate than humans in doing this task (not just for cats and dogs but more generally). This is supervised learning.

Clustering here takes a different approach. We use agnostic clustering techniques that have no previous knowledge of dogs and cats but sort the images in clusters according to different mathematical rules and algorithmic strategies. At the end if the method is well designed it will sort the pictures in two groups and, when we look at them post fact, we realise one cluster is dogs and one is cats.

The real promise is that the method can discover things we did not expect. If for example we have unknown animals and ask the ML to sort them, it might find a new species we did not know before.

Outside this simple example, we applied the techniques of clustering and unsupervised learning to sort observations of space processes obtained from space missions or computer simulations. In this chapter, we presented different applications that were shown to be especially useful:

- Using self organising maps to sort different types of plasmas. Here is an example, where we do not prejudge what are the different categories of plasmas, but we can post facto interpret what are the different types discovered by the clustering technique.

- Using clustering methods to identify, which current lawyers are reconnecting without prejudging "what reconnection is". We rely on the nature of reconnection as an extreme event and design a ML tool to discover the likely cases.

– Clustering can then be used to do Sisyphus's work in analysing turbulence. Rather than looking at generic statistical properties, ML clustering can identify every single feature in the myriad of features present. It can then analyse each and every one of the structure and correlate it with the environment to find when and where certain mechanisms such as energy dissipations are.

ACRONYM

ACE NASA's Advanced Composition Explorer

AI Artificial Intelligence

AIC Akaike Information Criterion

AIDA Artificial Intelligence for Data Analysis

AML Unsupervised Machine-Learning

API Application Programming Interface

AR Aspect Ratio

BIC Bayesian Information Criterion

BMU Best Matching Unit

CWI Centrum Wiskunde & Informatica

CNRS Centre national de la recherche scientifique

CS Current Sheet

CTIM Coupled Thermosphere Ionosphere Model

CUDA Compute Unified Device Architecture

DBSCAN Density-Based Spatial Clustering of Applications with Noise

EDR Electron-diffusion layer

GGCM Geospace General Circulation Model

GMM Gaussian Mixture Model

GSE Geocentric Solar Equatorial

IDL Interface Description Language

IDR Ion-Diffusion Layer

LHS left hand side

MHD Magnetohydrodynamics

ML Machine Learning

MMS Magnetospheric Multiscale Mission

PIC Particle-In-Cell

RCM Rice Convection Model

RHS right hand side

SOM Self-Organizing Map

THEMIS Time History of Events and Macroscale Interactions during Substorms

UDM Unified Distance Matrix

UTC Coordinated Universal Time

GLOSSARY

AIDA: a project initiated with funding from Horizon 2020 by the European Commission, which uses some of the most advanced data analysis techniques to analyze the data available from space missions and from space simulations.

AIDApy: a new tool developed by AIDA project in python, which will make the use of AI simpler for the space scientist and even the general public.

MATLAB: is a proprietary multi-paradigm programming language and numeric computing environment developed by MathWorks. MATLAB allows matrix manipulations, plotting of functions and data, implementation of algorithms, creation of user interfaces, and interfacing with programs written in other language

OMNIweb: a discontinued web browser that was developed and marketed by The Omni Group exclusively for Apple's macOS operating system.

iPIC3D: a three-dimensional parallel Particle-in-Cell (PIC) code, whereas PIC method refers to a technique used to solve a certain class of partial differential. equations

MMS mission of NASA: a robotic space mission to study the Earth's magnetosphere, using four identical spacecraft flying in a tetrahedral formation.

OpenGGCM: a program designed to be used as a research tool – as well as a prototype – for space weather forecasting models. It is considered a "Community Model".

MagnetoHydroDynamics: also called magneto-fluid dynamics or hydromagnetics, is the study of the magnetic properties and behaviour of electrically conducting fluids. Examples of such magnetofluids include plasmas, liquid metals, salt water, and electrolytes.

Compute Unified Device Architecture: or CUDA is a parallel computing platform and application programming interface (API) that allows software to use certain types of graphics processing units (GPUs) for general purpose processing, an approach called general-purpose computing on GPUs (GPGPU). CUDA is a software layer that gives direct access to the GPU's virtual instruction set and parallel computational elements, for the execution of compute kernels.

DBSCAN: a density-based clustering non-parametric algorithm: given a set of points in some space, it groups together points that are closely packed together (points with many nearby neighbors), marking as outliers points that lie alone in low-density regions (whose nearest neighbors are too far away). DBSCAN is one of the most common clustering algorithms and also most cited in scientific literature.

Unsupervised Machine-Learning: or AML, uses machine learning algorithms to analyze and cluster unlabeled datasets. These algorithms discover hidden patterns or data groupings without the need for human intervention.

FURTHER READING

Bishop, Christopher M., and Nasser M. Nasrabadi. (2006) "Pattern recognition and machine learning". Vol. 4. No. 4. New York: *Springer.*

Büchner, Jörg, ed. (2023) "Space and Astrophysical Plasma Simulation: Methods, Algorithms, and Applications". *Springer Nature.*

Burrell, Angeline G., et al. (2018) "Snakes on a spaceship—An overview of Python in heliophysics". *Journal of Geophysical Research: Space Physics* 123.12: 10–384.

Camporeale, Enrico, Simon Wing, and Jay Johnson, eds. (2018) "Machine learning techniques for space weather". *Elsevier.*

Goodfellow, Ian, Yoshua Bengio, and Aaron Courville. (2016) "Deep learning". *MIT press.*

Lapenta, G. (2012) "Particle simulations of space weather". *Journal of Computational Physics*, 231(3), 795–821.

Lapenta, G., Pierrard, V., Keppens, R., Markidis, S., Poedts, S., Šebek, O., ... & Borremans, K. (2013) "SWIFF: Space weather integrated forecasting framework". *Journal of Space Weather and Space Climate*, 3, A05.

Materassi, Massimo, et al., eds. (2019) "The dynamical ionosphere: A systems approach to ionospheric irregularity". *Elsevier.*

McGranaghan, R. M., et al. (2022) "Heliophysics Discovery Tools for the 21st Century: Data Science and Machine Learning Structures and Recommendations for 2020-2050". *arXiv preprint arXiv:2212.13325.*

Parks, G. K. (2019). "Physics of space plasmas: an introduction". *CRC Press.*

Schrijver, Carolus J., et al. (2015) "Understanding space weather to shield society: A global road map for 2015–2025 commissioned by COSPAR and ILWS". *Advances in Space Research* 55.12: 2745-2807.

BIBLIOGRAPHY

[1] M. E. Innocenti, J. Amaya, J. Raeder, R. Dupuis, B. Ferdousi, and G. Lapenta. Unsupervised classification of simulated magnetospheric regions. *Annales Geophysicae*, 39(5):861–881, 2021.

[2] Sophia K'ohne, Elisabetta Boella, and Maria Elena Innocenti. Unsupervised classification of fully kinetic simulations of plasmoid instability using self-organizing maps (soms). *Journal of Plasma Physics*, 89(3):895890301, 2023.

[3] Jorge Amaya, Romain Dupuis, Maria Elena Innocenti, and Giovanni Lapenta. Visualizing and interpreting unsupervised solar wind classifications. *Frontiers in Astronomy and Space Sciences*, 7:66, 2020.

[4] C. P. Escoubet, M. Fehringer, and M. Goldstein. Introductionthe cluster mission. *Annales Geophysicae*, 19(10/12):1197–1200, 2001.

[5] Vassilis Angelopoulos. The themis mission. In *The THEMIS mission*, pages 5–34. Springer, 2009.

[6] JL Burch, TE Moore, RB Torbert, and BL Giles. Magnetospheric multiscale overview and science objectives. *Space Science Reviews*, 199(1-4):5–21, 2016.

[7] D da Silva, A Barrie, J Shuster, C Schiff, R Attie, DJ Gershman, and B Giles. Automatic region identification over the mms orbit by partitioning nt space. *arXiv preprint arXiv:2003.08822*, 2020.

[8] Matthew R. Argall, Colin R. Small, Samantha Piatt, Liam Breen, Marek Petrik, Kim Kokkonen, Julie Barnum, Kristopher Larsen, Frederick D. Wilder, Mitsuo Oka, William R. Paterson, Roy B. Torbert, Robert E. Ergun, Tai Phan, Barbara L. Giles, and James L. Burch. Mms sitl ground loop: Automating the burst data selection process. *Frontiers in Astronomy and Space Sciences*, 7:54, 2020.

[9] Hugo Breuillard, Romain Dupuis, Alessandro Retino, Olivier Le Contel, Jorge Amaya, and Giovanni Lapenta. Automatic classification of plasma regions in near-earth space with supervised machine learning: Application to magnetospheric multi scale 2016–2019 observations. *Frontiers in Astronomy and Space Sciences*, 7:55, 2020.

[10] Vyacheslav Olshevsky, Yuri V Khotyaintsev, Ahmad Lalti, Andrey Divin, Gian Luca Delzanno, Sven Anderzén, Pawel Herman, Steven WD Chien, Levon Avanov, Andrew P Dimmock, et al. Automated classification of plasma regions using 3d particle energy distributions. *Journal of Geophysical Research: Space Physics*, 126(10):e2021JA029620, 2021.

[11] Gautier Nguyen, Nicolas Aunai, Bayane Michotte de Welle, Alexis Jeandet, Benoit Lavraud, and Dominique Fontaine. Massive multimission statistical study and analytical modeling of the earth's magnetopause: 1. a gradient boosting based automatic detection of near-earth regions. *Journal of Geophysical Research: Space Physics*, 127(1):e2021JA029773, 2022.

[12] A. Lalti, Yu. V. Khotyaintsev, A. P. Dimmock, A. Johlander, D. B. Graham, and V. Olshevsky. A database of mms bow shock crossings compiled using machine learning. *Journal of Geophysical Research: Space Physics*, 127(8):e2022JA030454, 2022. e2022JA030454 2022JA030454.

[13] J. Raeder. Global Magnetohydrodynamics – A Tutorial. In J. B′uchner, C. T. Dum, and M. Scholer, editors, *Space Plasma Simulation*. Springer Verlag, Berlin Heidelberg New York, 2003.

[14] Joachim Raeder, Yongli Wang, and Timothy J. Fuller-Rowell. Geomagnetic storm simulation with a coupled magnetosphere-ionosphere-thermosphere model. In *Geophysical Monograph Series*, pages 377–384. American Geophysical Union, mar 2013.

[15] Banafsheh Ferdousi and Joachim Raeder. Signal propagation time from the magnetotail to the ionosphere: Openggcm simulation. *Journal of Geophysical Research: Space Physics*, 121(7):6549–6561, 2016.

[16] Dimitrios Millas, Maria Elena Innocenti, Brecht Laperre, Joachim Raeder, Stefaan Poedts, and Giovanni Lapenta. Domain of influence analysis: Implications for data assimilation in space weather forecasting. *Frontiers in Astronomy and Space Sciences*, 7:73, 2020.

[17] NF Loureiro and DA Uzdensky. Magnetic reconnection: from the sweet–parker model to stochastic plasmoid chains. *Plasma Physics and Controlled Fusion*, 58(1):014021, 2015.

[18] F Pucci, M Velli, C Shi, KAP Singh, A Tenerani, F Alladio, F Ambrosino, P Buratti, W Fox, J Jara-Almonte, et al. Onset of fast magnetic reconnection and particle energization in laboratory and space plasmas. *Journal of Plasma Physics*, 86(6), 2020.

[19] JF Drake, M Swisdak, H Che, and MA Shay. Electron acceleration from contracting magnetic islands during reconnection. *Nature*, 443(7111):553–556, 2006.

[20] Fan Guo, Xiaocan Li, Hui Li, William Daughton, Bing Zhang, Nicole Lloyd-Ronning, Yi-Hsin Liu, Haocheng Zhang, and Wei Deng. Efficient production of high-energy nonthermal particles during magnetic reconnection in a magnetically dominated ion–electron plasma. *The Astrophysical Journal Letters*, 818(1):L9, 2016.

[21] Emanuele Cazzola, Maria Elena Innocenti, Martin V Goldman, David L Newman, Stefano Markidis, and Giovanni Lapenta. On the electron agyrotropy during rapid asymmetric magnetic island coalescence in presence of a guide field. *Geophysical Research Letters*, 43(15):7840–7849, 2016.

[22] Xiaocan Li, Fan Guo, Hui Li, and Gang Li. Particle acceleration during magnetic reconnection in a low-beta plasma. *The Astrophysical Journal*, 843(1):21, 2017.

[23] G. Lapenta, D. Gonzalez-Herrero, and E. Boella. Multiple-scale kinetic simulations with the energy conserving semi-implicit particle in cell method. *Journal of Plasma Physics*, 83:705830205, 2017.

[24] G. Lapenta. Exactly energy conserving semi-implicit particle in cell formulation. *Journal of Computational Physics*, 334:349–366, 2017.

[25] D. Gonzalez-Herrero, E. Boella, and G. Lapenta. Performance analysis and implementation details of the Energy Conserving Semi-Implicit Method code (ECsim). *Computer Physics Communications*, 229:162–169, 2018.

[26] Teuvo Kohonen. Self-organized formation of topologically correct feature maps. *Biological cybernetics*, 43(1):59–69, 1982.

[27] Thomas Villmann and Jens Christian Claussen. Magnification control in self-organizing maps and neural gas. *Neural Computation*, 18(2):446–469, 2006.

[28] Teuvo Kohonen et al. Matlab implementations and applications of the self-organizing map. *Unigrafia Oy, Helsinki, Finland*, 2, 2014.

[29] Stuart Lloyd. Least squares quantization in pcm. *IEEE transactions on information theory*, 28(2):129–137, 1982.

[30] F. Pedregosa, G. Varoquaux, A. Gramfort, V. Michel, B. Thirion, O. Grisel, M. Blondel, P. Prettenhofer, R. Weiss, V. Dubourg, J. Vanderplas, A. Passos, D. Cournapeau, M. Brucher, M. Perrot, and E. Duchesnay. Scikit-learn: Machine learning in Python. *Journal of Machine Learning Research*, 12:2825–2830, 2011. `https://scikit-learn.org/stable/`, last accessed: Mar 29, 2023, version: 1.0.2.

[31] Jonathon Shlens. A tutorial on principal component analysis. *arXiv preprint arXiv:1404.1100*, 2014.

[32] Giuseppe Vettigli. Minisom: minimalistic and numpy-based implementation of the self organizing map, 2018. GitHub.[Online]. Available: https://github.com/JustGlowing/minisom/.

[33] Matteo Mistri. Cuda-som, 2018. `https://github.com/mistrello96/CUDA-SOM/releases`, last accessed: Mar 22, 2023 (bugfix in src/utility_functions.cu in line 49: change "i¡nELements;" to "j¡nElements;").

[34] V. Satopaa, J. Albrecht, D. Irwin, and B. Raghavan. Finding a "kneedle" in a haystack: Detecting knee points in system behavior. In *2011 31st International Conference on Distributed Computing Systems Workshops*, pages 166–171, 2011.

[35] A. Hu, M. Sisti, F. Finelli, F. Califano, J. Dargent, M. Faganello, E. Camporeale, and J. Teunissen. Identifying Magnetic Reconnection in 2D Hybrid Vlasov Maxwell Simulations with Convolutional Neural Networks. *The Astrophysical Journal*, 900(1):86, September 2020.

[36] Manuela Sisti, Francesco Finelli, Giorgio Pedrazzi, Matteo Faganello, Francesco Califano, and Francesca Delli Ponti. Detecting reconnection events in kinetic vlasov hybrid simulations using clustering techniques. *The Astrophysical Journal*, 908(1):107, 2021.

[37] Seiji Zenitani, Michael Hesse, Alex Klimas, and Masha Kuznetsova. New Measure of the Dissipation Region in Collisionless Magnetic Reconnection. *Physical Review Letters*, 106(19):195003, May 2011.

[38] Francesco Califano, Silvio Sergio Cerri, Matteo Faganello, Dimitri Laveder, Manuela Sisti, and Matthew W. Kunz. Electron-only reconnection in plasma turbulence. *Frontiers in Physics*, 8:317, September 2020.

[39] Vladimir Zhdankin, Dmitri A. Uzdensky, Jean C. Perez, and Stanislav Boldyrev. Statistical Analysis of Current Sheets in Three-dimensional Magnetohydrodynamic Turbulence. *The Astrophysical Journal*, 771(2):124, July 2013.

[40] Hefin Rhys. *Machine Learning with R, the tidyverse, and mlr*. Simon and Schuster, 2020.

[41] Fouad Sahraoui, Lina Hadid, and Shiyong Huang. Magnetohydrodynamic and kinetic scale turbulence in the near-earth space plasmas:

a (short) biased review. *Reviews of Modern Plasma Physics*, 4:1–33, 2020.

[42] Francesco Pucci, Sergio Servidio, Luca Sorriso-Valvo, Vyacheslav Olshevsky, WH Matthaeus, Francesco Malara, MV Goldman, DL Newman, and Giovanni Lapenta. Properties of turbulence in the reconnection exhaust: numerical simulations compared with observations. *The Astrophysical Journal*, 841(1):60, 2017.

[43] S. Markidis, G. Lapenta, and Rizwan-uddin. Multi-scale simulations of plasma with iPIC3D. *Mathematics and Computers and Simulation*, 80:1509–1519, 2010.

[44] Giovanni Lapenta, Stefano Markidis, Martin V Goldman, and David L Newman. Secondary reconnection sites in reconnection-generated flux ropes and reconnection fronts. *Nature Physics*, 11(8):690–695, 2015.

[45] Giovanni Lapenta, Martin Goldman, David L Newman, and Stefan Eriksson. Formation and reconnection of electron scale current layers in the turbulent outflows of a primary reconnection site. *The Astrophysical Journal*, 940(2):187, 2022.

[46] E. G. Harris. On a plasma sheath separating regions of oppositely directed magnetic field. *Il Nuovo Cimento (1955-1965)*, 23:115–121, January 1962.

[47] Giovanni Lapenta, Martin V Goldman, David L Newman, and Stefano Markidis. Energy exchanges in reconnection outflows. *Plasma Physics and Controlled Fusion*, 59(1):014019, 2016.

[48] G Lapenta, F Pucci, MV Goldman, and DL Newman. Local regimes of turbulence in 3d magnetic reconnection. *The Astrophysical Journal*, 888(2):104, 2020.

[49] Lora Price, M Swisdak, JF Drake, JL Burch, PA Cassak, and RE Ergun. Turbulence in three-dimensional simulations of magnetopause reconnection. *Journal of Geophysical Research: Space Physics*, 122(11), 2017.

[50] Giovanni Lapenta, Mostafa El Alaoui, Jean Berchem, and Raymond Walker. Multiscale mhd-kinetic pic study of energy fluxes caused by reconnection. *Journal of Geophysical Research: Space Physics*, 125(3):no–no, 2020.

[51] Giovanni Lapenta. Detecting reconnection sites using the lorentz transformations for electromagnetic fields. *The Astrophysical Journal*, 911(2):147, 2021.

[52] M Zhou, HY Man, XH Deng, Y Pang, Y Khotyaintsev, G Lapenta, YY Yi, ZH Zhong, and WQ Ma. Observations of secondary magnetic reconnection in the turbulent reconnection outflow. *Geophysical Research Letters*, 48(4):e2020GL091215, 2021.

[53] Dieter Biskamp. *Magnetohydrodynamic turbulence*. Cambridge University Press, 2003.

[54] Martin Ester, Hans-Peter Kriegel, Jörg Sander, and Xiaowei Xu. A density-based algorithm for discovering clusters in large spatial databases with noise. In *Proceedings of the Second International Conference on Knowledge Discovery and Data Mining*, pages 226–231, 1996.

[55] MV Goldman, DL Newman, and Giovanni Lapenta. What can we learn about magnetotail reconnection from 2d pic harris-sheet simulations? *Space Science Reviews*, 199(1-4):651–688, 2016.

[56] Jack Scudder and William Daughton. "illuminating" electron diffusion regions of collisionless magnetic reconnection using electron agyrotropy. *Journal of Geophysical Research: Space Physics*, 113(A6), 2008.

[57] Seiji Zenitani, Michael Hesse, Alex Klimas, and Masha Kuznetsova. New measure of the dissipation region in collisionless magnetic reconnection. *Physical review letters*, 106(19):195003, 2011.

[58] TD Phan, Jonathan P Eastwood, MA Shay, JF Drake, BU Ö Sonnerup, Masaki Fujimoto, PA Cassak, M Øieroset, JL Burch, RB Torbert, et al. Electron magnetic reconnection without ion coupling in earth's turbulent magnetosheath. *Nature*, 557(7704):202–206, 2018.

[59] J Eastwood Stawarz, Jonathan P Eastwood, TD Phan, IL Gingell, MA Shay, JL Burch, RE Ergun, BL Giles, DJ Gershman, Olivier Le Contel, et al. Properties of the turbulence associated with electron-only magnetic reconnection in earth's magnetosheath. *The Astrophysical Journal Letters*, 877(2):L37, 2019.

[60] P Sharma Pyakurel, MA Shay, TD Phan, WH Matthaeus, JF Drake, JM TenBarge, CC Haggerty, KG Klein, PA Cassak, TN Parashar,

et al. Transition from ion-coupled to electron-only reconnection: Basic physics and implications for plasma turbulence. *Physics of Plasmas*, 26(8):082307, 2019.

[61] Francesco Califano, Silvio Sergio Cerri, Matteo Faganello, Dimitri Laveder, Manuela Sisti, and Matthew W Kunz. Electron-only reconnection in plasma turbulence. *Frontiers in Physics*, 8:317, 2020.

[62] Giuseppe Arrò, Francesco Califano, and Giovanni Lapenta. Statistical properties of turbulent fluctuations associated with electron-only magnetic reconnection. *Astronomy & Astrophysics*, 642:A45, 2020.

[63] Cristian Vega, Vadim Roytershteyn, Gian Luca Delzanno, and Stanislav Boldyrev. Electron-only reconnection in kinetic-alfvén turbulence. *The Astrophysical Journal Letters*, 893(1):L10, 2020.

[64] Md Mostofa Ali Patwary, Suren Byna, Nadathur Rajagopalan Satish, Narayanan Sundaram, Zarija Lukić, Vadim Roytershteyn, Michael J Anderson, Yushu Yao, Pradeep Dubey, et al. Bd-cats: big data clustering at trillion particle scale. In *SC'15: Proceedings of the International Conference for High Performance Computing, Networking, Storage and Analysis*, pages 1–12. IEEE, 2015.

[65] JL Burch, RB Torbert, TD Phan, L-J Chen, TE Moore, RE Ergun, JP Eastwood, DJ Gershman, PA Cassak, MR Argall, et al. Electron-scale measurements of magnetic reconnection in space. *Science*, 352(6290):aaf2939, 2016.

[66] Nicolas Aunai, Michael Hesse, and Maria Kuznetsova. Electron nongyrotropy in the context of collisionless magnetic reconnection. *Physics of Plasmas*, 20(9):092903, 2013.

[67] M Swisdak. Quantifying gyrotropy in magnetic reconnection. *Geophysical Research Letters*, 43(1):43–49, 2016.

[68] MP Pulupa, SD Bale, C Salem, and K Horaites. Spin-modulated spacecraft floating potential: Observations and effects on electron moments. *Journal of Geophysical Research: Space Physics*, 119(2):647–657, 2014.

[69] Lynn B. Wilson III, Li-Jen Chen, Shan Wang, Steven J. Schwartz, Drew L. Turner, Michael L. Stevens, Justin C. Kasper, Adnane Osmane, Damiano Caprioli, Stuart D. Bale, Marc P. Pulupa, Chadi S. Salem, and

Katherine A. Goodrich. Electron energy partition across interplanetary shocks. i. methodology and data product. *The Astrophysical Journal Supplement Series*, 243(1):8, jul 2019.

[70] Christopher M. Bishop. *Pattern Recognition and Machine Learning (Information Science and Statistics)*. Springer-Verlag New York, Inc., Secaucus, NJ, USA, 2006.

[71] JP Eastwood, MV Goldman, H Hietala, DL Newman, R Mistry, and Giovanni Lapenta. Ion reflection and acceleration near magnetotail dipolarization fronts associated with magnetic reconnection. *Journal of Geophysical Research: Space Physics*, 120(1):511–525, 2015.

[72] J Birn, JF Drake, MA Shay, BN Rogers, RE Denton, M Hesse, M Kuznetsova, ZW Ma, A Bhattacharjee, A Otto, et al. Geospace environmental modeling (gem) magnetic reconnection challenge. *Journal of Geophysical Research: Space Physics*, 106(A3):3715–3719, 2001.

[73] Pin Wu, MA Shay, TD Phan, M Oieroset, and M Oka. Effect of inflow density on ion diffusion region of magnetic reconnection: Particle-in-cell simulations. *Physics of Plasmas*, 18(11):111204, 2011.

[74] Stefano Markidis, Giovanni Lapenta, et al. Multi-scale simulations of plasma with ipic3d. *Mathematics and Computers in Simulation*, 80(7):1509–1519, 2010.

[75] Maria Elena Innocenti, Alec Johnson, Stefano Markidis, Jorge Amaya, Jan Deca, Vyacheslav Olshevsky, and Giovanni Lapenta. Progress towards physics based space weather forecasting with exascale computing. *Advances in Engineering Software*, 111:3–17, 2017.

IV

Legal Perspectives

Harnessing Artificial Intelligence Technologies for Sustainable Space Missions: Legal Perspectives

Steven Freeland

Western Sydney University School of Law, Sydney, Australia

Anne-Sophie Martin

Sapienza University of Rome, Rome, Italy

CONTENTS

8.1   Artificial Intelligence and Sustainability: An Outer Space
      Perspective 274
8.2   Legal Issues on the Use of Artificial Intelligence 279
      8.2.1   Opportunities and Challenges under International
              Space Law 279
      8.2.2   National Policy and Regulatory Frameworks linking to
              AI and Sustainability 282
8.3   Legal Considerations for the Safety of Space Operations 289
8.4   Fostering International Cooperation for Space Sustainability
      and Artificial Intelligence 292
8.5   Promoting Scientific Research in Space Missions with AI
      Components 294

DOI: 10.1201/9781003366386-8

8.6   Concluding Remarks 295
Bibliography 298

A RTIFICIAL INTELLIGENCE (AI) represents an opportunity, but also a challenge, for the future of space activities and its legal framework. Expanding connectivity and symbiotic interaction between humans and intelligent machines raises significant questions for the rule of law, including the liability regime in case of damage arising from actions undertaken through the utilization of increasingly advanced AI in space missions. Artificial Intelligence technologies also encompass a series of possibilities for space sustainability. We describe the legal issues brought by this technology, and then focus specifically on AI systems for space operations, which entail questions about how these interact with existing legal concepts and technical standards. We address the role of AI for space sustainability, especially in relation to how it can support the implementation of the 2019 UNCOPUOS Guidelines for the Long-term Sustainability of Space Activities. We also discuss how space law is relevant and applicable to AI use in the context of sustainable space operations, including autonomous action to avoid collisions in orbit and space traffic management to limit orbital congestion.

8.1   ARTIFICIAL INTELLIGENCE AND SUSTAINABILITY: AN OUTER SPACE PERSPECTIVE

Artificial intelligence (AI) represents a crucial component for the conduct of space missions. Research and innovation in this field have made great strides in recent years, giving us the ability to solve problems faster than traditional computing could ever enable. Advances in AI have allowed us to make progress in all kinds of disciplines – and these are not limited to applications on Earth.

One of the most commonly cited benefits of AI is automation, which has significant impacts across multiple industries. Automation is used to improve productivity and increase production, allowing for more efficient use of materials and reduced lead times.[1] AI in automation can be a tool for optimizing

---

[1]McKinsey & Company, *AI, Automation, and the Future of Work*, McKinsey Global Institute, June 2018: `https://www.mckinsey.com/featured-insights/future-of-work/ai-automation-and-the-future-of-work-ten-things-to-solve-for`
(last accessed 15 September 2022); IBM, *Intelligent Automation*, 5 March 2021: `https://www.ibm.com/cloud/learn/intelligent-automation` (last accessed 15

resources to ensure availability for more critical or complex tasks. We benefit from AI and machine learning (ML) by using these technologies to reduce human error, handle repetitive jobs, and maintain business continuity. Similarly, AI and machine learning technologies positively impact our support of complex missions that define the future of exploration and safety of operations in outer space.

An 'AI system' refers to a system that is either software-based or embedded in hardware devices, and which displays behaviour simulating intelligence by *inter alia* collecting and processing data, analysing and interpreting its environment, and by taking action, with some degree of autonomy, to achieve specific goals.[2] In other words, AI operates through machines that are programmed to think and act like humans.

Current space exploration missions use AI and machine learning capabilities in many areas of space operations[3], including mission planning and operations, data collection, autonomous navigation and manoeuvring, and spacecraft maintenance, thus contributing to the sustainability of space activities.

The positive impacts on space missions arising from the use of AI are also illustrated by its contribution to the sustainability of space activities, particularly relating to the mitigation of space debris[4], as well as the prevention of on-orbit break-ups, avoidance of on-orbit collisions with other space objects [1, p. 21], and the sharing of information on space objects and orbital events. In these respects, AI will be helpful to reduce the risk of conjunctions in space. For some satellites in LEO, for example, hundreds of alerts might

---

September 2022); Simplilearn, *How AI and Automation Are Changing the Nature of Work*, 14 September 2022: https://www.simplilearn.com/how-ai-and-automation-are-changing-the-nature-of-work-article (last accessed 15 September 2022).

[2] European Parliament, Report-A9-00012021, Report on artificial intelligence: questions of interpretation and application of international law insofar as the EU is affected in the areas of civil and military uses and of state authority outside the scope of criminal justice, P9_TA(2021)0009, 20 January 2021, para. 1: https://www.europarl.europa.eu/doceo/document/TA-9-2021-0009_EN.html (last accessed 9 August 2022); Definition of 'artificial intelligence', Cambridge Dictionary: https://dictionary.cambridge.org/fr/dictionnaire/anglais/artificial-intelligence (last accessed 9 August 2022).

[3] SpaceNews, *The Use of AI in Space Systems: Opportunities for Mission Improvement*, 26 March 2021: https://spacenews.com/op-ed-the-use-of-ai-in-space-systems-opportunities-for-mission-improvement/ (last accessed 15 September 2022).

[4] See the Inter-Agency Space Debris Coordination Committee (IADC) Space Debris Mitigation Guidelines adopted in 2002 (updated in June 2021) and the Space Debris Mitigation Guidelines of the Committee on the Peaceful Uses of Outer Space adopted in 2010.

be distributed in the form of 'conjunction data messages'[5]. For instance, ESA needs to perform on average more than one collision avoidance manoeuvre per satellite per year, the vast majority due to risks associated with space debris.[6]

Artificial Intelligence emerges as a valuable technological tool for the advancement and realization of a range of space activities [2, pp. 235–253]. Ambitious plans and missions with AI components in the context of space activities are being conceived by space agencies, governments, IGOs and industry around the globe. Indeed, the nature of the space and the satellite industry presents a compelling case for the use of AI[7]. Everything about the industry increasingly requires machine intelligence and assistance to help facilitate functions involving the launch, operation, maintenance, control, repair and overall success of sophisticated commercial (or other) missions.

Various aspects of AI potentially seem well suited to a number of 'traditional' space applications involving communications and data analysis, in particular considering the large amount of information that it is possible to collect through space technology, as well as data transfer[8]. Artificial intelligence consequently appears to be a powerful and essential tool in the interpretation of satellite data, potentially mixed with other digital sources[9]. In addition to the positive impact that AI represents for 'traditional' space activities, it could also constitute a significant component for the next generation of space programmes and applications, such as on-orbit servicing

---

[5]See European Space Agency, *Reentry and Collision Avoidance*: https://www.esa.int/ Safety_Security/Space_Debris/Reentry_and_collision_avoidance (last accessed 20 April 2020): "Benefiting from a data-sharing agreement with US Strategic Command, ESA uses Conjunction Data Messages provided by the US Joint Space Operations Center (JSpOC) together with ESA's own orbit data, to analyse all close approaches ('potential conjunctions') of a given satellite ('target') with any of the catalogued objects"

[6]European Space Agency, *AI Challenged to Stave Off Collisions in Space*: https://www.esa.int/Enabling_Support/Space_Engineering_Technology/ AI_challenged_to_stave_off_collisions_in_space (last accessed 9 August 2022).

[7]Hogan Lovells, *Artificial Intelligence and your space business: A guide for smart navigation of the challenges ahead*, February 2018, 4 ss: https://www.hoganlovells.com/~/ media/ai-article-space-09nl.pdf (last accessed 9 August 2022)

[8]*Idem.*

[9]see J.P. Darnis, X. Pasco, P. Wohrer, *Space and the Future of Europe as a Global Actor: EO as a Key Security Aspect*, IAI-FRS, February 2020, 19 ss [3]; Emerj, *AI Applications for Satellite Imagery and Satellite Data*, 17 May 2019: https://emerj.com/ai-sector-overviews/ai-applications-for-satellite-imagery-and-data/ (last accessed 19 September 2022).

[4, pp. 196–222], space traffic management intended to avoid collision in orbit[10], removing space debris[11], and deep space exploration [6].

Artificial Intelligence could additionally play a crucial role to enhance space situational awareness[12] and guide decision-making in the support for, and design of spacecraft systems. It can positively change the way satellites are able to interact with other space-related assets. In the near future, AI technology will enable fully autonomous satellite operations, enabling constellations of hundreds or thousands of satellites to operate in a completely autonomous way.[13] These satellites stay in continual communication with the swarm and modify their configuration to complete the mission even if a particular satellite(s) is lost. Clearly, it will be necessary to implement appropriate policy, law and regulation that adequately address this major development.

Additional complexities will arise with respect to new satellites engaged in innovative operations, such as mission extension vehicles that require continued relocation amidst a field of other satellites. Moreover, through the use of sensors, satellites will be able to detect obstacles and react in time to avoid an impact, which will potentially enhance the accuracy and effectiveness of space situational awareness [7, pp. 1–11] and space traffic management manoeuvres.

In sum, there is potentially a promising future for the utilisation of intelligent space-related objects. They could be used in rovers and probes for evaluating real-time data, as on-orbit servicing vehicles, for the monitoring of environmental or humanitarian disasters through imagery analysis, for strengthening planetary protection, and for supporting navigation and space debris remediation. Furthermore, space exploration, resource extraction and future space activities could be carried out by avatar-robots that are controlled by humans on Earth or in orbit around the Moon or perhaps even

---

[10]European Space Agency website, *Automating Collision Avoidance*, October 2019: `https://www.esa.int/Safety_Security/Space_Debris/Automating_collision_avoidance` (last accessed 9 August 2022); see also S. Hobe, *Space Law*, Nomos, 2019, 216 ss. [5].

[11]The Lead, *Getting Junk Out Space*, 18 March 2020: `http://theleadsouthaustralia.com.au/industries/space/getting-junk-out-of-space/` (last accessed 9 August 2022) "… machine learning to tackle the threat of space junk wrecking new satellites". "… machine learning to tackle the threat of space junk wrecking new satellites".

[12]M. Dabbah, *AI Offers New Approach to Space Situational Awareness*, ROOM Space Journal of Asgardia, 2(16), 2018: `https://room.eu.com/article/ai-offers-new-approach-to-space-situational-awareness` (last accessed 9 August 2022).

[13]DeepAI. Space, *How Swarms of Small Satellites Could Revolutionize Space Exploration*, 19 February, 2020: `https://deepai.space/how-swarms-of-small-satellites-could-revolutionize-space-exploration/` (last accessed 9 August 2022).

Mars.[14] Space activities in the future will represent a complex human-machine partnership very much supported and conducted by the utilisation of AI.

While the use of AI technology supports the realisation of space missions and has beneficial impacts for the sustainability of certain space programmes, it also poses challenges for the legal framework. The use of AI in space missions is subject to international law and also to the relevant governance principles that currently apply to outer space activities[15], including:

(i) The freedom to engage in peaceful exploration and use of outer space;

(ii) That space activities must be carried out in accordance with international law;

(iii) The principle of cooperation and mutual assistance, and due regard to the corresponding interests of other States in the conduct of space activities;

(iv) That States are internationally responsible for national activities in outer space;

(v) That States are to authorize and continuously supervise activities conducted by national non-governmental entities; and

(vi) That a launching State(s) is internationally liable to other States or their nationals for damage caused by a space object of the launching State(s).

This chapter briefly considers four key elements of the Long Terms Sustainability guidelines adopted by the Scientific and Technical Subcommittee (STSC) of the United Nations Committee on the Peaceful Uses of Outer Space (UNCOPUOS) in June 2019.[16] The guidelines deal with the following topics: (i) the policy and regulatory framework for space activities; (ii) the safety of space operations; (iii) international cooperation, capacity-building and awareness; and (iv) scientific and technical research and development.

At its sixty-second session in 2019, the Committee decided to establish a new Working Group on the topic under the STSC.[17] The Working Group aims to identify and study challenges, consider possible new guidelines for the long-term sustainability of space activities, share experiences, practices and lessons learned from voluntary national implementation of the already adopted guidelines, and raise awareness in particular among emerging space nations.[18]

---

[14]J.B. Sheldon, *Humanity's Virtual Presence in Space in the Quantum Age*, SpaceWatchGL Op-Ed: https://spacewatch.global/2020/03/spacewatchgl-op-ed-humanitys-virtual-presence-in-space-in-the-quantum-age/ (last accessed 9 August 2022).

[15]see A.S. Martin, S. Freeland, *Artificial Intelligence – A Challenging Realm for Regulating Space Activities*, XLV Annals of Air and Space Law, 2020 [8, pp. 298–301]; see also R.S. Jakhu, *Rule of Law Vital for Humanity's Sustainability and Survival*, ROOM Space Journal of Asgardia, Winter 2019/20#4(22) [9, pp. 14–19].

[16]UN Doc. A/74/20 (21 June 2019), at 163 and Annex II.

[17]*Ibid*, at 165.

[18]UN Doc. A/AC.105/C.1/2022/CRP.13 (7 February 2022), at 6.

8.2   LEGAL ISSUES ON THE USE OF ARTIFICIAL INTELLIGENCE

8.2.1   Opportunities and Challenges under International Space Law

Artificial Intelligence can impact international legal paradigms both directly and indirectly. It can lead to the establishment of 'possible' new legal entities or the enabling of new behaviour. Indirectly, AI can modify the values of States in their interactions with international law [10, pp. 1–29].

The utilisation of AI raises international law issues related to transparency, human control and responsibility and liability. The fact that technology contains ever more AI components requires a new approach to the applicable legal framework [11, pp. 74–79]. The legal 'status' of AI is still an open issue.[19] However, these systems must be developed and used in accordance with applicable international law [14].

First, it is important to verify and to figure out how AI technologies arrive at decisions. The principles of non-discrimination and proportionality need to be respected, and questions of causality, liability and responsibility, as well as transparency, accountability and plausibility, need to be clarified to determine whether, and to what extent, a State can act with the support of AI-based systems having a certain degree of autonomy, without breaching obligations stemming from international law. Here, the issue of the accountability of private actors under international law may need to be enhanced, given the decision-making hegemony and control of certain private entities over the development of these technologies,[20] an element particularly relevant to the conduct of space activities, given that States are internationally responsible for national space activities carried out by non-governmental entities.

It is necessary to test, evaluate, certify, monitor and verify the systems based on clear legal norms as well as on appropriate safety and security provisions, so as to ensure that, during the entire life cycle of AI-enabled systems, in particular during the phases of human-machine interaction, the systems and their effects do not go beyond the intended limits and that at all times they are used in a manner consistent with applicable international law.[21]

---

[19]see K. Ziemianin, *Civil legal personality of artificial intelligence. Future or utopia?*, Internet Policy Review, 10(2), 2021 (doi.org/10.14763/2021.2.1544) [12]; see also A. Atabekov, O. Yastrebov, *Legal Status of Artificial Intelligence Across Countries: Legislation on the Move*, European Research Studies Journal, Vol. XXI, Issue 4, 2018, pp. 773-782 [13]

[20]European Parliament, Report-A9-0001/2021, Report on artificial intelligence: questions of interpretation and application of international law insofar as the EU is affected in the areas of civil and military uses and of state authority outside the scope of criminal justice, *op.cit.*, para.72.

[21]*Ibid.*

Hence, human responsibility and control are of paramount consideration, given the unresolved issue of responsibility in cases of 'machine error', and the fact that accountability cannot seamlessly be transferred to a machine.[22] This 'accountability gap' raises issues around causality and compensation, thus necessitating a responsible and transparent chain of human command and control. In most cases, the programme-maker of an AI device will be held responsible, assuming he or she can be identified, but the shortfall of 'transparency' of decisions made by AI programmes is a critical issue [11].

There also remain problems related to the accuracy of the information disseminated and any 'biased' actions undertaken by an AI system.[23] These elements are also relevant for developing policy and legal frameworks for situations where AI components and autonomy are to be integrated into future space missions. It is important to promote responsible innovation and use of such technologies, which calls for the implementation of relevant risk assessments and appropriate mitigation measures.

The United Nations Space Treaties of the 1960s–1970s referred to below were not drafted in contemplation of some of these new technologies. There is now a need to 'connect' the legal issues of new technologies with these space treaties.

In this context, it is relevant to take into consideration the 1967 *Treaty on Principles Governing the Activities of States in the Exploration and Use of Outer Space, Including the Moon and Other Celestial Bodies* (Outer Space Treaty or OST)[24], the 1972 *Convention on International Liability for Damage Caused by Space Objects* (Liability Convention or LIAB)[25], and the 1975

---

[22]see F. Santoni de Sio, G. Mecacci, *Four Responsibility Gaps with Artificial Intelligence: Why they Matter and How to Address them*, Philosophy & Technology, 34, 2021 [15, pp. 1057–1084]; see also M. Coeckelbergh, *Artificial Intelligence, Responsibility Attribution, and a Relational Justification of Explainability*, Science and Engineering Ethics, 26, 2020 [16, pp. 2051–2068].

[23]Harvard Business Review, *What Do We Do About the Biases in AI?*, 25 October 2019: https://hbr.org/2019/10/what-do-we-do-about-the-biases-in-ai (last accessed 15 September 2022); McKinsey Global Institute, *Tackling Bias in Artificial Intelligence (and in Humans)*, 6 June 2019: https://www.mckinsey.com/featured-insights/artificial-intelligence/tackling-bias-in-artificial-intelligence-and-in-humans (last accessed 15 September 2022).

[24]Treaty on Principles Governing the Activities of States in the Exploration and Use of Outer Space, Including the Moon and Other Celestial Bodies (signed 27 January 1967, entered into force 10 October 1967), 610 U.N.T.S. 205.

[25]Convention on International Liability for Damage Caused by Space Objects (signed 29 March 1972, entered into force 1 September 1972), 961 U.N.T.S. 187.

Convention on Registration of Objects Launched into Outer Space (Registration Convention or REG).[26]

According to Article VI of the Outer Space Treaty, States are internationally responsible for national space activities. In relation to space activities conducted by non-governmental entities, including those that will in the future incorporate the use of AI, these must have been authorized (which typically requires the issuing of an appropriate license under a national regulatory framework) and under the continuous supervision by the appropriate State.

In addition, the relevant State, agencies, organizations or private companies will have the responsibility to maintain the security of a spacecraft with AI components, and to ensure in particular that its automation, navigation and communications systems are not being hacked or otherwise compromised.

Indeed, the deployment of AI underscores certain security and transparency concerns, in particular in the case of dual-use satellites (technology used for civil and military purposes), noting of course that outer space is increasingly a dual-use area of operation. Most likely, human intervention will be required beyond the "first" programming stage. Consequently, States remain responsible for national space activities also in the case of AI chip. In the potential case of a fully autonomous satellite, this raises the important question regarding the removal of human judgment from the scenario.

Responsibility for actions and decisions taken by artificial agents can be secured by resorting to strict or objective liability schemes, which do not require human fault, or by transferring human fault to the programmer or supervisor [17, p. 6]. Within that context, the United Nations Space Treaties deal with the issue of liability as outlined below.

Article VII of the OST provides that the launching State(s) is internationally liable for damage caused by its space object. The Liability Convention specifies two liability regimes: (i) absolute liability for damage caused by its space object on the surface of the Earth or to aircraft in flight (Article II); and (ii) liability for fault if the damage is caused in outer space (Article III). Article V(1) of the LIAB specifies that, if there are two or more launching States, they are jointly and severally liable for any damage caused.

In any case, the launching State(s) – as defined in the LIAB – will be liable for damage caused by a relevant space object. However, the complex issue to consider is whether, and how, the regime of liability, as foreseen in these Treaties, is suitable in the context of AI components in space programmes. On

---

[26]Convention on Registration of Objects Launched into Outer Space (signed 14 January 1975, entered into force 15 September 1976), 1023 U.N.T.S. 15.

this issue, it seems appropriate that the degree of automation should be taken into account in the implementation of any future legal framework, although this will undoubtedly not be completely accurate (or appropriate) in every case.

If something fails with a fully autonomous satellite, for example, the question arises as to who is liable, in the case of damage involving personal injury or property, or in case of non-compliance with national and/or international rules and regulations. Significant (potential) liability may arise in the event of a satellite failure, collision, destruction, or cyber-security incident, and the legal circumstance as to how to resolve any claims – and indeed, against whom the claims should be made – will need to be clarified.

8.2.2 National Policy and Regulatory Frameworks linking to AI and Sustainability

A first element considered by the LTS Guidelines is the development of policy and regulatory frameworks at the national level. States should consider the adoption, revision and amendment (as appropriate) of national regulatory frameworks for outer space activities (Guideline A.1). They should consider elements of sustainability (A.2), as well as the supervision of their national space activities (A.3), and the equitable, rational and efficient use of the radio frequency spectrum and the various orbital regions used by the relevant satellite (A.4).

Over recent years, a general trend can be observed towards the implementation of sustainability elements within national space legislation in the authorization procedure as well as in the framework concerning the obligation to monitor and control over space objects.[27] As part of this trend, States are

---

[27] Australia – Space (Launches and Returns) Act 2018, Article 34, 46G, 46L/ Space (Launches and Returns) (General) Rules 2019, Articles 47(a), 50 (g, j, i-vi), 53, 54 (debris mitigation strategy) and environment plan (Article 91); Austria – Austrian Outer Space Act, 6 December 2011, Sections 4 and 5/ Outer Space Regulation No. 36/2015, Sections 4, 5, 6; Belgium – Law of 17 September 2005 on the Activities of Launching, Flight Operation or Guidance of Space Objects, consolidated text as revised by the Law of 1 December 2013, Articles 8, par. 2/ 14, par. 2.8, 16/Royal Decree 19 March 2008, Article 7; Denmark – Outer Space Act no. 409 of 11 May 2016, Part 3, para. 6 (1,4,5), Part 6, para.8/Executive Order no. 552 of 2016, Part. 4 ; Finland – Act on Space Activities (2018), Chapter 2 - Section 5, Section 10 + Decree 74/2018, Section 3; France – Loi sur les opérations spatiales n°2008-518 du 3 juin 2008 Article 5/Arrêté du 31 mars 2011 relatif à la réglementation technique en application du décret n° 2009-643 du 9 juin 2009 relatif aux autorisations délivrées en application de la loi n° 2008-518 du 3 juin 2008 relative aux opérations spatiales Articles 21, 22, 26, 43, 46, 49 modifié par Arrêté du 23 février 2022 relatif à la composition des trois parties du dossier mentionné à l'article 1er du décret n° 2009-643 du 9 juin 2009 relatif aux

incorporating provisions on sustainability (for example, mitigation of space debris, protection of the environment) in their national legislation.

However, there is currently no existing national space law specifically regulating the use of artificial intelligence, although some policy instruments [8], such as the *2018 Montreal Declaration for a Responsible Development of Artificial Intelligence* and the *2018 Toronto Declaration*,[28] specify ethical

---

autorisations délivrées en application de la loi n° 2008-518 du 3 juin 2008 modifiée relative aux opérations spatiales Article 8/Arrêté du 23 février 2022 modifiant l'arrêté du 31 mars 2011 relatif à la réglementation technique en application du décret n°2009-643 du 9 juin 2009 relatif aux autorisations délivrées en application de loi n°2008-518 du 3 juin 2008 relative aux opérations spatiales; Greece – Law 4508/2017 on "Authorization of space activities -Registration in the National Register of Space Objects – Establishment of a Greek Space Organization and other provisions", Article 4, 5, 6, 12; Japan – Act on Launching of Spacecraft, etc. and Control of Spacecraft (Act No. 76 of 16 November 2016), Articles 22, 23 para. 2, 26; Luxembourg – Loi du 15 décembre 2020 portant sur les activités spatiales, Articles 9 (2), 15; Netherlands, Rules concerning Space Activities, 2007, Section 10; Nigeria – 2015 Regulations on Licensing and Supervision of Space Activities, Part I Section 9 and Section 10; Portugal – Decree-Law no. 16/2019 of 22 January, Legal regime of access to and exercise of space activities, Article 7 (1) (C), Article 7(3), Article 9 (2) (a) and (d)/Regulation no. 697/2019 of 5 September, Regulation on access to and exercise of space activities, Article 12-20, 39-42; Saudi Arabia – Federal Law No. 12/2019 of 19 December 2019, Regulation of the Space Sector, Articles 14, 17,19, 31, 36; United Kingdom – Outer Space Act 1986, Articles 5(2)(b), (e), i-ii/ Space Industry Act 2018, Articles 2, 26 para. 13; United States of America – 47 CFR§25.11, 47CFR§25.283, 14 CFR§415.203.

[28] The Toronto Declaration: Protecting the Right to Equality and Non-Discrimination in Machine Learning Systems: https://www.accessnow.org/cms/assets/uploads/ 2018/08/The-Toronto-Declaration_ENG_08-2018.pdf (last accessed 9 August 2022). It calls on governments and companies to ensure that machine learning applications respect the principles of equality and non-discrimination. It articulates those human rights norms that the public and private sector should meet so as to ensure that algorithms are applied equally and fairly; see Human Rights Watch, The Toronto Declaration, July 3, 2018: https://www.hrw.org/news/2018/07/03/toronto-declaration-protecting-rights-equality-and-non-discrimination-machine (last accessed 9 August 2022).

guidelines.[29] States such as France[30], Germany[31], Finland[32], Australia[33] and others are preparing national AI strategies and policies in this area, albeit at different stages of the development cycle.

Similarly, international intergovernmental organisations (IGOs) such as the European Space Agency (ESA)[34], the Organization for Economic

---

[29]The Montreal Declaration for the Responsible Development of Artificial Intelligence: https://www.canasean.com/the-montreal-declaration-for-the-responsible-development-of-artificial-intelligence-launched/ (last accessed 9 August 2022). The key objective of the Montreal Declaration is to identify the ethical principles and values applicable to the fields of digital technology and AI that promote the fundamental interests of people and groups. These include well-being, respect for autonomy, protection of privacy and intimacy, solidarity, democratic participation, equity, diversity inclusion, prudence, responsibility and sustainable development.

[30]C. Villani et al., *For A Meaningful Artificial Intelligence, Towards a French and European Strategy*, French Parliamentary Mission, March 2018: https://www.aiforhumanity.fr/pdfs/MissionVillani_Report_ENG-VF.pdf (last accessed 9 August 2022) [18]; see also Stratégie nationale pour l'intelligence artificielle (SNIA): https://www.intelligence-artificielle.gouv.fr/fr (last accessed 14 September 2022). La Stratégie, lancée en 2018, poursuit trois objectifs majeurs: renforcer l'attractivité des talents et des investissements, diffuser l'IA et la donnée dans l'économie et promouvoir un modèle éthique de l'IA.

[31]Federal Ministry of Transport and Digital Infrastructure, *Ethics Commission – Automated and Connected Driving*, June 2017: https://www.bmvi.de/SharedDocs/EN/publications/report-ethics-commission-automated-and-connected-driving.pdf?__blob=publicationFile (last accessed 9 August 2022).

[32]Ministry of Economic Affairs and Employment of Finland, *Work in the age of artificial intelligence: Four Perspectives on the Economy*, Employment, Skills and Ethics, 2018: http://julkaisut.valtioneuvosto.fi/bitstream/handle/10024/160980/TEMjul_21_2018_Work_in_the_age.pdf (last accessed 9 August 2022).

[33]Australian Government, Department of Industry, Science, Energy and Resources, *AI Ethics Framework*: https://www.industry.gov.au/data-and-publications/building-australias-artificial-intelligence-capability/ai-ethics-framework/ai-ethics-principles (last accessed 9 August 2022).

[34]European Space Agency, *Towards a European Artificial Intelligence for Earth Observation (AI4EO) R&I Agenda*, September 2018: https://eo4society.esa.int/wp-content/uploads/2018/09/ai4eo_v1.0.pdf (last accessed 9 August 2022).

Cooperation and Development (OECD)[35], and the United Nations Educational, Scientific and Cultural Organization (UNESCO)[36], are each developing frameworks without significant coordination with other IGOs.

The implementation of the principles that constitute such policy frameworks will most likely be significant in guiding the applicable legal rules that will apply to the use of AI, including as part of a space activity. From a legal point of view, AI not only operates within the domain of national law, but also involves the realm of regional and international law.

The European Union (EU) has also played an active role in establishing AI policy principles, including its 2019 Guidelines on Ethics in Artificial Intelligence[37]. In 2022, the European Commission published an overarching regulatory framework proposal titled "the Artificial Intelligence Act"[38]. The proposal focuses on the risks created by AI, with applications sorted into categories of minimal risk, limited risk, high risk, or unacceptable risk.

---

[35]In May 2019, the OECD adopted its Principles on Artificial Intelligence, the first international standards agreed by governments for the responsible stewardship of trustworthy AI: `https://www.oecd.org/going-digital/ai/principles/` (last accessed 9 August 2022). These Principles include concrete recommendations for public policy and strategy. The general scope of the Principles ensures they can be applied to AI developments around the world. Following on from this, in February 2020, the OECD launched the AI Policy Observatory with the aim of helping policymakers implement the AI Principles; see also OECD moves forward on developing guidelines for artificial intelligence (AI): `https://www.oecd.org/going-digital/ai/oecd-moves-forward-on-developing-guidelines-for-artificial-intelligence.htm` (last accessed 9 August 2022).

[36]UNESCO Doc, *Preliminary Study on the Technical and Legal Aspects relating to the Desirability of a Standard-Setting Instrument on the Ethics of Artificial Intelligence* (206 EX/42), Paris, 21 March 2019.

[37]EU Guidelines on Ethics in Artificial Intelligence: Context and Implementation, September 2019 (EPRS_BRI(2019)640163). The key elements of these guidelines are: Human agency and oversight – users should be able to understand and interact with AI systems to a satisfactory degree and there should always be human oversight; Artificial Intelligence algorithms should be secure, reliable and sufficiently robust to deal with errors or inconsistencies during all life-cycle phases of an AI system; Citizens should have full control over their own data and AI systems should be designed to guarantee privacy and data protection; Artificial Intelligence systems should be documented and traceable such that the principle of transparency is paramount, so as to ensure that AI is not unacceptably skewed; Mechanisms should be put in place to ensure responsibility and liability for AI systems; `https://www.europarl.europa.eu/RegData/etudes/BRIE/2019/640163/EPRS_BRI(2019)640163_EN.pdf` (last accessed 9 August 2022).

[38]Proposal for a Regulation of the European Parliament and of the Council laying down harmonised rules on artificial intelligence and amending certain Union legislative Acts, COM(2021 206final, 2021/0106 (COD), 21 April 2021; The European Union Artificial Intelligence Act (AI Act) is a proposed European law on artificial intelligence: `https://artificialintelligenceact.eu/` (last accessed 15 September 2022).

Depending on an application's designated risk level, there will be corresponding government action or obligations. So far, the proposed obligations focus on enhancing the security, transparency, and accountability of AI applications through human oversight and ongoing monitoring. Specifically, companies will be required to register stand-alone high-risk AI systems, such as remote biometric identification systems, into an EU database.

If the proposal is approved, the earliest date for compliance would be the second half of 2024[39]. In addition, the prior adopted EU General Data Protection Regulation (GDPR)[40] already has implications for AI technology. Indeed, Article 22 of the GDPR prohibits decisions based solely on automated processes that produce legal consequences or similar effects for individuals, unless the programme obtains users' explicit consent or meets other specific conditions.

In the United States, there has been a fragmented approach to AI regulation thus far, with various states enacting their own AI laws.[41] A number of the regulations adopted concern the establishment of commissions to determine the manner in which state agencies may use AI technology and to study the potential impacts of AI on the workforce and consumers. Some ongoing initiatives go beyond this and address the accountability and transparency of an AI system when it processes and makes decisions based on consumer data.

In February 2020, the Department of Defense adopted Ethical Principles for Artificial Intelligence[42]. In January 2021, U.S. Congress enacted the National AI Initiative Act, creating the National AI Initiative that provides "an

---

[39]The National Law Review, *AI Regulation: Where do China, the EU and the US Stand Today?*, Vol. XII, 257, 14 September 2022: https://www.natlawreview.com/article/ai-regulation-where-do-china-eu-and-us-stand-today (last accessed 14 September 2022).

[40]Regulation (EU) 2016/679 of the European Parliament and of the Council of 27 April 2016 on the protection of natural persons with regard to the processing of personal data and on the free movement of such data: https://eur-lex.europa.eu/legal-content/EN/TXT/PDF/?uri=CELEX:02016R0679-20160504&from=EN (last accessed 14 September 2022).

[41]National Conference of State Legislatures, Legislation Related to Artificial Intelligence: https://www.ncsl.org/research/telecommunications-and-information-technology/2020-legislation-related-to-artificial-intelligence.aspx (last accessed 14 September 2022); The National Law Review, *AI Regulation: Where do China, the EU and the US Stand Today?*, op.cit.

[42]US Department of Defense, *DoD Adopts Ethical Principles for Artificial Intelligence*, 24 February 2020: https://www.defense.gov/Newsroom/Releases/Release/Article/2091996/dod-adopts-ethical-principles-for-artificial-intelligence/ (last accessed 9 August 2022); see also Congressional Research Service, *Artificial Intelligence and National Security*, November 2019, 42: https://fas.org/sgp/crs/natsec/R45178.pdf (last accessed 9 August 2022).

overarching framework to strengthen and coordinate AI research, development, demonstration, and education activities across all U.S. Departments and Agencies".[43]

In 2022, the Congress adopted the Algorithmic Accountability Act [44]. In responding to reports that AI systems can lead to biased and discriminatory results, the legislation requires the Federal Trade Commission (FTC) to develop regulations that require "covered entities", including companies meeting certain criteria, to conduct impact assessments when using automated decision-making procedures [45]. This would specifically include those derived from AI or machine learning.[46]

Policy instruments on AI can be read as a first step toward the adoption of a national regulatory framework. There appears to be the general will to regulate the use of AI elements given the legal issues that the technology raises. The next step is to introduce regulations that address sustainability and AI elements in national space legislation. The relevant elements for consideration include regulating the use of AI by taking into account space activities in terms of transparency, control, data protection, reliable operation, specific insurance for AI components, liability (foreseen unintended consequences), and the State's level of responsibility (and that of the developer).

Another element considered in the LTS guidelines is the registration of space objects (Guideline A.5 "Enhance the practice of registering space objects"). In particular, it is interesting to examine the registration of space objects with AI component [17, pp. 6–7].

According to Article VIII of the OST, the State of registry "shall retain jurisdiction and control over [its] space object". This applies equally to space objects that incorporate AI components, and is particularly relevant in the

---

[43]Divison E – National Artificial Intelligence Initiative Act of 2020: `https://www.congress.gov/116/crpt/hrpt617/CRPT-116hrpt617.pdf#page=1210` (last accessed 14 Sptember 2022). The mission of the National AI Initiative is to ensure continued US leadership in AI research and development, lead the world in the development and use of trustworthy AI in the public and private sectors, and prepare the present and future US workforce for the integration of AI systems across all sectors of the economy and society.

[44]H.R. 6580 – Algorithmic Accountability Act of 2022: `https://www.congress.gov/117/bills/hr6580/BILLS-117hr6580ih.pdf` (last accessed 11 April 2023). The Act addresses growing public concerns about the widespread use of automated decision systems (ADS). It requires companies to assess the impacts of the automated systems they use and sell.

[45]H.R. 6580 – Algorithmic Accountability Act of 2022; section 3

[46]The National Law Review, *AI Regulation: Where do China, the EU and the US Stand Today?*, Vol.XII, 257, 14 September 2022: `https://www.natlawreview.com/article/ai-regulation-where-do-china-eu-and-us-stand-today` (last accessed 14 September 2022).

case of unpredictable actions stemming from the space object. The function and purpose of a register, and the notion of jurisdiction and control within the space-related paradigm, must also be examined in the context of the 1975 Convention on Registration of Objects Launched into Outer Space (Registration Convention or REG). In particular, Article II(1) of that instrument states that, "[w]hen a space object is launched into Earth orbit or beyond, the launching State shall register the space object by means of an entry in an appropriate registry which it shall maintain (...)". Paragraph 2 provides that, "[w]here there are two or more launching States in respect of any such space object, they shall jointly determine which one of them shall register the object (...)". As a result, a launching State will be required to register a space object with AI components.

The precise application of concepts such as the State of registry and the Launching State(s) will be relevant in the development of a workable and consistent legal framework for AI use in space activities, although further research may conclude that it might be appropriate to alter their existing scope depending on the precise circumstances. Furthermore, according to Article IV (1) of the REG: "Each State of registry shall furnish to the Secretary-General of the United Nations, as soon as practicable, the following information concerning each space object carried on its registry: (a) Name of launching State or States; (b) An appropriate designator of the space object or its registration number; (c) Date and territory or location of launch; (d) Basic orbital parameters [...]; (e) General function of the space object."

Notably, the additional possibility to provide such details to the United Nations Secretary-General (in practice, this will be to UNOOSA) over and above those details set out in Article IV (1) of the REG, may already exist by way of a voluntary notification pursuant to Article IV (2), which specifies that "[e]ach State of registry may provide to the Secretary-General of the United Nations with additional information concerning a space object carried on its registry".

All of these elements are relevant in determining the characteristics of a space object with AI components in order to strengthen transparency and trust in the relevant space activities. One other possibility, for example, might be the development of a 'special' registry,[47] at the national level, that specifies the unique features of spacecraft with on-board AI elements, as envisaged in the EU proposal.

The use of AI in future space missions therefore challenges the current legal and policy frameworks in many ways. If AI can be used to support the

---

[47]*Idem.* [17]

sustainability of space activities, it can also be used as a tool for interference, hacking or satellite destruction. Transparency in terms of AI utilization is of utmost importance. Even if artificial agents produce useful and reliable results, it must be openly explainable as to how these results are generated.

As noted above, some policy frameworks and legislation for the use of AI in space missions are being developed by States and organisations. These instruments might provide further guidance as to the future direction of the relevant law pertaining to the use AI and can take many forms.

8.3 LEGAL CONSIDERATIONS FOR THE SAFETY OF SPACE OPERATIONS

Another element of the LTS Guidelines is the safety of space operations. Artificial Intelligence plays a crucial role to enhance the safety of space operations, in particular in the context of space traffic management [19].

In addition, the Outer Space Treaty contains relevant elements related to the safety of space operations which can support the implementation of the LTS guidelines. For example, Article IX of the OST is particularly relevant, providing that "States [...] shall be guided by the principle of cooperation and mutual assistance and shall conduct all their activities in outer space [...] with due regard to the corresponding interests of all other States [...]. States [...] shall pursue studies of outer space [...] and conduct exploration of them so as to avoid their harmful contamination [...]. If a State [...] has reason to believe that an activity or experiment planned by it or its nationals in outer space [...] would cause potentially harmful interference with activities of other States [...], it shall undertake appropriate international consultations before proceeding with any such activity or experiment."

Article XI of the OST provides that States "agree to inform the Secretary-General of the United Nations as well as the public and the international scientific community, to the greatest extent feasible and practicable, of the nature, conduct, locations and results" of their space activities. The sharing of such information is necessary to avoid collision in orbit and for space traffic management purposes.

Artificial Intelligence is useful in mission planning, mission operations, data collection, autonomous navigation and manoeuvring, as well as on-orbit spacecraft maintenance.[48] In mission planning, the use of AI allows the

---

[48]Artificial Intelligence and the Future of Space Exploration, 1 April 2021: https://ai-solutions.com/newsroom/about-us/news-multimedia/artificial-intelligence-and-the-future-of-space-exploration/ (last accessed 14 September 2022).

spacecraft to be programmed in such a way that it can determine for itself how to "intelligently" perform a command for a specific function based on its past experiences and environment.

Artificial Intelligence is used for autonomous operations for space missions.[49] The use of AI and machine learning provides for safety-critical missions assessment and operational risk analysis evaluation. These technologies enable risk mitigation systems to process large amounts of data based on indicators from nominal actions and previous situations where anomalies have occurred.

Over the past decade, the volume of data collected from Earth-observing spacecraft, deep space probes, and planetary rovers has increased significantly, making the functioning of ground infrastructures used to distribute and deliver information to multiple end users more complex. The ability to optimize the large amount of data collected from scientific missions and to analyze this data using AI automation has a positive impacts on how data is processed and transmitted to ground station and end users[50]. The use of AI in data collection allows for a determination of which data sets are important to send to the relevant ground segments for processing.

In addition, ground operation communication capabilities do not always allow for continuous human monitoring and operations of the spacecraft. Artificial Intelligence contributes to the implementation of LTS Guidelines; B.2 – "Improve accuracy of orbital data on space objects and enhance the practice and utility of sharing orbital information on space objects"; B.3 – "Promote the collection, sharing and dissemination of space debris monitoring

---

[49]European Space Agency, AIKO: *Artificial Intelligence for Autonomous Space Missions*, 14 September 2018: `https://www.esa.int/Applications/Technology_Transfer/AIKO_Artificial_Intelligence_for_Autonomous_Space_Missions` (last accessed 15 September 2022).

[50]The Conversation, Five ways artificial intelligence Can Help Space Exploration, 25 January 2021: `https://theconversation.com/five-ways-artificial-intelligence-can-help-space-exploration-153664` (last accessed 14 September 2022). "For the sheer volume of data received, AI has been very effective in processing it smartly. It's been used to estimate 'heat storage in urban areas' (`https://www.sciencedirect.com/science/article/abs/pii/S0034425720304983`) and to combine 'meteorological data with satellite imagery' (`https://onlinelibrary.wiley.com/doi/abs/10.1002/er.6055`) for wind speed estimation. AI has also helped with 'solar radiation estimation' (`https://www.sciencedirect.com/science/article/abs/pii/S0038092X19303068`) using geostationary satellite data, among many other applications"; GeospatialWorld, *How Artificial Intelligence Is Advancing Space Efforts*, 10 August 2021: `https://www.geospatialworld.net/blogs/how-artificial-intelligence-is-advancing-space-efforts/` (last accessed 14 September 2022); see also N.H. Thanh Dang et al., *Artificial Intelligence in Data and Big Data Processing*, Springer, 2022. [20]

information" (States and international intergovernmental organizations are encouraged to develop and use relevant technologies for the measurement, monitoring and characterization of the orbital and physical properties of space debris"); B.4 – "Perform conjunction assessment during all orbital phases of controlled flight" (Conjunction assessment has to be performed for all spacecraft capable of adjusting trajectories during orbital phases of controlled flight for current and planned spacecraft trajectories.); and B.8 – "Design and operation of space objects regardless of their physical and operational characteristics" (States and international intergovernmental organizations are encouraged to promote design approaches that increase the traceability of space objects, regardless of their physical and operational characteristics, including small-size space objects, and those that are difficult to track throughout their orbital lifetime, as well as facilitate the accurate and precise determination of their position in orbit. Such design solutions could include the use of appropriate on-board technology).

Autonomy in navigation allows for spacecraft to travel through interstellar space and perform complex procedures necessary for arriving at and landing on distant celestial bodies [21, pp. 1–21]. Human intervention to control a spacecraft cannot occur in real-time from the ground due to the time lag/latency with respect to communications in deep space. Since spacecraft cannot wait for commands in given situations that require immediate determinations, using AI and machine learning facilitates the guiding and manoeuvring of a spacecraft while freeing up human resources for other mission activities.[51]

Moreover, the risk of collisions between human-made objects in space is growing. With the increase in space debris and the growing trend towards the launch of mega-constellations, the situation is expected to become far more challenging. Therefore, another area where machine learning is being tested is in space situational awareness and the prediction of conjunction events,[52] where two or more spacecrafts may arrive at the same point and time. Ideally, by identifying conjunction events requiring corrective actions, specific data on events and operator responses can be captured to feed into a machine learning algorithm in order to identify and minimize future events.

---

[51]AI Accelerator Institute, *AI in Space Exploration*: https://www.aiacceleratorinstitute.com/ai-in-space-exploration/ (last accessed 14 September 2022).

[52]Space, *Artificial Intelligence Is Learning How to Dodge Space Junk in Orbit*, 29 April 2021: https://www.space.com/AI-autonomous-space-debris-avoidance-esa (last accessed 14 September 2022).

Artificial Intelligence also contributes to the implementation of LTS Guidelines; B.1 – "Provide updated contact information and share information on space objects and orbital events" (States and international intergovernmental organizations have to establish appropriate means to enable timely coordination to reduce the probability of and/or to facilitate effective responses to orbital collisions, orbital break-ups and other events that might increase the probability of accidental collisions"; and B.9 – "Take measures to address risks associated with the uncontrolled re-entry of space objects".

Lastly, AI will be very helpful in the maintenance of a functioning mission. Maintenance mission technology driven by AI and machine learning technologies includes autonomous, real-time relative navigation systems, advanced robotic arms that precisely execute assignments, and powerful tools to perform servicing tasks.

8.4 FOSTERING INTERNATIONAL COOPERATION FOR SPACE SUSTAINABILITY AND ARTIFICIAL INTELLIGENCE

A third element for the sustainability of space activities is international cooperation. Indeed, international cooperation is an essential element for developing AI technology and ensuring a trusted environment,[53] especially in the space field where it represents a key aspect.[54] The space legal framework contains provisions related to the promotion of international cooperation. The OST's Preamble provides that: "desiring to contribute to broad international cooperation in the scientific as well as the legal aspects of the exploration and use of outer space for peaceful purposes" and "believing that such cooperation will contribute to the development of mutual understanding [...]".

Article I of the OST states that "[...] there shall be freedom of scientific investigation in outer space, including the Moon and other celestial bodies, and States shall facilitate and encourage international cooperation in such investigation".

---

[53] Strengthening International Cooperation on AI, Report, Brookings, October 2021: https://www.brookings.edu/wp-content/uploads/2021/10/Strengthening-International-Cooperation-AI_Oct21.pdf (last accessed 14 September 2022); E. Noor, *Artificial Intelligence: the Case for International Cooperation*, The Survival Editors' Blog, 2nd January 2020: https://www.iiss.org/blogs/survival-blog/2019/12/ai-international-cooperation (last accessed 14 September 2022).

[54] UNGA Res. 51/122, *Declaration on International Cooperation in the Exploration and Use of Outer Space for the Benefit and in the Interest of All States, Taking into Particular Account the Needs of Developing Countries* (13 December 1996).

Article III of the OST provides that "States Parties shall carry on activities in the exploration and use of outer space [...] in the interest of maintaining international peace and security and promoting international cooperation [...]".

Article X of the OST requires that "in order to promote international cooperation in the exploration and use of outer space [...] the States Parties to the Treaty shall consider on a basis of equality any requests by other States to the Treaty to be afforded an opportunity to observe the flight of space objects launched by those States".

Article XI of the OST highlights the fact that "in order to promote international cooperation in the peaceful exploration and use of outer space, States Parties conducting activities in outer space, agree to inform the Secretary-General of the UN as well as the public and the international scientific community, to the greatest extent feasible and practicable, of the nature, conduct, locations and results of such activities".

The space legal framework thus supports the implementation of various of the LTS Guidelines, in particular C.1 – "Promote and facilitate international cooperation in support of the long-term sustainability of outer space activities" and C.2 – "Share experience related to the long-term sustainability of outer space activities and develop new procedures, as appropriate, for information exchange". Furthermore, AI components are valuable tools allowing the observation of a space asset and operational control over it.

Cooperation is a core element in the international legal framework as well as of the LTS Guidelines. Cooperation involves the sharing of experience which is valuable also in the field of AI in order to enhance the sustainability of space activities. International cooperation also involves the public, private and academic sectors, and may include *inter alia* the exchange of scientific knowledge, technology and equipment for space activities on an equitable and mutually acceptable basis. These elements are important in the context of space sustainability and the use of AI.

Moreover, States and international intergovernmental organizations are invited to share, as mutually agreed, experiences, expertise and information relating to the long-term sustainability of outer space activities, including with non-governmental entities, and develop and adopt procedures to facilitate the compilation and effective dissemination of information on the ways and means of enhancing the long-term sustainability of space activities.

The experiences and expertise acquired by those engaged in space activities should be regarded as instrumental in the development of effective measures to enhance the long-term sustainability of outer space activities.

8.5 PROMOTING SCIENTIFIC RESEARCH IN SPACE MISSIONS WITH AI COMPONENTS

Another element of the LTS Guidelines deals with "scientific and technical research and development". The International space legal framework addresses the importance of scientific research in the conduct of space activities. Research is crucial in the field of AI[55] and sustainability.

Article I of the OST provides that: "the exploration and use of outer space, including the Moon and other celestial bodies, shall be carried out for the benefit and in the interests of all countries, irrespective of their degree of economic or scientific development, and shall be the province of all mankind. [...]There shall be freedom of scientific investigation in outer space, including the Moon and other celestial bodies, and States shall facilitate and encourage international cooperation in such investigation".

The Moon Agreement[56] deals also with scientific cooperation and released scientific results. Article 5 specifies that "States shall inform the Secretary-General of the United Nations as well as the public and the international scientific community, to the greatest extent feasible and practicable, of their activities concerned with the exploration and use of the Moon. [...] while information on the results of each mission, including scientific results, shall be furnish upon completion of the mission". Article 6 provides that "there shall be freedom of scientific investigation on the Moon... [...] States agree on the desirability of exchanging scientific and other personnel on expeditions to or installations on the Moon to the greatest extent feasible and practicable". Lastly, Article 7 para. 3 underscores the fact that "States shall report to other States Parties and to the Secretary-General concerning areas of the Moon having special scientific interest...".

The international space legal framework underpins scientific research and, according to LTS Guideline D.1, it is necessary to develop ways to support sustainable exploration and use of outer space. Artificial Intelligence has a central role here through appropriate space technologies, processes and services, and other initiatives for the sustainable exploration and use of outer space, including celestial bodies and, pursuant to Guideline D.2, "States and international intergovernmental organizations should investigate the necessity

---

[55]see Y. Xu et al., *Artificial Intelligence : A Powerful Paradigm for Scientific Research*, The Innovation, 2(4), 2021 [22]; see also The Royal Society, *The AI Revolution in Scientific Research*, 7 August 2019: https://royalsociety.org/-/media/policy/projects/ai-and-society/AI-revolution-in-science.pdf?la=en-GB&hash=5240F21B56364A00053538A0BC29FF5F (last accessed 23 September 2022).

[56]Agreement Governing the Activities of States on the Moon and Other Celestial Bodies (signed 18 December 1979, entered into force 11 July 1984), 1363 U.N.T.S. 3.

and feasibility of possible new measures, including technological solutions, and consider implementation thereof, in order to address the evolution of and manage the space debris population in the long term".

Investigation of new measures could include *inter alia* methods for the extension of operational lifetime, novel techniques to prevent collision with and among debris and objects with no means of changing their trajectory[57], advanced measures for spacecraft passivation and post-mission disposal and designs to enhance the disintegration of space systems during uncontrolled atmospheric re-entry. In all of these cases, AI has a role to play.

8.6 CONCLUDING REMARKS

Artificial Intelligence capabilities are making significant impacts in the space industry by creating efficiencies in mission planning and operations and providing scientists with the ability to further explore the far reaches of space. While automation of tasks paves the way for the use of AI, the ability for spacecraft to become fully cognitive machines, capable of making critical decisions based on their current environment, and without reliance on ground systems to perform essential functions, will create more time for humans to spend on other valuable and complex research activities.

Emerging technologies with AI components are becoming ever more present in our modern society within the realm of civil, commercial, scientific and military activities. The capacity of intelligent objects to give more autonomy to novel technology will likely offer an even more expansive future for space missions through 'intelligent' space assets, which can be deployed in various space activities. Although AI platforms will take time to mature, the technology is becoming a reality and could soon become standard operating practice in helping space actors, from companies to governmental agencies, to conduct their space activities.

However, one important issue (among many) is how much autonomy, if any, intelligent space objects should have, and what decisions necessitate ongoing human oversight, despite the crucial role machine intelligence will play in the functioning and operation of space projects.

Whilst the international space legal framework contains elements to support the development of AI technologies, that will benefit for the sustainability of space programmes, the use of artificial intelligence raises legal issues in terms related to *inter alia* responsibility, liability and transparency.

---

[57]European Space Agency (ESA), *Mitigation Space Debris Generation*: `https://www.esa.int/Space_Safety/Space_Debris/Mitigating_space_debris_generation` (last accessed 19 September 2022).

There is a pressing need to develop a more comprehensive and practical legal framework in the field of AI, complementing a sustainability framework, that will adequately address not only these legal issues, but other concerns related to ethics, decision making, accountability, societal norms and security-related issues. There is also a need of harmonization and standardization, in order to make the future possibilities for space missions clear and secure.

The growing reliance on autonomous technologies may require a new perspective on many traditional concepts regulating space activities. These technical developments may shape and transform the existing body of legal rules, regulations and practices that apply to space activities. The imperative is to progress these discussions quickly, but also in a way that delas with the myriad complexities that have and will no doubt arise.

ACRONYM

AI Artificial Intelligence

ESA European Space Agency

EU European Union

FTC Federal Trade Commission

GEO Geostationary Earth Orbit

GRDP General Data Protection Regulation

IADC Inter-Agency Space Debris Coordination Committee

IGO International intergovernmental organisations

LEO Low Earth Orbit Region

LTS Long-Term Sustainability

NASA National Aeronautics and Space Administration

OECD Organization for Economic Cooperation and Development

STSC Scientific and Technical Subcommittee

UNCOPUOS United Nations Committee on the Peaceful Uses of Outer Space

UNESCO United Nations Educational, Scientific and Cultural Organization

UNOOSA United Nations Office for Outer Space Affairs

GLOSSARY

Geostationary Earth Orbit (GEO): 'Earth orbit (…) whose orbital period is equal to the Earth's sidereal period. The altitude of this unique circular orbit is close to 35,786 km' (IADC Space Debris Mitigation Guidelines, section 3.3.2.).

IADC Guidelines: Inter-Agency Space Debris Coordination Committee Space Debris Mitigation Guidelines (2002).

Liability Convention: Convention on International Liability for Damage Caused by Space Objects (1972).

Long-Term Sustainability Guidelines: Guidelines for the Long-Term Sustainability of Outer Space Activities of the United Nations Committee on the Peaceful Uses of Outer Space (2019).

Low Earth Orbit Region (LEO): 'Spherical region that extends from the Earth's surface up to an altitude of 2,000 km' (IADC Space Debris Mitigation Guidelines, section 3.3.2.).

Montreal Declaration: Montreal Declaration for a Responsible Development of Artficial Intelligence (2018).

Moon Agreement: Agreement Governing the Activities of States on the Moon and Other Celestial Bodies (1979).

Outer Space Treaty: Treaty on Principles Governing the Activities of States in the Exploration and Use of Outer Space, including the Moon and Other Celestial Bodies (1967).

Registration Convention: Convention on Registration of Objects Launched into Outer Space (1975).

Rescue Agreement: Agreement on the Rescue of Astronauts, the Return of Astronauts and the Return of Objects Launched into Outer Space (1968).

Toronto Declaration: Toronto Declaration: Protecting the Rights to Equality and Non-Discrimination in Machine Learning Systems (2018)

UNCOPUOS (Mitigation) Guidelines: Space Debris Mitigation Guidelines of the United Nations Committee on the Peaceful Uses of Outer Space (2007).

FURTHER READING

Peter Martinez. (2023) "Implementing the Long-Term Sustainability Guidelines: What's Next?". *Air&Space Law*, 48: 41-58.

Custers, B. and Fosch-Villaronga, E. (2022) "Law and Artificial Intelligence: Regulating AI and Applying AI in Legal Practice." Berlin: *Springer*.

Masson-Zwaan, T. and Hofmann, M. (2019) "Introduction to Space Law." The Hague: *Wolters Kluwer*.

Lyall, F. and Larsen, P.B. (2018) "Space Law A Treatise." New York: *Routledge*.

Jakhu, R.S. and Dempsey, P.S. (2017) "Routledge Handbook of Space Law." New York: *Routledge*.

von der Dunk, F. and Tronchetti, F. (2015) "Handbook of Space Law." Cheltenham: *Edward Elgar Publishing*.

Viikari, L. (2008) "The Environmental Element in Space Law: Assessing the Present and Chartering the Future." Leiden: *Martinus Nijhoff*.

BIBLIOGRAPHY

[1] Carmen Pardini and Luciano Anselmo. Evaluating the impact of space activities in low earth orbit. *Acta Astronautica*, 184:11–22, 2021.

[2] Daniela Girimonte and Dario Izzo. *Artificial Intelligence for Space Applications*, pages 235–253. Springer London, London, 2007.

[3] Jean-Pierre Darnis, Xavier Pasco, and Paul Wohrer. Space and the future of europe as a global actor: Eo as a key security aspect, 2020.

[4] Anne-Sophie Martin and Steven Freeland. Exploring the legal challenges of future on-orbit servicing missions and proximity operations. *J. Space L.*, 43:196, 2019.

[5] Stephan Hobe. *Space law*. Nomos Verlag, 2019.

[6] George S Robinson and Rita M Lauria. Legal rights and accountability of cyberpresence: a void in space law/astrolaw jurisprudence. *Robinson, G. and Lauria, R., ANNALS OF AIR AND SPACE LAW*, 28:311–326, 2003.

[7] GD Kyriakopoulos, P Pazartzis, A Koskina, and C Bourcha. Artificial intelligence and space situational awareness: Data processing and sharing in debris-crowded areas. In *Proceedings of the 8th European Conference on Space Debris*, pages 20–23, 2021.

[8] Anne Sophie Martin and Steven R Freeland. Artificial intelligence– a challenging realm for regulating space activities. *Annals of Air and Space Law*, 45:275–306, 2020.

[9] RS Jakhu. Rule of law vital for humanity's sustainability and survival. *ROOM–Space Journal of Asgardia, Winter*, 20(4):22, 2019.

[10] Matthijs M Maas. International law does not compute: Artificial intelligence and the development, displacement or destruction of the global legal order. *Melbourne Journal of International Law*, 20(1):1–29, 2019.

[11] Steven Freeland and Anne-Sophie Martin. Ai in space: a legal perspective. *ROOM–Space Journal of Asgardia, Spring*, pages 74–79, 2021.

[12] Karolina Ziemianin. Civil legal personality of artificial intelligence: Future or utopia? *Internet Policy Review*, 10(2):1–22, 2021.

[13] Atabek Atabekov and Oleg Yastrebov. Legal status of artificial intelligence across countries: Legislation on the move. *European Research Studies*, 21(4):773–782, 2018.

[14] Chae-min Yi and Jaemin Lee. *Artificial Intelligence and International Law*. Springer Nature, 2022.

[15] Filippo Santoni de Sio and Giulio Mecacci. Four responsibility gaps with artificial intelligence: Why they matter and how to address them. *Philosophy & Technology*, 34:1057–1084, 2021.

[16] Mark Coeckelbergh. Artificial intelligence, responsibility attribution, and a relational justification of explainability. *Science and engineering ethics*, 26(4):2051–2068, 2020.

[17] Anne-Sophie Martin and Steven Freeland. The advent of artificial intelligence in space activities: New legal challenges. *Space Policy*, 55:101408, 2021.

[18] Cédric Villani, Yann Bonnet, Bertrand Rondepierre, et al. *For a meaningful artificial intelligence: Towards a French and European strategy*. Conseil national du numérique, 2018.

[19] Massimiliano Vasile, Víctor Rodríguez-Fernández, Romain Serra, David Camacho, and Annalisa Riccardi. Artificial intelligence in support to space traffic management. 2018.

[20] Ngoc Hoang Thanh Dang. *Artificial Intelligence in Data and Big Data Processing: Proceedings of ICABDE 2021*, volume 124. Springer Nature, 2022.

[21] Antonia Russo and Gianluca Lax. Using artificial intelligence for space challenges: A survey. *Applied Sciences*, 12(10), 2022.

[22] Yongjun Xu, Xin Liu, Xin Cao, Changping Huang, Enke Liu, Sen Qian, Xingchen Liu, Yanjun Wu, Fengliang Dong, Cheng-Wei Qiu, et al. Artificial intelligence: A powerful paradigm for scientific research. *The Innovation*, 2(4):100179, 2021.

V

Market Perspectives

CHAPTER 9

Future-ready space missions enabled by end-to-end AI adoption

Lorenzo Feruglio, Alessandro Benetton, Mattia Varile, Davide Vittori, Ilaria Bloise, Riccardo Maderna, Christian Cardenio, Paolo Madonia, Francesco Rossi, Federica Paganelli Azza, Pietro De Marchi, Luca Romanelli, Matteo Stoisa, Luca Manca, Gianluca Campagna, and Armando La Rocca

AIKO S.r.l., Torino, Italy

CONTENTS

9.1   The Space Domain Context 304
    9.1.1   The Space for Earth Market untapped potential; Earth
        Observation use case 304
    9.1.2   The Space for Space Market; autonomy as enabler for
        an economy of scale 308
9.2   The Artificial Intelligence Impact 310
    9.2.1   Edge Autonomy 311
    9.2.2   Continuous Intelligence 335
9.3   Overcome the Penetration Barrier 346
    9.3.1   Product Usability and User Experience 347
    9.3.2   AI-based Products Maintainability and Scalability 349
9.4   Conclusion 351
Bibliography 356

THE demand for space-based services, ranging from to established Earth Observation and Telecommunications verticals to the rising In-Orbit Servicing market, is accelerating, providing lucrative business cases for

DOI: 10.1201/9781003366386-9

industry players. As a consequence, not only is the number of satellites in Earth's orbits growing, but so is the complexity of the mission itself. This is made evident by the increasingly popular constellation architectures and multi-payload satellite configurations. In such a dynamic backdrop, the need for autonomous spacecraft that rapidly act upon unexpected events has never been greater. The traditional spacecraft management systems with multiple decision-making layers centered around human command are today technologically obsolete and inadequate. The satellite cannot remain a non-intelligent platform that executes pre-defined commands only, as such an approach leads to deep inefficiencies from operating the satellite to transmitting and processing the acquired data. Instead, the satellite must become an intelligent system that structures its tasks independently from the intervention of ground control to maximize the mission outcome. In this sense, using AI for the platform's self-assessment, self-healing, and enhanced manageability is becoming crucial to ensure mission scalability and profitability. Indeed, the next generation of space missions will see software as the differentiating asset, while the hardware will remain only a standardized commodity. Thanks to recent advancements in computational hardware, even small satellites can now be equipped with impressive computational capabilities. Edge computers can effortlessly multitask, managing multiple applications simultaneously and supporting a wide range of popular networking protocols. Therefore, space companies can transport relevant applications from their data centers and into the operational environment. In this chapter, we explore how AI-based technologies will support the next generation of space missions, from the design of the space asset to the delivery of the service to the end customer.

9.1 THE SPACE DOMAIN CONTEXT

The following section will provide an overview of the space ecosystem, highlighting major scalability issues, bottlenecks, and technological gaps to be addressed.

9.1.1 The Space for Earth Market untapped potential; Earth Observation use case

In 2019, roughly 95% of the estimated $366 billion in revenue earned in the space sector was from the space-for-earth economy: goods or services produced in space for use on Earth. Earth Observation (EO) is part of this economy.

Figure 9.1   Earth's surface as seen from LEO satellite for Earth Observation. Rendering produced by AIKO Virtual Environments Department.

Despite booming consistently, it is facing the challenges of harsh competition in a specific overcrowded environment characterized by scarce resources. The ultimate goal of the market linked to the EO sector is shifting from acquiring and supplying images of the Earth's surface to generating actionable data and insights related to the environment that surrounds us and in which we live. This growing demand for geospatial analytical services will expand the sector into B2B and even B2C markets, enabling its economic sustainability on a large scale, targeting an $8 billion market within the upcoming decade [1].

EO private space companies are fighting a war based on resolution delivery, global coverage, and daily revisit. Differentiation becomes essential to overcome competitors. We all know about the disruptions in the launcher industry and the smart manufacturing sector, and the payload and space segment is following suit; hardware, in general, is becoming a commodity in the space industry. On the other hand, the entire operation pipeline is lagging. Prominent constellations of satellites, each with multiple instruments tasked to provide more in-depth insight at a very high frequency, will require innovative operations concepts. The software will be the core asset of this evolving competition.

B2B and B2C market segments perceive Value-Added Services as more relevant concerning the services providing basic Earth imagery (in the best-case scenario, such services offer analysis-ready products, but many others provide raw data that must be processed and made available by third parties in the middle of the operational stack). A lower predisposition to data processing characterizes these markets; customers in these segments need to integrate insights and actionable information in their decision-making pipelines and want to obtain this information at a low cost and with low latency. This trend represents a paradigm shift from single-purchase to ongoing subscription of actionable and easy-to-use intelligence data and insights.

The space industry is at the foundation of the entire EO operating stack. Those companies own and operate satellites and constellations to acquire the most considerable possible amount of daily Earth imagery. In the last years, the reduction of launch and manufacturing costs, combined with the miniaturization of remote sensing technology, has shifted the focus of the competition, in this precise segment, to constellation architectures characterized by high numbers of small satellites (up to Cubesat of size of a shoe box) capable of providing a massive amount of data with high revisit values. The trend is evidently to give the most outstanding amount of data possible at the lowest price with a wide heterogeneity of resolution and acquisition bands.

Mid-layer players in the EO stack aggregate raw or (best case scenario) Analysis-Ready Data (ARD) of the various space-based providers and make them available to end-users through cloud platforms and simplified APIs. This step in the stack is fundamental to efficiently provide data to the end-user for further processing (institution, universities, research department) or to different software service providers that will generate relevant insights queried by target customers in multiple verticals. The infrastructural layer and the space-based one responsible for the provision of the heterogeneous imagery data sources constitute the abstraction of the EO Big Data Ecosystem.

There is a significant issue associated with this rapidly growing ecosystem. The fragmentation of data sources, the heterogeneity of the data, and the excessive volume, not necessarily proportionate to the value of the acquired data, could drive the entire sector towards a situation where it is becoming data-rich but insight poor. EO companies accumulate vast stores of data they have yet to learn what to do with and have no hope of providing actionable insights. This issue is identified as Big Data Overload; too much data is available, which leads to difficulties in extracting useful information and insights, with a high probability that the overall system is affected by a substantial untapped potential.

Additionally, space companies started providing the capabilities to the end-user to task satellites and constellations to obtain precisely the data they want. This scenario can be represented through a feedback loop that originates from the customer and is interrupted when he obtains the information or data of the required quality he wants. What if the tasked acquisition is so covered by clouds as to be useless? What if the acquired frame is not actionable for the customer? What if multiple end-user requests produce planning conflicts that could be resolved in a non-optimal way regarding the data quality provided (the most relevant image could be the one not acquired for reasons of priority)? This approach can lead to inefficiencies in the service, long iterations between the customer and the provider for single acquisition activity, and, in the end, higher costs and possible missed revenues.

Many players are also changing their business models and moving towards providing vertically integrated intelligence services through synergistic whole-product offerings established through partnerships and acquisitions of allies operating at different levels of the EO stack. Although this approach significantly improves the efficiency of the insight production pipeline (companies have control over the entire stack, from acquisition to final delivery to the end-user), the inefficiencies linked to the disproportion between the volume of acquired data and the actionability of the same remain predominant. Consider a nominal EO scenario where a polar-orbiting satellite travels at 7.8 km/s. It will be in the range of a high data-rate downlinking station for less than 10 minutes, during which it must downlink all of the collected data since the last pass. Operators must ensure they do not capture more data than they can send in one pass. Moreover, downlinking/processing data comes at a cost. It is evident that not guaranteeing the maximum possible ratio between the volume of downloaded data and the value of the same in terms of actionability represents a significant inefficiency in terms of untapped service potential for the EO company.

For those space companies that base their business models on providing raw or ARD data directly to end-users or platforms and marketplace providers, maximizing the value of the volume of data they enter into the Big Data Ecosystem becomes a consistent opportunity for differentiation from competitors. Furthermore, having data already analyzed, contextualized, and whose actionability is provided as metadata accompanying the image itself becomes an added value for those who have to aggregate the data and make it available to users who will then have to process it to obtain relevant information.

To solve the Big Data overload issue and improve their business models, EO companies should transform the big data they generate into smart data by operating classification to reduce the information leading to operational or strategic decisions. Making big data smart requires structuring and prioritizing data to define a more contained set, which offers value and on which third companies or end-users can act. Switching to smart data means optimizing volume with value and actionability. In this sense, it is necessary to work upstream of the value chain, where the information is acquired, enabling data processing functions directly onboard the satellite and allowing a preliminary data classification based on actionability and data relevance.

9.1.2 The Space for Space Market; autonomy as enabler for an economy of scale

The space race implemented in recent decades has increased the number of functional activities, whose past actions and future growth scenarios amplify two critical issues: the management of orbiting objects (orbital congestion, dangerous debris) and the volume of activities to be coordinated by the ground control centers (i.e., ground station). These topics, despite some differences, affect both low orbits (Low Earth Orbit – LEO) and geostationary (Geostationary Earth Orbit – GEO) and are reflected on both conventional and miniaturized satellites (e.g., CubeSat, Nanosat).

Moreover, the space logistics market is a rapidly growing industry becoming increasingly relevant in the modern space sector. As humanity continues to explore and exploit space, the need for transportation and logistics services will only continue to increase. A critical aspect of this market is in-orbit servicing, which involves maintaining and repairing satellites and other spacecraft while in orbit. A further shift must be considered when accounting for space operations: while a significant reduction in launch and space segment costs is underway, ground segment costs do not scale with satellite size, as ground stations used for missions, both in LEO/GEO and deep space, have incredibly high running costs. The study of navigation and control techniques for spacecraft proximity operations is framed in this context. In this context, a wide range of rapidly growing activities and services are rising under the title of In-Orbit Servicing (IOS), which includes: life-extension, refueling, de-orbiting, Active Debris Removal, and repairing. Science has already triggered a significant increase in efforts in developing these operations, which are enabled by new innovative engineering automation standards, including AI.

In-orbit servicing is critical for the long-term sustainability of space operations. Satellites and other spacecraft have a limited lifespan and can experience various problems in orbit, including malfunctions and collisions. In the past, when a satellite failed or reached the end of its operational life, it was left to orbit until it eventually burned up in the atmosphere. However, with in-orbit servicing, it is possible to repair and maintain these spacecraft, extending their operational lifetimes and reducing the amount of space debris. In-orbit servicing also has important implications for space exploration. As we continue to send spacecraft to more distant locations, the ability to repair and maintain these vehicles becomes even more critical. In the past, a single malfunction could spell the end of a mission, but with in-orbit servicing, it may be possible to fix problems and continue the mission. The market for in-orbit servicing and, in general, for in-orbit assembly and manufacturing is expected to grow significantly in the coming years. As more and more satellites are launched into orbit, the need for maintenance and repair services will continue to increase. Several companies are already operating in space for the space market, offering various services from satellite inspection to refueling and repair. These companies are also exploring new technologies, such as robotic arms and drones, that can perform maintenance tasks in space. As the industry continues to grow and develop, we can expect to see even more innovative solutions to the challenges of those kinds of operations.

In recent years, robotic autonomy has become a crucial enabler for space exploration and exploitation. The ability of robots to operate independently in space has not only made space missions safer and more efficient. Still, it has also enabled us to achieve scientific and economic goals that were previously impossible. One of the critical advantages of robotic autonomy is the ability to operate in environments that are hostile or dangerous to humans. Space is an example of such an environment, where humans face various challenges that make it difficult to work effectively. For example, the lack of gravity, extreme temperatures, and high radiation levels can make it difficult for humans to operate for extended periods. However, robots can be designed to withstand these challenges and operate for long periods without human intervention. Another advantage of robotic autonomy is that it allows for faster and more efficient space exploration. Traditional space exploration methods involve sending a spacecraft to a specific location and then waiting for data to be transmitted back to Earth. This process can take weeks or even months to complete. However, with autonomous robots, data can be collected in real-time, and decisions can be made on the fly. This allows for much more efficient use of resources and can lead to faster and more meaningful scientific discoveries. Robotic autonomy is also critical for the development of space infrastructure. To establish a permanent human presence in space, we need to be able to build and maintain infrastructure such as space stations, habitats, and communication networks.

Commercial exploitation of space could also be impacted. As the cost of launching payloads into space continues to decrease, the commercial potential of space is becoming increasingly attractive. However, many tasks required for space exploitation, such as mining and manufacturing, are extremely difficult for humans. By leveraging spacecraft autonomy, companies can

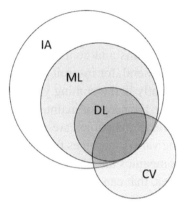

Figure 9.2 Diagram exploiting relations between Artificial Intelligence (AI), Machine Learning (ML), Deep Learning (DL), and Computer Vision (CV).

significantly increase the efficiency and safety of their operations, making space exploitation more economically viable.

Overall, autonomy will become an essential enabler for space exploration and exploitation. It unlocks the ability to operate in hostile environments, collects data in real-time, builds and maintains infrastructure, and supports commercial operations making it a critical technology for the future of space. As we continue to push the boundaries of what is possible in space, we expect even more tremendous advances in robotic autonomy and its application in this exciting field.

9.2 THE ARTIFICIAL INTELLIGENCE IMPACT

The adoption of AI in the space domain is set to impact how we explore and utilize space profoundly. The potential benefits of AI adoption in the space domain are vast, from improved space exploration and satellite operations to increased space-based research and development and improved space traffic management. As AI technology advances, we can expect to see even more significant benefits in the future. This chapter will explore the potential impact of AI adoption in the future of the space domain.

In principle, two main domains can be identified while presenting the future impact of AI and automation technology on the space sector: *Edge Autonomy* to extend the operational capabilities of space-based services and *Continuous Intelligence* for augmenting the scalability and the effectiveness of the same. Those two lines will embrace a broader range of commercial use cases in the space environment domain.

9.2.1   Edge Autonomy

Edge computing will enable use cases within many industrial and commercial ecosystems, such as drones, robots, autonomous cars, AR/VR, home automation, and security systems. In short, edge computing is cloud principles applied near the user's location and consumption. Moving large data sets across the network, such as to the centralized cloud and back, takes time and creates performance issues and latency, hindering real-time use cases around the ecosystems mentioned above. The platform-generated content will dramatically spawn at larger scales considering the space economy expansion. This scenario will lead in the following decades to bandwidth saturation and latency. Many mission concepts and space applications will then become impossible to operate. On the other hand, edge computing reduces network costs, avoids bandwidth constraints, reduces transmission delays, limit service errors, and better control sensitive data transfers. Moreover, edge computing allows more reactive and proactive devices to make operational decisions in autonomous platforms' increasingly generalized mission environment.

Even small satellites can now be equipped with high computational capabilities thanks to the integration of edge computers, including commercial ones, capable of running modern operating systems, usually based upon the Linux kernel. Given this technological advance, edge computers can effortlessly multitask, managing multiple applications simultaneously, supporting a wide range of popular networking protocols, etc. In general, edge computers today can run almost anything a server computer in a data center can. So, space companies can now move appropriate types of application code out of their data centers and into the operational environment.

Edge computer's powerful hardware can perform highly complex tasks, like machine inferencing. They can analyze data locally and send lower volume, higher value data to the center. They also often enable multiple tenants to run applications simultaneously on the same hardware. The management of edge devices will have to be autonomous, given the scale, volume, and workload requirements. In this sense, using AI for the platform's self-assessment, self-heal, and manageability is mandatory to enhance human capabilities. It is crucial to embed state-of-the-art Machine-Learning algorithms on edge to detect, predict, and even autonomously fix possible problems and issues. The paradigm shift will require preventing issues before they occur and not reacting to them. Autonomous platform operations and self-healing capabilities will create a new generation of independent platforms and critical support to enable advanced mission concepts, operationalize new technologies and use cases, and repurpose human and technology assets.

Autonomy means being independent or able to control and govern oneself by learning and adapting to the environment. Globally, a shift is taking place in several industrial verticals. This transformation will move the entire industrial chain from automation to autonomy.

Although the space environment does not represent the avant-garde in this sense, its evolution, which is already dramatically characterizing its growth, will bring a greater demand for autonomy to optimize the various systems' performance. Applied to the industrial domain, autonomy outclasses its automation counterpart by adding layer cognitive capabilities and intelligent sensing to the various platforms and system-level. All is done to anticipate and adapt to unforeseen circumstances and opportunities, removing the need for continuous human intervention. Adopting autonomy will provide advanced capabilities in several application areas, from maintenance, planning, and scheduling, to operations. In this sense, system autonomy will apply to predictive maintenance or issues detection and recommend both human and robotic platforms for optimal strategies and remedial actions, if not acting proactively. Companies will extensively achieve their business objectives, monitor their technological assets, and conduct automated operations with only higher-level strategic control. In the following years, the space domain shall aim for an automation level where most assets operate autonomously and are synchronized to optimize production, safety, and maintenance. Identifying four main benefits of adopting autonomy at a large industrial scale is possible.

Adaptability

Thanks to the evolution of sensors and data processing and the appearance of powerful data fusion algorithms, it is now possible for an autonomous platform to navigate and operate, relying on elements already present in the environment. This technological advance makes it much easier to react quickly and proact to unforeseen events and uncertainties in the operational environment. Given the innovative nature of such approaches, a smooth and controlled transition will be beneficial to let customers and operators appreciate the added value of this advanced technology without incurring risks.

Optimization of costs and productivity

The adoption of autonomy and automation in general humans to refocus their activities on higher value-added and strategic tasks. In the context of a commercial telecom mission, the added value resides in the strategic decisions related to providing better-quality service to the customer. Understand users' preferences depending on the usage data and modify the service to meet customers' needs. Automating repetitive and low value-added tasks, such as the nominal operations and the self-management and self-assessment capabilities of the platforms, makes sense in the context of reduced margins and increased workload necessities to perform these tasks.

Safety

Adopting autonomy will allow for achieving higher levels of safety, especially regarding the environment of low earth orbit. Themes of Space Situational Awareness (SSA) and Space Traffic Management(STM) are now hot topics; also, considering the growing number of satellites in orbit, the risk of collisions is constantly increasing. This scenario represents a risk for many strategic assets operating in this environment without considering the numerous human-crewed missions already foreseen in the coming years. Being able to rely on autonomous systems capable of reacting in real-time to possible risks is essential to maintain a safer and more sustainable environment over time.

Flexibility

Integrating AI on edge to provide autonomy capabilities guarantees an unprecedented level of generalization to the system. Satellites and rovers may perform more low- and medium-level tasks in significantly different operational contexts. Furthermore, combining such autonomy capabilities in federated systems and architectures further increases the functional capabilities of the individual elements and the architecture as a whole, leading to systems capable of performing unprecedented activities and objectives.

9.2.1.1 Satellite Autonomy

As per NASA's definition [2], autonomy is "the ability of a system to achieve goals while operating independently of external control." Autonomous robotics is expected to be one of the key pillars in the future evolution of the space sector. Artificial intelligence (AI) applications are already prevalent in many industries and research fields, and the space industry is poised to be the next frontier for AI.

During the last few years, various autonomous features and automated system-level abilities have been shown and employed in spacecraft operations. This aspect is especially true in the case of threshold analysis of telemetry data for Fault Detection, Isolation, and Recovery (FDIR) purposes, which has been now in adoption for many decades. However, spacecraft are still primarily dependent on ground-based systems to evaluate circumstances and decide on the next steps, utilizing pre-written command sequences. Indeed, while we experienced exceptional advancements in hardware and spacecraft miniaturization (which boosted the recent growth of the space economy), little has changed concerning flight software and how operations are carried out. The commercial space race and the increased public interest in space-related activities are now driving the development of autonomous platforms that can overcome operational constraints and break the last barriers that prevent the adoption of autonomous technologies. Indeed, the space environment presents several challenges for robotic systems, from low Earth orbit up to deep space, and automated or autonomous approaches have often been hard to be implemented with sufficient safety or with low enough costs. To date, we can identify three significant aspects that will benefit from adopting increasingly autonomous space systems: responsivity, platform complexity, and operations scalability.

Responsivity

Historically, satellites, spacecraft, constellations, rovers, and landers have all faced uncertain mission environments and unexpected events, and their ability to efficiently react, adjust, and explore is still limited. In low Earth orbit, latency, scattered communication, and limited communication windows represent a significant bottleneck for the success of scientific and commercial spacecraft missions since any decision or action has to pass through the control of the ground segment, which may sometimes be be out of reach for several hours.

Platform Complexity and Mission Efficiency

As technology advances and commercial interest in the space economy grows, satellite platforms will become increasingly complex, with more capabilities and advanced subsystems. As platforms become increasingly powerful and capable, the way we use them must follow suit to ensure that we are exploiting the full potential of the hardware and software that composes the platform.

Operations Scalability

Thanks to the miniaturization of satellite platforms, we have experienced a surge in the number of satellites operating in Earth's orbit in the last ten years. At a current value of above 4000 units, this number is set to grow exponentially over the upcoming decade, primarily due to the launch of several constellations of satellites. The current approach to operations -heavily reliant on ground control and human decisions- is not equipped to scale along with this increase in operating assets.

9.2.1.1.1 orbital_OLIVER, the State-of-the-Art Application for Satellite Autonomy These problems will affect commercial spacecraft operations and institutional missions [3], and the search for their solutions has been marked as a priority. Over the next decade, spacecraft autonomy is expected to be a significant enabler for missions of different natures, including EO, telecommunications, defense, space exploration, and even crewed missions [4]. In light of this, integrating AI-based architectures is crucial to increase the operational capabilities of robotics systems in the space domain while reducing the workload on the ground.

In this scenario, orbital_OLIVER [5], an onboard software application developed by AIKO for enabling autonomous mission operations services, represents one of the most advanced solutions in the space domain for what concerns satellite autonomy. The software application relies on symbolic AI and Machine Learning as its main building blocks. orbital_OLIVER streamlines spacecraft operations and augment mission performance by making satellites less dependent on ground control. This tool has been specifically intended for use by spacecraft owners and operators that need to improve mission performance and increase scalability by reducing the human workload on the ground.

In the canonical approach to space operations, the satellite periodically receives a schedule generated by the ground segment. That schedule will be valid until contact with the ground is established again and the task list is updated. While this approach has been excellent since the start of space exploration, the advancements in data processing -specifically, onboard data processing- may make this an obsolete flow. In a mission equipped with onboard data processing capabilities and autonomy tools, the flow of the operations would start from the combined action of the sensing and reasoning modules. Those components allow the software to ingest the platform telemetry and monitor the spacecraft's health in near-real time, ensuring faster response to anomalies. Thanks to this, it will not be necessary to downlink the full telemetry and analyze them on Earth. Still, the operators will be able to select and downlink only the most relevant data packets to monitor on the ground. Furthermore, with onboard autonomy, if the sensing-reasoning components detect a specific event, an onboard planner would be triggered to adapt the mission schedule almost in real-time to take advantage of unforeseen opportunities during the last schedule instantiation.

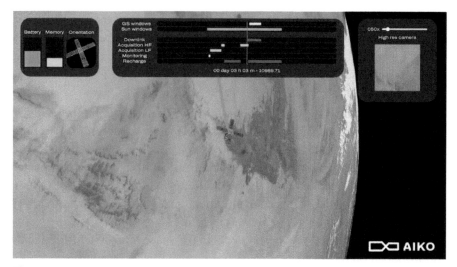

Figure 9.3   orbital_OLIVER capabilities demonstration in a simulated Earth Observation scenario.

orbital_OLIVER technology is based on a modular architecture and a three-stage autonomous pipeline that covers:

- platform and environment sensing;

- reasoning and goal definition;

- planning and scheduling.

This breakdown adapts to the most generalized abstraction of an autonomous system [6]. In orbital_OLIVER, the abstraction scheme is mapped onto a modular structure composed of three independent (but interlinked) components: the sensing module, the reasoning module (reasoner), and the planning module (planner). With this approach, the satellite is capable of:

- perceiving its surroundings and its condition through onboard data processing;

- using the acquired data to reason on the best goals to pursue or to trigger an emergency procedure in the case of an anomaly;

- re-planning the mission schedule if the goals generated at the second step differ from the directions provided by the ground segment or if an anomaly has been detected.

orbital_OLIVER is a clear example of enabling a higher satellite autonomy by combining symbolic AI and Learning approaches. Machine Learning (ML)-based inference engines are dedicated to real-time telemetry data processing. Thanks to this, the classic FDIR approach (based on threshold analysis) could be improved by having a tool for monitoring the platform's health more efficiently and preemptively. Indeed, by using ML, the sensing module integrated into orbital_OLIVER enables both health diagnostics functionalities (i.e., to investigate failures and understand the root cause), and prognostics, allowing the prediction of potential failures before they occur, thus extending the lifetime of the platform. Before the mission, the engine is trained on historical data (if available) and synthetic data for greater model robustness. The ML model can be retrained during the mission using actual operational data from the platform. Then it can be uploaded again to update the inference engine onboard and deliver a more accurate telemetry analysis.

orbital_OLIVER includes a dedicated reasoning module to provide explanations (when needed) for the information collected by analyzing telemetry streams and payload data products and devise the optimal goals to pursue based on those information and higher-level mission goals. Symbolic AI technology makes magic. Scenario and system-level information are stored in a knowledge base to solve problems that usually require a human expert, thus preserving its knowledge in a database. An inference engine is applied to the knowledge base to derive information from already-known facts. Lower-level information is queried during inference until a known fact is encountered, thus reconstructing the existing system and mission knowledge state. In light of this, the reasoning module takes as input the information extracted from the telemetry and payload data, placing it in the mission context by using the specific knowledge defined in the design phase to generate an optimal high-level goal that must be pursued by the spacecraft.

Finally, the autonomy loop is closed onboard by AI-powered planning techniques that enable near-real-time mission timeline adjustment to guarantee optimal mission outcomes even in uncertain conditions or unforeseen events. In orbital_OLIVER, a dedicated planning module encompasses those capabilities. It is tasked with the last activity in the pipeline: generating an actual mission plan based on the goals defined by the reasoning module. Once a goal has been identified, it is divided into smaller tasks. This breakdown is defined as well within the mission knowledge. Then, the planning module processes this list, associating each task with a specific execution time based on: task duration, task priority, onboard resources utilization, and task precedences. To produce the schedule, the module adopts an algorithm based on linear integer programming, which optimizes the allocation of tasks within the satellite's time horizon.

At the time of writing, the major milestone ahead in orbital_OLIVER's roadmap is completing the in-orbit validation currently in progress through the ESA InCubed+ program. This tool validation is expected to occur between Q2/Q3-2023. Preliminary in-orbit tests have already been accomplished with promising results, confirming expectations from a functional and performance point of view. In 2023, satellites' onboard autonomy is expected to become a reality thanks to integrating technologies based on artificial intelligence, which will only become more effective. AI-enabled spacecraft autonomy will improve spacecraft performance, allowing for faster response, better platform health monitoring, and exploiting mission opportunities that would go unnoticed in the ground-based mission operation framework. As a byproduct of the enhanced independence from the ground, spacecraft autonomy will reduce the workload on the mission control centers, providing a scalable solution that can be applied to handle the tens of thousands of satellites that will orbit Earth by the end of this decade.

9.2.1.2 Onboard Payload Processing

Space-based services for Earth Observation are critical for a wide range of applications, including weather forecasting, climate modeling, disaster response, natural resource management, infrastructure monitoring, surveillance, and constant monitoring. However, the vast amount of data generated by satellite EO missions poses significant challenges for data storage, transmission, and processing. Onboard processing has emerged as a promising approach for addressing these challenges by enabling data to be processed and analyzed on edge (the satellite) before being transmitted to the ground.

One of the main advantages of onboard processing in satellite EO is improved data efficiency. By processing and analyzing data on the satellite itself, only relevant information needs to be transmitted to the ground, reducing the amount of data that must be stored and transmitted. This approach can also lead to significant cost savings and increased operational efficiency, especially for missions that generate large amounts of data, such as high-resolution imaging or atmospheric sensing. In addition to improving data efficiency, onboard processing can e real-time data analysis and decision-making. The satellite can respond to changing conditions in real time by processing data onboard, providing more timely and accurate information for latency-sensitive applications. Real-time data analysis can also enable the satellite to make decisions and adjust its behavior in response to changing conditions, which can increase the autonomy and flexibility of the satellite.

Several relevant use cases demonstrate the potential of onboard processing in satellite EO. One example is onboard processing for cloud detection and analysis in high-resolution imaging. The satellite can identify and classify different clouds in real-time by processing and analyzing data onboard, enabling more accurate weather forecasting and climate modeling. Another use case is onboard processing for vegetation monitoring and analysis, which can provide real-time information on the health and growth of crops and forests, enabling more efficient and sustainable natural resource management.

Despite the potential advantages of onboard processing in satellite EO, the current status quo is that most data processing is still performed on the ground due to various factors, including limitations in onboard computing power and memory, the need for advanced data processing algorithms, and the high cost of developing and launching onboard processing hardware. However, technological advancements, such as developing more efficient processors and using Machine Learning algorithms for data processing, are envisioned enablers for the broader adoption of onboard processing in satellite EO. In conclusion, onboard processing can transform how we process and analyze data from satellite EO missions, enabling real-time data analysis, improved data efficiency, and increased autonomy and flexibility for the satellite. While there are still challenges to overcome, the potential benefits of onboard processing make it a promising area of research for the future of satellite EO. Technological advancements, such as developing more efficient processors and Machine Learning algorithms, are envisioned enablers for the broader adoption of onboard processing in satellite EO. The following paragraphs provide examples of Machine Learning-enabled EO payload processing applications.

9.2.1.2.1 Cloud Detection The presence of clouds can make images taken by EO satellites in VIS or NIR useless. An onboard pre-filtering of data stored onboard can thus significantly increase the effectiveness of EO missions. A set of techniques have been exploited in the last few years to overcome this issue. Most of them use threshold or rule-based image analysis. While computationally efficient, those approaches come at the expense of lower accuracy, which implies lower generalization capabilities. More recently, however, the advent of ML and Deep Learning (DL) has disrupted the way images are processed and features are extracted.

In this context, AIKO is at the forefront of innovation, boasting state-of-the-art solutions. For cloud segmentation, AIKO proposes a DL-based segmentation software application called cloudy_CHARLES [7], capable of classifying pixels in a satellite image with high fidelity and constrained

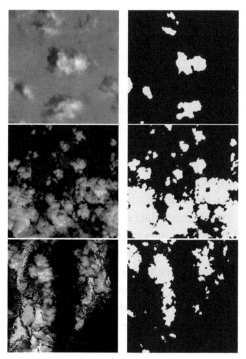

Figure 9.4 Cloudy Images form Sentinel-2 Dataset (left) and relative cloud masks produced by cloud_CHARLES (right); average Intersection over Union grater than 95%.

computational costs. The algorithms have been tested and validated on various hardware platforms (COTS, space-graded) and have already been validated in orbit.

9.2.1.2.2 Object Detection Leveraging DL and Convolutional Neural Network (CNN) techniques for onboard object detection in Earth Observation has significant advantages over traditional image processing methods. DL and CNN techniques are particularly effective in detecting and localizing objects in complex images.

They are ideal for Earth Observation use cases where many objects could be relevant in a scene. By leveraging these techniques for onboard object detection, LEO satellites can detect and classify objects with high accuracy, providing more detailed and reliable information for various applications. One advantage of leveraging DL and CNN techniques for onboard object detection is increased accuracy. DL and CNN techniques are trained on large datasets, which enables them to learn complex features and patterns in images that may not be apparent to traditional image processing techniques, resulting in more accurate object detection and classification and providing more reliable and detailed information for a range of Earth Observation applications. DL and CNN techniques also enable the detection of objects at different scales and orientations, making them suitable for applications where objects may be small or partially occluded by other objects.

The use of DL-based detection capabilities onboard the satellite enables:

- real-time response to events (i.e., fire detection, ship detection), where warnings are sent to the ground once an event is detected (reduced latency, lower downlink bandwidth needed);

- continuous satellite monitoring of a particular area acting to events in real-time (i.e., adjusting attitude for continuous monitoring);

- crop-to-the-target to reduce the data to downlink by cropping the payload product to the actual target/s to track.

9.2.1.2.3 Change Detection Onboard change detection performed by LEO satellites using Deep Learning techniques has significant implications for Earth Observation use cases. Change detection is identifying and localizing changes in a scene between two or more images, essential for many applications, such as monitoring land use changes, detecting natural disasters, and tracking the spread of wildfires. By performing change detection onboard the satellite, the need for ground-based processing is reduced, enabling real-time analysis and decision-making. Time-critical applications (disaster response, where timely and accurate information can be critical for saving lives and mitigating damage) could primarily benefit from onboard real-time change detection capabilities.

One of the main benefits of leveraging DL techniques for onboard change detection in Earth Observation is improved accuracy. DL-based approaches are particularly effective in detecting and localizing changes in complex images, making them ideal for Earth Observation use cases with many changes of interest in a scene. By leveraging these techniques for onboard change detection, LEO satellites can detect and classify changes accurately, providing more detailed and reliable information for various applications. DL and CNN techniques also enable the detection of changes in different scales and orientations, making them suitable for applications where changes may be small or occur over a large area.

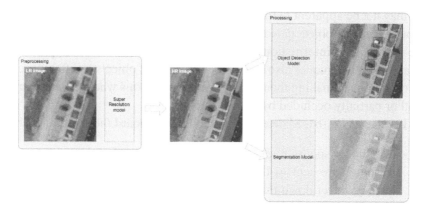

Figure 9.5 Diagram highlighting the Super-Resolution pipeline applied to an object detection scenario.

9.2.1.2.4 Super-Resolution Deep Learning-based super-resolution has the potential to revolutionize the way we process and analyze images in space. Super-resolution increases an image's resolution beyond the original image's limits, essential for many applications, such as high-resolution imaging and atmospheric sensing. A comparative study by [8] targeting satellite imagery suggests a significant increase in object detection accuracy. By performing super-resolution onboard the satellite, the need for ground-based processing is reduced, enabling real-time analysis and decision-making. This aspect can be essential in time-critical applications, such as disaster response or military surveillance, where timely and accurate information can be critical for decision-making. DL algorithms can learn to extract more detailed and accurate information from low-resolution images, creating high-resolution images with finer details and more accurate information. Onboard super-resolution can improve satellite imagery's accuracy and reliability for various applications, such as weather forecasting, climate modeling, disaster response, and military surveillance.

There are a bunch of relevant DL-based approaches to be considered. First, Single Image Super Resolution (SISR) exploits spatial correlation in a single image to try to recover the high-resolution version of it. The problem with this application is the amount of information available in a single image, which is quite limited as some information may be lost in the low-resolution formation process. Generally, the low-resolution image is created by simply downscaling, blurring, or other operations that emulate camera image acquisition. The main advantages of such an approach are: i) no need for supplementary low spatial resolution images within the exact location and ii) augmented generalization capabilities for different scenarios and spacecraft. However, SISR also presents a significant drawback because less information would be available for high-resolution image reconstruction. Interesting approaches are presented in [9] and in [10], where Generative Adversarial Networks (GANs) are used to accomplish SISR tasks.

Another relevant approach to Super Resolution is represented by Multi-Image Super Resolution (MISR), which uses multiple images of the same scene, taking full advantage of their complementary information, to create one single high-resolution image with high spatial frequency details. This is usually the case for remote sensing applications, where multiple images of the same scene can typically be acquired by a spacecraft during multiple orbits, by multiple satellites imaging the same scene at different times, or may be obtained simultaneously with different sensors. The increasing spatial and spectral resolution of onboard instruments capable of generating high amounts of data challenge the compression algorithms to meet the available downlink bandwidth. As a result, this often reduces the availability of high-resolution products. The main advantage of this approach is represented by the availability of information to recover the super-resolution image, but considering that it could not be easy to obtain an adequate number of low-resolution images from the same scene. A relevant MISR approach is presented in [11].

9.2.1.2.5 Synthetic Aperture Radar Data Processing Synthetic Aperture Radar (SAR) systems widely provide high-resolution images that reach a centimeter scale regardless of illumination and weather conditions. Such images are independent of the day-and-night cycle because the SAR's sensors carry its illumination. Also, they are independent of the weather as microwaves penetrate clouds and storms with low deterioration. Moreover, radar frequencies interact with matter differently from optical ones, so SAR images provide complementary information to those collected with optical sensors. The primary example of a SAR mission is Sentinel-1, which comprises a constellation of two polar-orbiting satellites performing C-band SAR imaging to provide a continuous radar mapping of the Earth. Unfortunately, on 23 December 2021, Sentinel-1B experienced an anomaly that prevented it from delivering radar data.

Products obtained through SAR are used for various applications, including geoscience, climate research, environmental monitoring, 2-D, 3-D, and 4-D (space and time) Earth mapping, and even planetary exploration. The data availability encourages AI solutions for EO. However, AI solutions are nearly unexplored for SAR data utilization. The onboard SAR data processing is one of the most promising areas of innovative research in the satellite onboard payload data processing field. It can open up new opportunities and enhance the EO services model of the European market. By advancing onboard SAR processing capabilities, EO data producers can generate actionable information directly on the satellite. This technology will allow the feedback loop to be closed on the payload products in nearly real-time, drastically reducing the latency in response to potential information obtainable from the SAR image produced by the satellite. In such a way, SAR platforms will become central to emergency response, Intelligence, and alerting applications, possibly coordinating with other satellites in the constellation, capable of collecting further information through heterogeneous payloads.

It must be noted that, differently from optical payload data products, SAR data acquired onboard the spacecraft consists of raw signals nearly uninterpretable to the human eye. The so-called focusing procedure must produce a recognizable image and enable further data processing. The state-of-the-art in the SAR focusing application is represented by traditional algorithms which work in the frequency domain exploiting fast convolutions based on the Fast Fourier Transform (FFT). The main ones are Range Doppler Algorithm (RDA), Chirp Scaling Algorithm (CSA), Omega-K Algorithm (ωKA), and Spectral Analysis Algorithm (SPECAN). The RDA is the oldest but is still primarily used as it represents one of the best trade-offs between accuracy, efficiency, and generality. This is a demanding computational task currently performed on the ground in dedicated processing facilities, hindering real-time surveillance capabilities.

In this scenario, AIKO is carrying out various research activities to demonstrate the applicability of approaches based on the use of Dee Learning models to enable the focusing of raw SAR data directly onboard the satellite. Specifically, the focusing algorithm transforms the Level-0 raw signal into Level-1 Single-Look-Complex (SLC) data. It consists of a two-layers hybrid architecture: a traditional FFT algorithm for range processing and a Deep Neural Network (DNN) trained to solve the azimuth processing task, which provides scalability and modularity benefits. Finally, because the focusing process must remain functional to onboard data processing activities, the pipeline must be completed by a further DL model for an object or feature detection. Given the particularity of the SAR data, the ability to focus the raw data at a sufficiently high quality, such as allowing further processing to obtain actionable information, remains the primary technological enabler for onboard processing applications for this specific payload class.

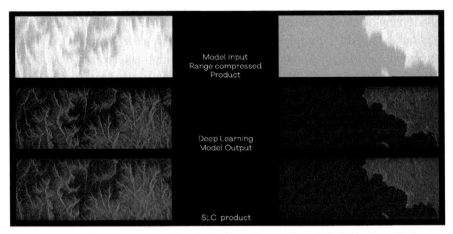

Figure 9.6   Results of SAR data focusing using DL-based techniques; AIKO R&D project. A comparison with SLC products on Sentinel-1IW data products is shown.

9.2.1.3   Autonomous Guidance, Navigation, and Control

9.2.1.3.1   Station-Keeping Strategies powered by Onboard AI   In LEO, satellites and spacecraft must maintain their position and orientation relative to each other and the Earth's surface. This aspect is known as station-keeping, critical for the successful operation of many important space missions, such as Earth observation, communications, and navigation. Station-keeping is achieved through propulsion systems that adjust the spacecraft's velocity and direction. However, maintaining a stable orbit in LEO is a complex and ongoing process that requires constant monitoring and adjustments to ensure that the spacecraft remains in its desired position and orientation. An autonomous Station Keeping architecture can be implemented using AI-based techniques to enable higher autonomy levels. Specifically, ECSS (European Cooperation for Space Standardization) standards define an E4 level of autonomy concerning the capability of a spacecraft to execute goal-oriented mission operations directly onboard. Applying this concept to SK operations, it is possible to derive two high-level objectives: i) the spacecraft shall be capable of maintaining the nominal operational orbit while minimizing the fuel (high-level goal), and ii) no thresholds or triggering events shall be defined apriori.

Many references could be found in the literature regarding automated approaches to spacecraft station keeping in LEO, GEO, and even Deep Space missions [12–14]. However, AI-based approaches would augment and optimize the performances of the approaches implemented thanks to the capability of adopting a data-driven approach that can generalize the operational context without the need for complex scenario modeling. In recent years, Deep Reinforcement Learning has emerged as a promising approach for solving complex control problems in various fields, including robotics and transportation. In the context of LEO station-keeping, Deep Reinforcement Learning (DRL) can potentially improve the efficiency and accuracy of spacecraft control, which can have significant implications for the success of space missions. Firstly, DRL offers several advantages over traditional control methods for spacecraft station-keeping in LEO. DRL is a type of Machine Learning algorithm that enables an agent (in this case, a spacecraft) to learn how to perform a task through trial and error without the need for explicit programming, meaning that the spacecraft can adapt its behavior to changing conditions in real-time, which is essential for maintaining a stable orbit in LEO. DRL approaches have been shown to outperform traditional control methods in various control problems, including autonomous driving and robotics, suggesting that they could also be effective for spacecraft station-keeping in LEO.

However, several challenges are also associated with using DRL for LEO station-keeping. One of the main challenges is the complexity of the control problem. Various factors influence a spacecraft's behavior in LEO, including gravitational and atmospheric forces, solar radiation pressure, and perturbations from other celestial bodies. To effectively control a spacecraft in this environment using DRL, the algorithm must learn how to respond to these different factors and adjust its behavior accordingly. This approach requires extensive data, computational power, and advanced control and navigation techniques. Despite these challenges, there are several potential applications of DRL approaches for LEO station-keeping. One application is to improve the efficiency of propulsion systems. By using DRL to optimize the spacecraft's trajectory and minimize fuel consumption, it may be possible to extend the lifespan of the spacecraft and reduce the cost of space missions. Another application is to enhance the autonomy of spacecraft. By using DRL to enable spacecraft to make decisions and adapt their behavior to changing conditions in real-time, it may be possible to reduce the need for ground-based control and increase the operational flexibility of space missions. In conclusion, the impact of DRL approaches for LEO station-keeping is a promising area of research that can potentially transform how we control spacecraft in space. While several challenges must be overcome, the potential advantages of using DRL for LEO station-keeping are significant, and the development of advanced control and navigation techniques will be essential for realizing the full potential of this approach.

9.2.1.3.2 Orbit Transfers using Deep Reinforcement Learning Space missions often require the transfer of spacecraft from one orbit to another. This process, known as orbit transfer, can be achieved through low-thrust propulsion systems, which provide continuous and low thrust levels over an extended time. Low-thrust orbit transfers in LEO are necessary for a wide range of space missions, including satellite deployment, repositioning, and deorbiting. However, achieving the desired orbit transfer in LEO using low-thrust propulsion systems is a challenging task that requires careful planning and execution and advanced control and navigation techniques to ensure that the spacecraft reaches its target orbit accurately and efficiently.

Recently, there has been a growing interest in using low-thrust propulsion systems for orbit transfers in LEO. These systems offer advantages over traditional high-thrust systems, such as lower fuel consumption and longer lifetime. However, low-thrust propulsion also poses new challenges for autonomous Guidance and Control (G&C) systems [15]. The state-of-the-art in G&C for low-thrust orbit transfers in LEO can be broadly divided into model-based and model-free methods [16–18]. Model-based approaches involve mathematical models representing the system's dynamics and generating optimal control inputs. On the other hand, model-free approaches involve learning from experience and do not rely on explicit mathematical models. These approaches involve formulating the problem as a mathematical optimization. The goal is to find the control inputs that minimize a performance metric such as fuel consumption or Time of Flight. Different optimization algorithms are then used to find the optimal solutions, such as Sequential Quadratic Programming, Sliding Mode Control, Model Predictive Control, Genetic Algorithms, and Particle Swarm Optimization.

Another method that has gained attention in recent years is the use of AI techniques, such as Reinforcement Learning (RL) as shown in Fig. 9.7 [19].

These approaches have been used to address the challenges of G&C for low-thrust orbit transfers in LEO by learning from data and adapting to changing conditions. In particular, DRL has gained attention as a powerful alternative to classical algorithms for finding optimal controllers for nonlinear and possibly stochastic dynamics that are unknown or highly uncertain [20]. The Markov Decision Process (MDP) [21] is a valuable framework for modeling sequential decision problems to achieve long-term goals. Reinforcement Learning is a powerful technique for solving MDPs, in which the agent makes sequential decisions by continually interacting with the environment. The agent aims to discover an optimal policy that maximizes the cumulative reward. In RL, the mapping between states and actions is typically stored in a tabular form, which can be impractical, mainly when the state space is ample,

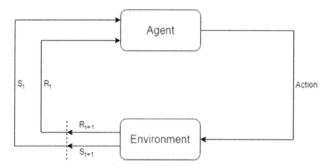

Figure 9.7   Diagram highlighting the interaction between an RL agent and the training environment.

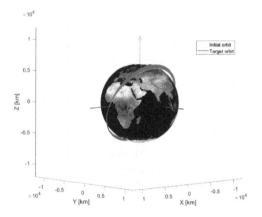

Figure 9.8   Autonomous Orbit Transfer using DRL-based techniques (3D representation). AIKO R&D project.

or the action space is continuous. With some form of function approximator – such as DNNs – model-free DRL can make intelligent sequential decisions in challenging environments.

In [22], the authors propose a model-free DRL algorithm for a spacecraft's autonomous maneuvering during an LEO orbit transfer. This method is particularly suitable for tackling the complex nonlinear dynamics of low-thrust G&C and allows for dealing with external perturbations. The paper presents an RL approach to autonomously maneuver a satellite during a low-thrust LEO orbit transfer. The goal is to reach a target orbit starting from nominal conditions while considering dynamic models of increasing complexity.

The problem is formalized as an MDP model, and a model-free deep RL algorithm is applied to solve two types of maneuvers. The effectiveness of

Figure 9.9   Autonomous Orbit Transfer using DRL-based techniques (realistic simulated scenario). AIKO R&D project.

RL agents in learning how to solve such optimization problems is demonstrated. Preliminary assessments aimed at testing agent robustness starting in non-nominal initial conditions are currently under evaluation, early results are promising, but their systematic analysis represents a future goal. Other than this, future works could focus on improving the realism of the scenario by introducing random noise to observation values and action effectiveness, as well as generalizing initial and target conditions, both from a noise point of view and aiming at generalized maneuvers.

9.2.1.3.3 Visual Navigation and AI   Visual Navigation (VN) is the problem of navigating an agent in an environment primarily using camera input. The agent is considered any device or physical system that integrates the vision sensor, such as a mobile robot, vehicle, drone, or satellite. VN in space can be applied in IOS operations, particularly for Rendezvous and Proximity Operations (RPO) activities for collaborative and non-collaborative targets. The vision sensor comprises one or more cameras, color (RGB) or grayscale. It can be integrated with other sensors such as IMU (Inertial Measurement Unit), LIDAR (Laser Imaging Detection and Ranging), and GPS (Global Positioning System). The multi-sensor approach can infer the agent's status (position, speed) more precisely, even if the setup is more expansive and complex. The VN is based on Computer Vision (CV) techniques and significantly benefits from Deep Learning and Machine Learning. The Vision requires complex

Figure 9.10   Three-dimensional mapping of the features obtained from the sequence of images, with triangulation and estimating the camera pose.

actions, such as spatial geometry and feature perception. These steps require a more complex aggregation from the most straightforward perception levels. The environmental perception of a target depends on the data the agents can integrate, both from a qualitative and quantitative point of view. Below, a set of the most relevant VN techniques are provided.

3D Reconstruction

This technique is possible when a relative movement exists between the agent, the environment, or a target. Given a sequence of ordered and unordered frames, two possible reconstructions can be obtained. The first is the so-called *Structure from Motion* (SfM), a technique used to estimate the scattered 3D dimensional structure, also called "point cloud," starting from the sequence of 2D images. The second reconstruction approach is based on the *Photogrammetry* method to obtain a dense 3D reconstruction, similar to SfM. In this case, it is also possible to integrate physics-based rendering characteristics that describe the target's albedo and color ranges with a detailed volumetric appearance.

Visual Odometry

The Visual Odometry methodology aims at estimating the movement of the agent within the environment in which it is inserted, in terms of relative or absolute position (depending on the input type), to determine the position and orientation in a defined space-time neighborhood (e.g., frame pair).

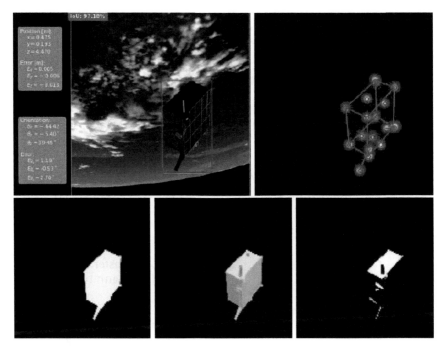

Figure 9.11   AIKO-NET V8 Object Detection network prediction for close-range satellite.

Simultaneous Localization and Mapping (SLAM)

An agent locates itself globally in an unknown context through a SLAM technique and simultaneously builds an incremental map of the surrounding environment without any prior information.

Pose Estimation

With Pose Estimation (PE) as shown in Fig. 9.11 [23], experts refer to a generic activity to identify a target and determine its position and orientation (6-DoF) to a coordinate system. If the agent is already familiar with an environment or target (i.e., after SfM or during SLAM), another PE subtype called re-localization is possible.

Hazard Detection and Avoidance (HDA)

HDA techniques are used to navigate and avoid predefined obstacles and dangers. It consists of visual recognition of one or more instances (i.e.,

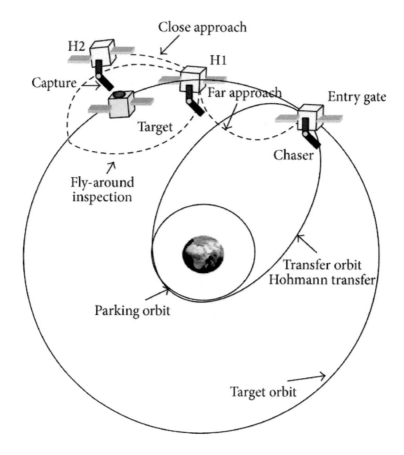

Figure 9.12   Rendezvous Operations Overview.

objects, models, elements of the terrain topography) or according to a list of fixed rules. Eventually, this domain can involve other CV and VN techniques such as, for example, the detection and segmentation of objects and their tracking, PE and GNC.

9.2.1.3.4 Close-Proximity Operations   Space missions often involve proximity operations, where a spacecraft must approach and interact with other objects (cooperative or uncooperative) in space, such as docking with a space station or conducting a rendezvous with a spacecraft. Proximity operations are complex and challenging control problems that require the spacecraft to navigate in a dynamic and uncertain environment while avoiding collisions and performing specific tasks.

In recent years, the use of Artificial Intelligence has emerged as a promising approach to enhance the autonomy of spacecraft during proximity operations. Specifically, DRL can be used for control and visual navigation to allow spacecraft to learn from its environment and adapt its behavior to achieve its objectives. The main advantage of augmenting the autonomy of spacecraft during proximity operations with AI is increased safety and reliability. Using DRL for control, the spacecraft can learn how to perform maneuvers optimized for safety and efficiency, reducing the risk of collisions and improving the mission's success rate. DRL approaches can also adapt to changing environmental conditions and uncertainties, such as unpredictable movements of the target object or variations in gravitational forces, which are challenging to model in advance. This flexibility allows the spacecraft to respond in real time and adjust its behavior accordingly, improving the mission's success rate and reducing the risk of failure.

9.2.1.3.5 Asteroid and Lunar Landing Deep reinforcement learning also has the potential to revolutionize the way we control spacecraft during lunar and asteroid landing missions. Like LEO station-keeping, landing on the Moon or an asteroid presents a complex control problem that requires the spacecraft to adjust its behavior in response to changing conditions. However, the challenges of landing on a celestial body are even more significant than those in LEO, as the spacecraft must contend with various additional factors, including variable gravitational forces, uneven terrain, and limited communication with ground-based control. DRL approaches have the potential to address these challenges by enabling spacecraft to learn how to respond to these factors and adjust their behavior accordingly in real-time, which could improve the accuracy and safety of landing missions.

In addition to enhancing spacecraft control during lunar and asteroid landings, DRL can be coupled with DL-based visual navigation techniques to detect and avoid potential hazards during the landing process. DL approaches are effective in object detection and image classification, which can identify hazardous terrain features, such as boulders or steep slopes, that may pose a risk to the spacecraft during landing. By coupling DRL with DL-based visual navigation, the spacecraft can learn how to avoid these hazards and adjust its trajectory accordingly, improving the safety and accuracy of landing missions. This approach has the potential to reduce the reliance on ground-based control during landing missions and increase the autonomy of spacecraft in space.

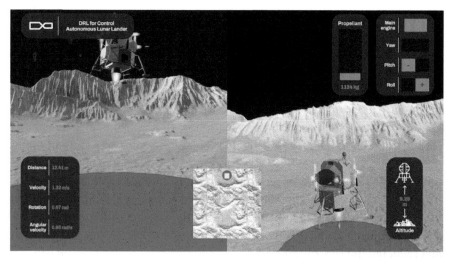

Figure 9.13 Autonomous Lunar Landing Scenario using DRL-based techniques. AIKO R&D project.

9.2.1.4 Closing Remarks

The classical approach to space operations typically involves ground-based mission control centers with highly trained human operators closely monitoring spacecraft operations and making decisions based on real-time data. As spacecraft become highly autonomous, however, there are limitations to continuing to rely solely on this approach. One major limitation is that as spacecraft will become capable of making decisions faster and more accurately than human operators, the traditional approach of relying on human operators to monitor and control spacecraft becomes less effective, as they cannot keep up with the speed and precision of autonomous systems. Additionally, relying solely on ground-based control of spacecraft can introduce communication delays and limitations, slowing decision-making and limiting the ability to respond to unexpected events. Another limitation is that spacecraft will require more sophisticated software and hardware. This aspect increases the systems' complexity and can introduce new failure modes and risks that are difficult to predict and mitigate. New approaches to spacecraft design, testing, and operations are needed to fully realize the benefits of autonomy, which may involve rethinking traditional mission control architectures and processes. This aspect will be extensively covered in the following section.

9.2.2 Continuous Intelligence

"Continuous intelligence is a design pattern in which real-time analytics are integrated into business operations, processing current and historical data to prescribe actions in response to business moments and other events."

Gartner IT Glossary

Adopting AI and automation at an industrial scale on the ground segment of a space mission or space-based service will result in implementing a Continuous Intelligence pattern for the space domain, thus, enabling future-ready operations for advanced autonomous robotic assets in orbit. Continuous Intelligence (CI) from all the mission data differs from another phrase to describe real-time, speed, or throughput. It is about frictionless cycle time to derive constant business value from all missions, robotic assets, and payload data.

CI is a modern data-driven approach to analytics that quickly gets to all the relevant data and accelerates the analysis. No matter how off the beaten track it is, how many data sources there are, or how vast the volumes are. Using CI, it is possible to automate the entire analysis and decision process to become continuous and frictionless. Customers can leverage information derived from real-time analytics to identify anomalies and scenario changes as they occur. Rather than being reactive, streaming analytics will let immediate corrective actions or seize opportunities that the customer might miss otherwise. CI is not a single piece of technology, but a strategic capability and a culture, although CI platforms integrate key tech elements. It draws on emerging technologies to deliver real-time alerts, recommendations, and automated fixes to operational and business problems. Through Continuous intelligence, a space company will eventually enable capabilities integrating key technology elements to deliver real-time guidance on operational problems.

The space market presents solutions that provide incremental efficiency at the lowest common denominator but no significant competitive advantage for the end customer. Those solutions fall in the wake of analytic silos, automation of inefficient decisions, and calcified operating systems. Deploying in production AI-based software systems will allow decision-makers to achieve a significant strategic advantage by integrating their data, operative and business decisions, and systems' operations, a connected ecosystem of applications that constantly talk, improve, and compound upon one another. The resulting architecture will provide different mission services, enabling high-level system autonomy distributed across the different mission segments. This pattern will maximize the mission outcome, simplify operations, and allow unprecedented system autonomy.

Specifically, Continuous intelligence use cases that are particularly effective combine real-time analytics of streaming data and analysis of historical data. The ability to continuously store and analyze this enormous amount of data is fundamental; autonomously connecting this information flow to the decision-making pipeline represents a distinctive business value for the company that operates the system. Data monitoring and analytics are constantly on guard, building a mission and system-related knowledge base.

The underlying analysis pipeline shall be supported by symbolic AI algorithms and Machine Learning models that can guarantee a high degree of generalization while maintaining a concrete understanding of the underlying physical models that describe the overall space system (ground and space segment). Such an approach can provide ground operators with tools for detecting and analyzing failure, for the preventive evaluation of the operating status of the onboard systems, and consequently for carrying out the practical root causes analysis on possible anomalies and malfunctions. In this scenario, operators do not have to wait until a failure occurs before they take action.

The field of spacecraft operations comprehends the operation, commanding, and monitoring of satellites and other space assets in orbit and preparing in-orbit activities for ongoing and upcoming satellite missions. Machine Learning shows excellent potential in spacecraft operations, especially in the upcoming mega-constellations. As the amount of data produced cannot be trivially monitored and analyzed manually by operators, continuous and automated Telemetry (TM) analysis and data mining powered by AI are mandatory for managing large numbers of spacecraft simultaneously. Overall, an AI-based system can support the operator in routine tasks by autonomously preparing data, detecting anomalies and unusual behavior, visualizing the TM data, and reporting interdependencies of different states which are not easily traceable.

Machine Learning could disrupt how operators analyze and detect novelties affecting their space assets; referring to FDIR systems, the last two blocks (isolation and recovery) have been tackled through different approaches. Specifically, there is the need to provide a valid strategy to prevent a fault from spreading to other subsystems and remedy harmful effects to ensure the mission's operability. AI-based algorithms can provide a way to identify cascading failures by modeling the spacecraft and the interdependence between its subsystems. For instance, a Markov process, coupled with algorithms to decide which actions to take to recover the fault, could represent a good solution for the above problem. Moreover could be considered to employ Bayesian networks, which infer, from a probabilistic standpoint, how the fault can spread further and how it can be isolated.

Finally, AI could broadly impact how recommendation systems and mission planning systems are integrated into a mission control center or in the ground segment of a space-based service to complete the intelligence pipeline on the ground. Those components play a crucial role in facilitating efficient and effective decision-making. A recommendation system is designed to suggest actions based on data analysis and machine learning algorithms. In the context of space operations, these systems can provide recommendations for optimal trajectories, propulsion adjustments, and communication protocols for spacecraft. On the other hand, mission planning systems are responsible for developing plans for various stages of a space mission, including launch, orbit insertion, and planetary exploration. These systems can take into account a wide range of factors, such as spacecraft capabilities, mission objectives, and environmental conditions to generate mission plans that are both feasible and effective. Ultimately, the importance of these systems lies in their ability to improve mission success rates, reduce risks, and increase efficiency in space operations.

The selected approaches will be presented in more detail in the following sections.

9.2.2.1 Telemetry Analysis

In recent years, research has investigated the adoption of techniques and models for FDIR in spacecraft systems. Models are applied at the component and subsystem levels, primarily focusing on the Attitude and Orbit Control System, aiming to detect and isolate reaction wheel, GPS, Star Tracker, and magnetometer faults. Machine Learning algorithms are used for pattern recognition, prediction, and anomaly detection in onboard telemetry. Research results are promising, reaching high accuracy for several classes of failures.

In addition, the possibility of neural networks to output probability distributions helps extract insights about the component behavior even when no fault is confidently detected. Reasoning algorithms could use these data to infer structured information, explanations, and eventually recovering strategies.

Generally, the area of anomaly detection comes with its complexities, such as the unknown nature of possible anomalies, heterogeneous anomaly classes, rarity, imbalance of anomalies, and diverse types of anomalies. In [24], authors identified six main challenges to which they try to map the different Deep Learning based approaches for anomaly detection, which can be summarized in: low anomaly detection recall rate, anomaly detection in high-dimensional

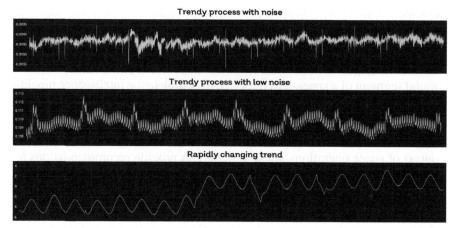

Figure 9.14 Forecasting-based anomaly detection approach. Screenshot from AIKO's Grafana dashboards.

and not independent data, data-efficient Learning of normality/abnormality, noise-resilient anomaly detection, detection of complex anomalies and anomaly explanation.

With particular attention on anomaly detection for space-related applications, [25] focuses on the comparison of anomaly detection and TM prediction between various techniques such as auto-regressed integrated moving average (ARIMA), Multi-Layer Perception (MLP), Recurrent Neural Networks (RNNs) with particular attention to Long Short-Term Memory (LSTM), Gated Recurring Units (GRUs), deep LSTM and deep GRU. When applied to three different TM data sets, good prediction results were achieved by ARIMA, but LSTM achieved the highest accuracy. Considering the execution time, the MLP was the fastest with the lowest accuracy. They show that using Neural Networks-based techniques such as LSTM is more efficient for long-term predictions, e.g., for communication satellites with a long lifetime.

Interesting work is also provided in [26]. The authors present a study investigating the detection of anomalies using LSTMs. They use a labeled data set from the Soil Moisture Active Passive satellite and the Mars Science Laboratory Rover Curiosity. Besides anomaly detection, the study focuses on TM prediction and introduces dynamic error thresholds. An issue that is often encountered in Deep Learning is a high rate of false positives. To counteract this specific issue, they adopted an anomaly-pruning approach. A threshold is placed based on historical data, and anomalies are re-classified as nominal based on the residual or distance towards the threshold.

To conclude, a brief look into Predictive Maintenance (PdM) techniques might be helpful. Within PdM, [27] extensively reviews AI-based methods. Besides knowledge-based approaches and traditional ML methods, their review focuses on Deep Learning. Methods such as Auto Encoders (AE) and RNN (especially LSTM and GRU) are analyzed with hybrid approaches that combine different DL approaches, specifically combinations of AE and LSTM, CNN and LSTM, and AE and CNN.

9.2.2.2 Root Cause Analysis

Suppose an anomaly is detected during routine operations. In that case, the root cause analysis involves finding the reason for lower operative modes (such as safe mode) and internal triggers issued by analyzing thousands of parameters and auxiliary data like event logs and telecommands sent.

A comprehensive review of the use of Deep Learning in fault analysis is presented in [28].In particular, it is shown that this is typically a two-step approach, which involves the detection of the occurring fault and then its diagnosis/classification. In the reviewed implementations, they show the progressive switch, during the research advancements, from MLP-based architectures, chosen because of their simplicity, flexibility, and reliability, to more powerful RNN-based architectures.

For what concerns more precisely the root cause analysis of anomalies during spacecraft operations, [29] proposes, also in this case, a two-step approach for diagnosing time-varying faults in a simulated CubeSat Attitude Determination and Control System (ADCS). First, single-class Support Vector Machines (SVMs) outputs (time series of fault signals) train an LSTM for fault isolation. Then, exploiting transfer learning, the SVMs are retrained with flight telemetry data, and the LSTM capabilities are tested on actual faults. A similar multi-step approach to diagnosing spacecraft faults' root causes based on SVMs is presented in [30]. First, Principal Component Analysis projects the input data onto a representational space with lower dimensionality. Then, a binary State Vector Machine (SVM) is used for fault detection. In the case of an affirmative response, nested multi-class SVMs are used to identify the root cause, obtaining a 97% classification accuracy on the test set (containing earth sensor, gyroscope, and infrared sensor faults).

Still, in spacecraft ADCS, [31] investigates a peculiar multi-classifier framework for fault detection and classification of satellite reaction wheels. An ensemble of classifiers such as Random Forest, SVM, Partial Least Square, and Naïve Bayes is employed to detect the fault's occurrence and subsequently identify its root cause. The results show that the ensemble of classifiers

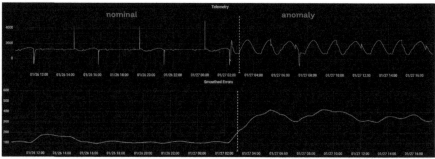

Fig: Forecasting-based anomaly detection

Figure 9.15   Spacecraft telemetry time series with different trends. Screenshot from AIKO's Grafana dashboards.

achieves better performance concerning the single classifiers' accuracy, sensitivity, and specificity. It is worth mentioning that concerning root cause analysis, far fewer resources are available concerning anomaly detection. Furthermore, most of the research done in this field focused on supervised techniques. However, this is not the typical spacecraft operational scenario, where only a few anomalies are generally available.

9.2.2.3   Diagnosis of Trends

The early detection and diagnosis of trends is a common task for AI. Trend detection of historical data and prediction of future timestamps is usually done for time series data. Several research directions on applying AI and DL are surveyed in different industries and branches. Regarding early detection and diagnosis of trends in space operations, a reason to rely on AI-based agents is that usual data visualization systems display a limited amount of telemetry data; thus, the operator cannot immediately identify trends that develop over time.

Statistical analysis can extract additional features and insights from historical telemetry data. A famous technique used in this context is the decomposition of time series. It is a statistical approach that deconstructs the time-series stream in multiple components to highlight patterns. This method decomposes the time series into three principal components.

- Trend. This component describes the general direction of the time series over a period (decreasing or increasing trend). It is calculated using a centered moving average of the data. The moving average is computed using a window length equal to the time-series frequency. The frequency is the observing repeating period (hours, months, years).

- Seasonality. By analyzing seasonality, finding the repeating pattern (repeating cycle) observed in a fixed, regular interval is possible. It is computed using seasonal averages repeated for the entire time series.

- Residuals. This irregular component consists of fluctuations, random deviations, or, eventually, the noise inside the time series. The remaining part of the original data is obtained by subtracting the trend and the seasonal component.

This analysis may be relevant when one aims to identify slow changes within the telemetries, such as degradation and aging factors, that can be useful to predict the remaining lifetime of a particular component. Sequence models represent the state-of-the-art for time-series forecasting. In [32], they provide an extensive overview of RNNs for diagnosing noisy time series, defining the fundamental trend concept. They formulate the trend estimation problem as a sequence-to-sequence prediction, considering three standard structures: vanilla RNN, LSTM, and GRU. The primary outcome is that the two latter architectures can be considered the best building blocks for trend detection in time series.

Despite being more complex, other suitable architectures for trend diagnosis from time series are Transformer Networks, particularly suited for forecasting complex dynamics. They differ from mainly other used models, such as LSTMs, for the working principle: within Transformers, data are treated as a whole, with dependencies learned by the internal self-attention mechanism. Moreover, Temporal Fusion Transformers (TFTs) can learn temporal relationships at different scales by combining recurrent layers for local processing and self-attention layers for long-term dependencies [33]. A further improvement is represented by Informer [34], a transformer-based architecture developed for long-sequence time-series forecasting with limited time complexity and memory usage.

9.2.2.4 Decision-Making and Recommendation Systems

MDPs are extensions of Markov chains: chains are stochastic models representable by a graph, where nodes are states and directed links represent the possibility of transitioning from a state to another; links are weighted with the transition probability, which in this way only depends on the previous state and not in the system history. Decision processes introduce actions, which allow choice in some transitions, and rewards, to score the result of actions and events. There are several methods to analyze an MDP, depending on whether the time horizon is finite or infinite. In the finite horizon case, the most popular approach is computing the optimal policy of action through backward induction based on the state reward. This approach provides a direct application to the use case of recovery from a system fault. The MDP graph represents the system states and the possible recovery actions, along with their effectiveness, formulated in terms of reward. Figure 6.1 shows an example of an MDP graph.

A Fault Tree (FT) is a logical and graphical diagram describing how basic equipment faults propagate through the system and eventually become a system-wide failure. They are obtained by performing a Fault Tree Analysis to analyze the risks related to safety and economically critical assets. Their graphical representations are intuitive and easy to understand. Some tree nodes represent units or subsystems potentially affected by a fault, and others stand as gates to regulate the fault spreading, such as AND or OR gates. FT can be extended with new features to Dynamic Fault Trees, which introduce temporal dependencies to redundancy concepts such as spare units, which can be included in the system and activated when a failure needs recovery. A further extension is represented by Non-deterministic Dynamic Fault Trees, which do not impose a fixed order on using spares and the occurring events in the system. An approach to FDIR is to turn a Non-deterministic Dynamic FT into a Markov Automaton, which offers a suitable way of computing the best non-deterministic decisions on spare activations. A recovery strategy can be synthesized as the definition of which spare has to be employed in which failure state of the system and can therefore guarantee optimal reliability at any time.

Knowledge-based Systems are a wide range of approaches that attempt to represent human-like knowledge in a machine-understandable structure and elaborate it to provide artificial intelligence skills. A typical example is expert systems, where the aim is to leverage knowledge to provide intelligent support or replace human agents in a specific domain, emulating the decision-making capability of a human expert. Such systems are composed of two core functions: an explicit representation of knowledge in the field of application, the knowledge base, and a reasoning algorithm operating on the knowledge base, the inference engine, which exploits data patterns and provides conclusions and guidelines for acting in the scenario. The knowledge base is often encoded as an ontology, a structured way of ordering concepts related to an environment; in other words, it is a representation vocabulary offering the means to build a model for a specific domain, detailed in objects appearing in the scenario, their qualities and mutual relations, behavior, and interactions. Usually, ontologies are layered, where different levels of granularity are described: the uppermost layers define the generic entities in the form of classes, which can be further specialized in underlying layers to domain-specific elements. Classes are instantiated when representing a concrete scenario and take values shaping the actual scenario.

The inference engine can reason on the assertions, relations, and properties at its disposal and calculate the likelihood of hypotheses, possible results of actions, and strategies to cope with tasks. The reasoning is based on general rules about the scenario entities and properties, guiding the engine to discover correlations and possible outcomes among all data available. Facts and connections between concepts are recorded, composing the reasoning baseline for the engine's operation. Rules in the field of application can be straightforwardly submitted by domain experts, thanks to the interface with ontology concepts, which provide a familiar representation of objects and events. Furthermore, the reasoning engine can elaborate current data to infer new rules or parameters to its own goals or verifications; for example, backward chain reasoning can start from desired goal states and trace back initial conditions that were unpredictable and unexploited. This approach also has a relevant advantage regarding AI explainability: the inference pattern can finally be exposed for human revision, highlighting the line of reasoning which brought the various outcomes.

Bayesian Networks (BNs) are a formalism for representing and reasoning under uncertainty in intelligent systems. In a BN, entities of interest are treated as random variables and represented as nodes in a network, connected by directed arcs indicating probabilistic dependencies between them. The network structure and conditional probability tables associated with each node provide a compact representation of the joint probability distribution of all variables. BNs are suitable for probabilistic inference by computing the posterior probabilities of variables from partial observations. This approach is similar to MDP in several aspects, but it differs as the nodes in a BN represent variables describing some system feature, not a state. At the same time, the links are strictly directed, and the graph is acyclic.

Dynamic BNs have already been employed in FDIR systems, appreciated for integrating the possibility of reasoning on system variables of different types and under uncertain circumstances. This approach provides a reactive response to failures and a preventive recovery strategy, where signs of potential faults may arise beforehand.

9.2.2.5 *Planning & Scheduling*

Planning and scheduling technologies are integral parts of space operations and closely related to the other aspects of the overall monitoring and control of space missions. The intended meaning of planning and scheduling in space operations often differs from their general meaning in the AI-automated planning and scheduling domain. Mission planning in space operations refers to all the activities related to defining an operational plan to operate and control a mission, including decision-making, generation of detailed schedules to be uploaded, plan execution, and satellite telemetry analysis. These can be performed manually, automated, or combined, usually called mixed-initiative planning.

Prominent constellations of satellites are becoming increasingly crucial for a wide range of applications, including telecommunications, remote sensing, and scientific research. However, a mission plan for a large constellation can be challenging due to overwhelming operational constraints, such as limited resources, operational efficiency, cost-effectiveness, conflicting end-users' requests and needs, and time constraints. Ultimately, the success of a large constellation mission depends on achieving its objectives. Optimizing the mission plan can help collect the necessary data, provide the required services, and achieve the desired outcomes. In summary, optimizing the mission plan for a large constellation is critical to ensure that the mission is successful and efficient and that the investment in the constellation is cost-effective. It requires careful consideration of operational constraints and the development of sophisticated optimization algorithms that consider various factors, including satellite orbits, coverage requirements, and ground station capabilities.

Symbolic AI approaches coupled with data-driven modeling could be deployed to raise the automation and optimality of mission planning and scheduling pipelines in such an endeavor. Those will be key technologies for many of the use cases described above. Automated planning in Artificial Intelligence is about decision-making to achieve a specific goal. An AI-based automated planning system differs from a traditional decision-supporting system because the input specifies an objective (what) rather than a 'how.' A model specifies "how" to achieve objectives and is then left to the planning process to devise 'how' to achieve that objective (i.e., with planned activities) coherently with the model. It is then a more general concept in terms of applicability. It supports a generic decisional process but is more specific in scope in the more general mission planning activity, being related only to the decision-making process.

The AI-based approach is agnostic concerning the specific mission use case to which it is applied because of the flexibility of the symbolic represent-ation of goals, constraints, logic, and parameters to be optimized. As the sys-tem is not designed to achieve (possibly parameterize) goals in each domain but to manipulate symbolic entities, the software deployment and test are sub-stantially independent of the specific mission. Timeline-based planning is his-torically a viable solution for mission planning-related applications because of the modeling primitives that provide extensive capabilities for managing resources and task allocation. Modeling is one of the most critical points for AI model-based applications, and to cope with modeling issues, the cognit-ive distance between modeling primitives made available by the system and the objects to be modeled has to be as small as possible. For this reason, the timeline-oriented paradigm has proved particularly suitable for space applic-ations because it is very close to how problems and constraints are naturally represented in space applications. Besides that, timeline-based systems allow the management of arbitrarily coded timelines, constraint models, and solving processes that are not easy or convenient to be modeled symbolically. Integ-rating learning techniques could significantly add value to the overall planning system concept.

Planning algorithms can be supported by integrating external (user-defined) modules. Those software components can supply inputs to the plan-ning problem, such as constraints, priorities, and activities to be included, thus increasing the automation level of the planning process and the ability to ex-plore what-if scenarios quickly and effectively. Moreover, custom models can be implemented to describe the behavior of specific procedures, subsystems, or components (resource consumption, priority computation, and long-term effects). These modules, powered by AI, can capture behaviors encountered in different missions and form a growing library of reusable modules that re-duces the tailoring effort of future missions.

For example, Deep Learning could provide a wide range of techniques to train models from available mission and platform data. Deep Learning mod-els can be exploited inside plug-in modules to encode and generate system constraints and model the behavior of specific subsystems or processes. Su-pervised approaches based on artificial neural networks are primarily envi-sioned to learn from historical planning data made available by the system or service owner. Furthermore, it would be possible to reduce model complex-ity and speed-up iterations by applying Reinforcement Learning approaches. This technique can provide solutions for automatic planning and scheduling or quick approximate plans on maneuvers changes to orient the solution search before computing a precise, exact solution.

9.3   OVERCOME THE PENETRATION BARRIER

In today's fast-paced and ever-changing technological landscape, it is common for new and innovative technologies to be introduced to the market. However, educating end-users on their use and benefits is essential with new technologies. This step is crucial for complex and cutting-edge technologies, where proper usage and comprehension can determine its success or failure. The case for introducing AI-powered applications in the space domain highly represents this issue.

One of the primary benefits of educating space operators, decision-makers, and space engineers, in general, is maximizing the adoption of AI technology in the space market value chain. Educating those users on advanced data-driven automation benefits and how to leverage and complement it correctly will increase their confidence in using it. They will become more familiar with the technology, its use cases, and how it can make their lives easier. With proper education and training, end-users will better understand this pioneering technology, leading to higher adoption rates. Space engineers will become more able to streamline their work processes, reduce errors, and improve the quality of their output, leading to time and cost savings, making the AI-powered applications more valuable to both the end-users and the company that develops them.

On the other hand, new technologies often bring new risks and potential hazards. Proper education and training can help to mitigate these risks and ensure the safe use of the technology. By educating end-users on the proper use and safety measures, companies can reduce the potential for accidents and errors when using the technology, ultimately leading to a safer and more secure work environment for everyone involved.

Innovative AI-based approaches in the whole space domain industry will potentially transform the market and improve space-based services' scalability and effectiveness over time. However, for their full potential to be realized, it is essential for organizations innovating in this field to educate end-users and take care of their needs and expectations. By doing so, companies can maximize technology adoption, increase efficiency and productivity, reduce risks, and improve customer satisfaction. Companies must prioritize user research, education and training, and continuous maintenance as AI technologies emerge to ensure their success in the marketplace.

The following sections will provide an overview of the most relevant aspects to be considered in this sense.

9.3.1 Product Usability and User Experience

Usability is the degree to which a software application can be used by its target users to achieve specific goals with effectiveness, efficiency, and satisfaction. It involves ease of use, learnability, memorability, and error prevention. A highly usable software application is more likely to be adopted and used by its target audience, leading to higher user satisfaction and improved business outcomes. As the use of AI continues to grow, so does the number of AI-based software applications such as chatbots and virtual assistants, or predictive analytics and Machine Learning algorithms. However, despite the potential benefits of AI, the success of an AI-based software application depends on more than just its technical capabilities. Usability is one of the crucial factors that can significantly impact the adoption and success of an AI-based software application. By ensuring that the application is easy to use and understand, users are more likely to accept and embrace the technology. A software application that is easy to use and understand reduces the user's cognitive load and allows them to focus on the task. By providing a positive user experience, users are likelier to continue using the application, leading to improved business outcomes.

When designing and developing AI-powered software applications with Machine Learning models at their core, taking care of the end-user experience is crucial. User Experience (UX) design must ensure accountability, reliability, trustworthiness, and explainability in these applications to maximize user satisfaction and acceptance. Reliability is critical in ML-powered applications as users rely on the software to make accurate predictions and decisions. UX design can impact reliability by ensuring that the software is intuitive and easy to use, reducing the potential for user error.

Trustworthiness is also an essential factor in the UX of AI-powered software applications, as users need to have confidence in the accuracy and reliability of the predictions and decisions made by the software. UX design can help establish trust by making the decision-making process transparent and explainable, enabling users to understand how and why the software makes certain decisions.

In conclusion, UX design is crucial in developing ML-powered software applications, focusing on reliability, trustworthiness, and explainability. By designing transparent, reliable, and trustworthy user-friendly software, developers can maximize user satisfaction and drive the success of AI technology.

9.3.1.1 Explainable AI

We have already described the enormous potential of AI-powered applications in space. However, AI must be transparent and explainable in safety-critical applications where the consequences of errors can be severe (i.e., maneuvering, commissioning phase, strategic decision-making). Explainable AI (XAI) refers to AI models that provide users with clear explanations of their decision-making process. XAI is a subfield of AI that aims to make Machine Learning models transparent and interpretable.

Explainability is a crucial aspect of UX in ML-powered applications, as users need to understand the reasoning behind the software's decisions. UX design can support explainability by providing precise and concise explanations of the model's outputs and offering interactive tools that enable users to explore the data and decision-making process. The goal of XAI is to provide users with an understanding of the decision-making process of the AI models. XAI techniques can be used to understand how the model works, identify biases or errors, and ensure that the model makes decisions based on relevant factors. In strategic applications, the transparency of the decision-making process is essential. For example, when assessing the root cause of a specific anomalous trend of a particular telemetry stream, the system's decision-making process must highlight how the AI system arrived at a particular decision, allowing operators and space engineers to make informed decisions about the specific space asset and the overall mission.

Safety-critical applications in the space domain are often subject to stringent regulations requiring transparent and explainable decision-making processes. By complying with these regulations, companies can ensure their products are safe and reliable, reducing the risk of accidents or other incidents. Furthermore, AI models are not perfect, and they can make errors. In the harsh space environment, these errors can have serious consequences. XAI can be used to debug the model and identify errors or biases that may impact the model's decisions. The AI model can become more accurate and reliable by identifying and correcting these errors.

Additionally, XAI can help to increase trust in the AI system. When an AI system's decision-making process is transparent and understandable, users are more likely to trust the system. This aspect is critical in applications that may affect the health status of a satellite and the overall availability of the service, where the consequences of an incorrect decision can be severe. An AI system can instill user confidence by explaining its decision-making process.

9.3.2   AI-based Products Maintainability and Scalability

Very few machine learning models get deployed [35]; according to a recent Gartner research, more than 85% of models developed do not reach deployment in the production environment [36].

As we said, Machine Learning models are becoming increasingly popular in today's businesses and organizations to solve complex problems, automate tasks and decision-making processes, and improve overall efficiency and productivity. However, as these models are deployed into production environments, they also bring new challenges. These challenges can arise from the difficulty in managing the lifecycle of these models, ensuring they are performing optimally and accurately, and maintaining the stability and reliability of the production environment. To overcome these challenges, organizations must rely on Machine Learning Operations (MLOps) procedures to manage their ML systems and models more efficiently and effectively.

It may be relatively simple to develop an ML model to analyze (detection, prediction, root-cause analysis) a stream of data (real-time and offline). Deploying the same model reliably while ensuring it maintains a qualitative performance on real, generalized data is far from obvious, resulting from being the main challenge of the industrialization and operationalization of ML-based products. In the space domain, the acceptance bar for innovative technology is extremely high, and the reliability of an innovative technology becomes a crucial aspect of being taken care of. So the necessity of MLOps to properly maintain over time AI-powered applications can be summarized as follows:

- Data Volume. The volume of data generated by the end-users in the production environment continues to grow, making it increasingly challenging to manage and extract meaningful insights. With MLOps, organizations can process and analyze large amounts of data consistently and reliably, reducing the risk of errors and ensuring the accuracy and quality of their models.

- Fine-Tuning of ML model parameters. The process of fine-tuning the parameters of ML models could be streamlined, ensuring that the models are optimized for their specific use cases and data. This process automation helps reduce the risk of manual errors and the time and effort required to fine-tune the model's parameters. In addition, MLOps allows organizations to automate the process of monitoring and evaluating the performance of their models, providing valuable insights into the effectiveness of the fine-tuning process. By automating these processes,

ML-based product providers can quickly identify areas where the model is performing suboptimally and adjust the parameters as needed, helping to optimize the model's performance.

- Keep track of the features the model works with. Feature engineering contributes largely to model accuracy; demand for this task, through a highly automated process, to data scientists and ML model developers is essential in maintaining the effectiveness of the ML-based final product.

- ML models monitoring and debugging. Anticipating any performance degradation of the analysis models underlying the product is essential to ensure it does not stop providing its added value to the end-user, becoming useless.

- Production environment data may change over time. Models rely on real-world data for predicting; as real-world data changes, so should the model. We must keep track of new data changes and ensure the model learns accordingly.

MLOps combines software engineering and data science practices that help the model provider automate and manage ML models' entire lifecycle, from development to deployment and ongoing maintenance. Those automated pipelines are designed to streamline the deployment and maintenance of ML models, reducing the time and effort required to manage them and ensuring that the models perform optimally. Organizations can make the most of their investment in ML, enabling them to leverage these models to their full potential. One of the key benefits of adopting MLOps procedures is that it increases the speed of deployment of their models. ML product providers can ensure their models are quickly and easily deployed into the production environment by automating the deployment process. This reduces the time required to get new models into production and enables organizations to respond more rapidly to changing business needs and demands.

MLOps also helps companies maintain their production environment's stability and reliability. By automating the monitoring and maintenance of the models, ML providers can ensure that the models are performing optimally and accurately, even in complex and dynamic environments. This aspect helps to reduce the risk of unexpected failures and improves the overall stability and reliability of the production environment. Another essential benefit of MLOps is that it helps maintain overall models' quality over time. ML model providers can ensure that their models meet the required accuracy and performance standards by automating the testing and validation process. This reduces the risk of models producing incorrect results and improving quality.

Definitely, MLOps also enables cost reduction in deploying and maintaining ML models. Users can make the most of their investment in ML and achieve a higher investment return. In conclusion, MLOps is essential to the ML model deployment process. It is becoming increasingly important as companies and institutions in space adopt ML models in their operational pipelines and workflows.

9.4 CONCLUSION

As highlighted at the beginning of this chapter, the demand for space-based services is accelerating. Consequently, the number of satellites in Earth's orbit and the mission's complexity are exploding. The need for autonomous spacecraft it has never been higher.

Automation is about relieving humans of repetitive tasks that intelligent machines can perform better in several orders of magnitude lower times. Human operators and decision-makers can use highly automated processes to focus on creative tasks with higher added value for the specific industrial context. Indeed, automation is a critical enabler in the 4.0 industrial digital transformation for the optimal management of complex, innovative, highly performant technologies. Companies that introduce automation technologies into their operating pipelines will optimize their workforce's processes and activities and create new business models.

Yet, the space sector is now struggling to find its way towards intelligent automated operations, implementing robotic automation workflows at best. Continuous data integration in such workflows is still out of sight due to multiple siloed information sources, the inability to manage diverse and heterogeneous data types, and the lack of innovative, helpful technology to extract relevant intelligence insights from data streams. Artificial Intelligence could be considered the technological enabler of the shift toward future-ready space missions characterized by enhanced autonomy and automation.

AI has been revolutionizing industries worldwide, and the space market is no exception. By utilizing AI in various space-related applications, we can improve efficiency, accuracy, and reliability and push the boundaries of what is possible in space exploration. AI-based technologies represent critical enablers in the transition aiming at reshaping how space systems operate; this includes the entire collection of tasks and routines required to manage space assets and value-adding payloads and support in-space human operations.

AI adoption in space missions could create a future where human creativity and machine intelligence work seamlessly on Earth and beyond. This approach will ultimately accelerate progress in space exploration by leveraging the power of AI to enable breakthroughs and discoveries. By combining human ingenuity with the incredible capabilities of machine intelligence, operators and engineers can achieve unprecedented levels of innovation and success in space.

ACRONYM

ω**KA** Omega-K Algorithm

ADCS Attitude Determination Control System

AE Auto Encoders

AI Artificial Intelligence

ARD Analysis-Ready Data

ARIMA Auto Regressive Integrated Moving Average

B2B Business to Business

B2C Business to Customer

BN Bayesian Network

CI Continuous Intelligence

CNN Convolutional Neural Network

CSA Chirp Scaling Algorithm

CV Computer Vision

DL Deep Learning

DNN Deep Neural Network

DRL Deep Reinforcement Learning

ECSS European Cooperation for Space Standardization

EO Earth Observation

FDIR Failure Detection Isolation and Recovery

FFT Fast Fourier Transform

FT Fault Tree

G&C Guidance & Control

GAN Generative Adversarial Network

GEO Geostationary Earth Orbit

GPS Global Positioning System

GRU Gated Recurring Unit

HDA Hazard Detection and Avoidance

IMU Inertial Measurement Unit

IOS In-Orbit Servicing

LEO Low-Earth Orbit

LSTM Long Short-Term Memory

MDP Markov Decision Process

MISR Multi Image Super Resolution

ML Machine Learning

MLOps Machine Learning Operations

MLP Multi-Layer Perception

PdM Predictive Maintenance

RDA Range Doppler Algorithm

RNN Recurrent Neural Network

RPO Rendezvous and Proximity Operations

SAR Synthetic Aperture Radar

SfM Structure from Motion

SISR Single Image Super Resolution

SLAM Simultaneous Localization and Mapping

SLC Single-Look-Complex

SPECAN Spectral Analysis Algorithm

SSA Space Situational Awareness

STM Space Traffic Management

SVM State Vector Machine

TFT Temporal Fusion Transformer

TM Telemetry Analysis

UX User Experience

VN Visual Navigation

XAI Explainable Artificial Intelligence

GLOSSARY

Auto Encoders: a type of neural network architecture used for unsupervised learning and dimensionality reduction tasks. They consist of an encoder network that maps input data to a lower-dimensional latent space and a decoder network that reconstructs the original data from the latent representation.

cloudy_CHARLES: an on-board data processing software – developed by AIKO S.r.l. – based on Machine Learning models. It identifies clouds in optical images moments after their acquisition, allowing them to keep only the most useful and clean frames.

Convolutional Neural Networks - CNNs: are a type of deep neural network widely used in image and video processing tasks. They excel at capturing spatial patterns and features within data, making them particularly effective in tasks such as image classification, object detection, and image segmentation. CNNs employ convolutional layers, which apply filters to input data to extract relevant features, followed by pooling layers that downsample the data. These networks have revolutionized computer vision by achieving state-of-the-art performance and enabling advancements in areas like autonomous vehicles, medical imaging, and facial recognition.

Deep Learning - DL: a subset of machine learning that utilizes artificial neural networks with multiple layers to learn and extract hierarchical representations of data. Deep learning has revolutionized various fields by achieving remarkable performance in tasks such as image and speech recognition.

Deep Reinforcement Learning - DRL: combines reinforcement learning and deep learning techniques to enable agents to learn sequential decision-making through trial and error. Key terms in this field include reinforcement learning, deep learning, agent, environment, Q-learning, Deep Q-Network (DQN), policy, exploration vs. exploitation, reward function, convolutional neural network (CNN), and policy gradient. These concepts form the foundation for understanding how deep reinforcement learning algorithms optimize decision-making processes in complex domains.

Explainable Artificial Intelligence - XAI: refers to the field of research and development focused on creating artificial intelligence systems that can provide understandable and transparent explanations for their decisions and actions.

Generative Adversarial Networks - GANs: is a type of deep learning model that consists of two interconnected neural networks, namely the generator and the discriminator, which are trained in a competitive manner. The generator network aims to produce synthetic data, such as images or text, that closely resemble real data from a given distribution, while the discriminator network tries to distinguish between the real and generated data. The two networks are trained simultaneously, with the generator improving its ability to produce realistic data as the discriminator becomes more skilled at differentiating between real and fake samples.

Machine Learning: is a branch of artificial intelligence that focuses on developing algorithms and models that enable computer systems to learn and make predictions or decisions without being explicitly programmed.

orbital_OLIVER: an AI-powered software – developed by AIKO S.r.l. – conceived for satellite providers and operators. It streamlines spacecraft operations and augments the capabilities of the satellite. orbital_OLIVER acts as a satellite operator onboard the spacecraft, optimizing the mission plan in real-time.

Reinforcement Learning - RL: is a type of machine learning where an agent learns to make sequential decisions through trial and error by interacting with an environment. The agent receives feedback in the form of rewards or penalties to guide its learning process.

FURTHER READING

LeCun, Y. (2022) "A path towards autonomous machine intelligence version 0.9. 2, 2022-06-27". *Open Review*, 62. Accessible on `https://openreview.net/pdf?id=BZ5a1r-kVsf`

Warren B. Powell (2022) "Reinforcement Learning and Stochastic Optimization: A unified framework for sequential decisions". *Wiley.* Accessible on `https://doi.org/10.1080/14697688.2022.2135456`

Huen, Chip (2022) "Designing Machine Learning Systems: An Iterative Process for Production-Ready Applications". *O'Reilly Media, Inc.*

Troesch, M., et al. (2020) "MEXEC: An Onboard Integrated Planning and Execution Approach for Spacecraft Commanding". *JPL Open Repository*, V1. Accessible on `https://hdl.handle.net/2014/53230`

Castano, R. et al. (2022) "Operations for Autonomous Spacecraft". In *2022 IEEE Aerospace Conference (AERO), IEEE*, pp. 1–20. Accessible on `https://doi.org/10.1109/AERO53065.2022.9843352`

Vaquero, Tiago S., et al. (2022) "A Knowledge Engineering Framework for Mission Operations of Increasingly Autonomous Spacecraft". *California Institute of Technology*. Accessible on `https://icaps22.icaps-conference.org/workshops/KEPS/KEPS-22_paper_7985.pdf`

Chien, S., Parjan, S., & Harrod, R. (2022) "Distributed Observation Allocation for a Large-Scale Constellation". *JPL, NASA*. Accessible on `https://ai.jpl.nasa.gov/public/documents/papers/Parjan_DistributedConstellations_OptLearnMAS_2022.pdf`

BIBLIOGRAPHY

[1] Euroconsult. "global market for commercial earth observation data and services to reach $7.9 billion by 2031". `https://www.euroconsult-ec.com/press-release/global-market-for-commercial-earth-observation-data-and-services-to-reach-7-9-billion-by-2031/`, 2022.

[2] NASA Autonomous Systems, Systems Capability Leadership Team. "autonomous systems taxonomy report", (2018), pp.1, (accessed 25.01.2023). https://ntrs.nasa.gov/citations/20180003082, 2018.

[3] NASA JPL. "strategic technologies report (2019)", pp.11–13. https://scienceandtechnology.jpl.nasa.gov/sites/default/files/documents/JPL_Strategic_Technologies_2019.pdf, 2019.

[4] Jeremy Frank. "NASA autonomous mission operations roadmap", nasa ames research center. https://www.nasa.gov/sites/default/files/files/J_Frank-Autonomous_Mission_Operations.pdf.

[5] AIKO S.r.l. "orbital_OLIVER white paper". https://www.datocms-assets.com/52867/1675272050-orbital_oliver_wp.pdf.

[6] Issa AD Nesnas, Lorraine M Fesq, and Richard A Volpe. Autonomy for space robots: Past, present, and future. *Current Robotics Reports*, 2(3):251–263, 2021.

[7] AIKO S.r.l. "cloudy_CHARLES product page". https://aikospace.com/products/cloudy-charles/, 2023.

[8] Jacob Shermeyer and Adam Van Etten. The effects of super-resolution on object detection performance in satellite imagery. In *Proceedings of the IEEE/CVF Conference on Computer Vision and Pattern Recognition (CVPR) Workshops*, 2019.

[9] Christian Ledig, Lucas Theis, Ferenc Huszar, Jose Caballero, Andrew Cunningham, Alejandro Acosta, Andrew Aitken, Alykhan Tejani, Johannes Totz, Zehan Wang, and Wenzhe Shi. Photo-realistic single image super-resolution using a generative adversarial network. In *Proceedings of the IEEE Conference on Computer Vision and Pattern Recognition (CVPR)*, July 2017.

[10] Xintao Wang, Ke Yu, Shixiang Wu, Jinjin Gu, Yihao Liu, Chao Dong, Yu Qiao, and Chen Change Loy. Esrgan: Enhanced super-resolution generative adversarial networks. In *Proceedings of the European Conference on Computer Vision (ECCV) Workshops*, September 2018.

[11] Soo Ye Kim, Jeongyeon Lim, Taeyoung Na, Munchurl Kim. "3CSRnet: Video super-resolution using 3D convolutional neural networks;

computer vision and patter recognition". `https://doi.org/10.48550/arXiv.1812.09079`, 2019.

[12] Luc Maisonobe; Pascal Parraud. Very low thrust station-keeping for low earth orbiting satellites. *Advances in Space Research*, 71(3):1558–1593, 2023.

[13] Avishai Weiss; Uroš V. Kalabić; Stefano Di Cairano. Station keeping and momentum management of low-thrust satellites using MPC. *Aerospace Science and Technology*, 76:229–241, 2018.

[14] Jérôme Thomassin, Sophie Laurens, and François Toussaint. Asteria: Autonomous collision risks management. *Acta Astronautica*, 200:599–611, 2022.

[15] Morante, D.; Sanjurjo Rivo, M.; Soler, M. A survey on low-thrust trajectory optimization approaches. *Aerospace*, 8(3):88, 2021.

[16] John T Betts. *Practical methods for optimal control and estimation using nonlinear programming*. SIAM, 2010.

[17] Anil V Rao. A survey of numerical methods for optimal control. *Advances in the Astronautical Sciences*, 135(1):497–528, 2009.

[18] Francesco Topputo and Chen Zhang. Survey of direct transcription for low-thrust space trajectory optimization with applications. In *Abstract and Applied Analysis*, volume 2014. Hindawi, 2014.

[19] Dario Izzo and Ekin Öztürk. Real-time guidance for low-thrust transfers using deep neural networks. *Journal of Guidance, Control, and Dynamics*, 44(2):315–327, 2021.

[20] D. S. Kolosa. A reinforcement learning approach to spacecraft trajectory optimization. *Journal of Guidance, Control, and Dynamics*, 44(2):315–327, 2019.

[21] Sutton, R. S., and Barto, A. G. "Reinforcement Learning: An introduction". `http: //incompleteideas.net/book/the-book-2nd.html`, 2nd ed., The MIT Press. 2018.

[22] L. Romanelli, M. Stoisa, F. Paganelli Azza, P. De Marchi, M. Varile. "on-board guidance and control for low-thrust orbit transfers", 17th international conference on space operations using Deep Reinforcement Learning. `https://spaceops2023.org/wp-content/uploads/2023/03/A4_SpaceOps_Digital-Program-Book_V8.1.pdf`.

[23] Bo Chen, Jiewei Cao, Alvaro Parra, and Tat-Jun Chin. Satellite pose estimation with deep landmark regression and nonlinear pose refinement. In *Proceedings of the IEEE/CVF International Conference on Computer Vision Workshops*, 2019.

[24] Guansong Pang, Chunhua Shen, Longbing Cao, and Anton Van Den Hengel. Deep learning for anomaly detection: A review. *ACM computing surveys (CSUR)*, 54(2):1–38, 2021.

[25] Sara K Ibrahim, Ayman Ahmed, M Amal Eldin Zeidan, and Ibrahim E Ziedan. Machine learning methods for spacecraft telemetry mining. *IEEE Transactions on Aerospace and Electronic Systems*, 55(4):1816–1827, 2018.

[26] Kyle Hundman, Valentino Constantinou, Christopher Laporte, Ian Colwell, and Tom Soderstrom. Detecting spacecraft anomalies using lstms and nonparametric dynamic thresholding. In *Proceedings of the 24th ACM SIGKDD international conference on knowledge discovery & data mining*, pages 387–395, 2018.

[27] Yongyi Ran, Xin Zhou, Pengfeng Lin, Yonggang Wen, and Ruilong Deng. A survey of predictive maintenance: Systems, purposes and approaches. *arXiv preprint arXiv:1912.07383*, 2019.

[28] Ahmad Azharuddin Azhari Mohd Amiruddin, Haslinda Zabiri, Syed Ali Ammar Taqvi, and Lemma Dendena Tufa. Neural network applications in fault diagnosis and detection: an overview of implementations in engineering-related systems. *Neural Computing and Applications*, 32:447–472, 2020.

[29] Justin Mansell and David Spencer. Data-driven fault detection and isolation for small spacecraft. 10 2019.

[30] Yu Gao, Tianshe Yang, Nan Xing, and Minqiang Xu. Fault detection and diagnosis for spacecraft using principal component analysis and support vector machines. In *2012 7th IEEE Conference on Industrial Electronics and Applications (ICIEA)*, pages 1984–1988. IEEE, 2012.

[31] Hasan Abbasi Nozari, Paolo Castaldi, Hamed Dehghan Banadaki, and Silvio Simani. Novel non-model-based fault detection and isolation of satellite reaction wheels based on a mixed-learning fusion framework. *Ifac-papersonline*, 52(12):194–199, 2019.

[32] Alexandre Miot and Gilles Drigout. An empirical study of Neural Networks for trend detection in time series. *SN Computer Science*, 1(6):347, 2020.

[33] Bryan Lim, Sercan Ö Arık, Nicolas Loeff, and Tomas Pfister. Temporal fusion transformers for interpretable multi-horizon time series forecasting. *International Journal of Forecasting*, 37(4):1748–1764, 2021.

[34] Haoyi Zhou, Shanghang Zhang, Jieqi Peng, Shuai Zhang, Jianxin Li, Hui Xiong, and Wancai Zhang. Informer: Beyond efficient transformer for long sequence time-series forecasting. In *Proceedings of the AAAI conference on artificial intelligence*, volume 35, pages 11106–11115, 2021.

[35] KDNuggets. "models are rarely deployed: An industry-wide failure in machine learning leadership"; KDNuggets web post. https://www.kdnuggets.com/2022/01/models-rarely-deployed-industrywide-failure-machine-learning-leadership.html.

[36] GARTNER. "gartner says nearly half of CIOs are planning to deploy artificial intelligence"; Gartner web post. https://www.gartner.com/en/newsroom/press-releases/2018-02-13-gartner-says-nearly-half-of-cios-are-planning-to-deploy-artificial-intelligence, 2018.

VI

Look Back from the Future

CHAPTER 10

Commercial Human Space Exploration Assisted by Artificial Intelligence

Anousheh Ansari

XPRIZE Foundation, California, USA

Jim Mainard

XPRIZE Foundation, California, USA

CONTENTS

10.1	Introduction	364
10.2	We pass the test!	365
10.3	To the Moon	370
10.4	Starlab	373
10.5	Big day	384
10.6	Real Intelligence	390
10.7	Epilogue	393

A S the cost of access to space drops while safety and frequency of manned space mission increases, the dream of making space accessible to the public is within reach. This means allowing those who would want to brave a rocket ride, the opportunity to experience space for a few hours, few days, or permanently establishing a home base in orbit or on the Moon or Mars. The rate of technological advancement and innovation in space has never been greater since the Yuri Gagarin, a colonel in soviet air force orbited our planet in 1961, becoming the first earthling to leave our planet. As the road is being

DOI: 10.1201/9781003366386-10

paved for more and more of us to venture into space, certain types of innovation would be necessary to make space accessible, equitable, safe, and enjoyable. Being in space would not be a mission, but an adventure destination or a way of life. The wide array of technologies in need of significant advancement includes material science, robotics, artificial intelligence, quantum chemistry, propulsion, and countless others. The solutions that will shape the future of humanity in space will include ways to protect human body and mind from the harsh environment and isolation of space. As we don't want to repeat our mistakes on earth, all solutions that would be implemented should have sustainability, accessibility and equity at the center. This means creating closed loop systems to sustain life in space independent of earth resources. We would not only tap into our planetary resources, but harnessing energy and resources of the Moon or the asteroids for Earth would be essential part of this explorations journey. As we make space our home, it needs to feel like home with art, music, recreation, and ways to enrich our experience and life. The best part is that, in micro gravity and low gravity of space, even the simplest thing would act differently than when entrenched in the gravity well of our planet. This would make for interesting creations that would forever change us. This book so far has explored the technological possibilities of AI and how it can play a role in paving the path for humans to become an interplanetary species. This chapter will focus on the use of Artificial Intelligence, not only to protect human health and body in space, but also feeding human mind to make the experience, one to remember and desire.

10.1 INTRODUCTION

The rate of technological advancement and innovation in space has never been greater since Yuri Gagarin, a colonel in soviet air force orbited our planet in 1961, becoming the first earthling to leave our planet. One of the powerful technologies leading the way is Artificial Intelligence (AI) but its effect is amplified by the wide array of other technologies that are also advancing at an exponential rate, including material science, robotics, quantum chemistry, propulsion, and countless others. AI combined with these fast-evolving technologies are finally making it possible for us to imagine a world where space is not a place just for a brave few to visit, but a place where we live, play, and build businesses.

Sometimes it is hard to imagine this future and how it would evolve and alter our way of life both on and off Earth. In this chapter we attempt to bring these possibilities to life through the story of a family's journey in distant future where many of the technologies you read about in previous chapters have become mainstream and have shaped humanity's future. It takes place at a time when safe and frequent manned space missions are the norm. This means the public have an opportunity to experience space for a few hours, a few days, or permanently make a home base in orbit, on the Moon or Mars. It focuses on how Artificial Intelligence is embedded in many things not only to protect human health and body in space, but also in feeding the human mind to make the experience one to remember and desire.

As we make space our home, it needs to feel like home with art, music, recreation, and ways to enrich our experience and life. As you read about the experiences of this fictional family, our hope is to transport you in time and space and help formulate important questions in your mind to impact the work you do today in shaping the future of humanity in space tomorrow.

10.2 WE PASS THE TEST!

As I woke from a much needed, yet not entirely-fulfilling sleep, I was disoriented. It took me a minute or so to realize where I was, and why I was waking up after just a few hours of sleep. I'd left the shades open, and the light was streaming down into my sleeping pod. Framed by the porthole, was my home, a beautiful blue marble unlike the gray dust of my temporary home. My entire existence up until a few weeks ago had been comfortably living on that blue beauty. Now I was here on the Moon with my parents and brother on what my Mom described as a workcation — at least for her. The rest of us were on an adventure, a vacation, and frankly, a life-altering experience. Nothing would be the same when or if we made it back to Earth.

Working in space was now accessible well beyond just the few brave souls that ventured out over the intervening decades, now some fifty years past the launch of the Artemis mission. Space travel had become commonplace enough that even space tourism was a growing business, and some chose to stay and colonize; tourists becoming extraterrestrial semi-permanent residents.

How did I come to be here? Well, that has much to do with my grandpa. He'd been recounting to my Mom, and later me, stories of his days in space, when space was the province of highly trained astronauts and daring explorers. I was captivated by his stories as was my mother. He was now in his late nineties, and while healthy and active, he was considered to be too old to safely make the trip. If it were up to him, he'd be here with us, but health assessments by doct-o-bots (DBs) would say otherwise. While excited for us, he was heartbroken to be left behind, even if it was only 3-6 months.

The DBs had been fed data and trained over decades to assess potential space travelers and determine their probability of space suitability. Predecessor DBs had used machine learning techniques initially developed at the turn of the millennium that later became Generative Pre-Trained Transformers (a.k.a. ChatGPT) introduced a few decades later. I studied them in middle school. It seems a narrow and limited learning solution, but my grandpa has often reminded me that hindsight is an exact science. Someday what we do today will be viewed similarly. Maybe, but at 17 years old, it's a bit hard for me to embrace that concept. Today, our modern systems use highly advanced artificial intelligence and, unlike prior systems, are highly adaptable to changing conditions and broad decision making. Design and construction of these systems is required coursework in high school.

One thing I've noticed here in space is that I have a more difficult time overcoming distraction; seems like something someone, not me, should study with the aid of a solver-bot. I've already gotten off-track, I was explaining how we got here.

My mother is a medical doctor and researcher. She is evaluating the physical and metaphysical aspects of living in space. Not everyone tolerates space as well. It can still be a lonely place, despite the work that has been done to make it more like home. For those born on Earth my mother says it will likely never feel like home. The next generation that has never lived on Earth will adapt and find comfort here. My Mom's company is the one that developed the gene therapy system that makes the human body resilient to radiation, one of the biggest dangers for those living outside the protection of Earth's atmosphere. The first settlers on the Moon had received an external tiny pump that injected subdermally, releasing the gene enhancing therapeutics based on data it would collect as the genes started changing. Eventually, the pumps would be dissolved, and the remains would exit the body. Based on the data collected, her company hopes to come up with a one-time therapy that would be personalized based on the formula the solver-bots would develop after analyzing the information collected. My Mom's assignment here was to do onsite interviews and collect tissue samples and evaluations of long staying residents, and the Company felt her being in space would help her finalize the development of the one-time therapy.

I remember when she first broached the opportunity with us. It was a Sunday family dinner with the four of us and Grandpa. Grandma had died several years prior, and Grandpa missed her dearly. Mom tested the waters telling us it was an opportunity for us all to experience what Grandpa had lived, but with far more safety and comfort. Grandpa spoke right up encouraging us to take the leap. Papa was more than a little concerned. Papa was an accomplished artist, and, frankly, the worrier in the family. My younger brother of three years resisted going because he didn't want to be away from his friends and teammates for so long. This was a six-month commitment, maybe longer. In the end, Mom and Grandpa won out, and the process of assessment and preparation began.

Getting cleared for a space trip for work or play was quite an undertaking. Demand was high to experience space, but due to Mom's work we were on the accelerated track and found ourselves two weeks later at the Virgin Galactic Space Port in Las Cruces, New Mexico. From the several floor-high walls of glass in the traveler space readiness assessment and preparation building (AKA the "flame pit", tribute to the many wash outs in space suitability), we could see the Organ Mountain range across the desertscape with its shards of red and orange, and Needle peak rising from the desert floor nearly 3,000 meters.

Assessment was broken into two tracks, physical and psychological. For most, the physical requirements were easier to achieve. If you were fit and healthy you had a very good chance of making it through this trial and most that didn't make the cut were filtered out by the DBs review of their medical records before arriving. Grandpa had received word that he was deemed "unsuitable for space travel" just one week ago. This wasn't a human decision, but rather a logical, practical decision by a data-driven, emotionless, enhanced neuro-mimicking autocrat. We had given over a great deal of decision making in our lives as a species to these autocrats, and most believed it was for the better, but these "beings" could not understand the emotional havoc they could wreak on the human psyche. Having been a decorated test pilot and astronaut for much of his adult life and one of the lunar astronauts on the first Artemis return mission to the Moon, we could see this hit hard at his emotional core. We tried to comfort him, and while logically he understood, emotionally he felt he had been stripped of a key aspect of his identity. He was no longer an astronaut or a husband, but he was still a father and grandfather as we reminded him frequently. Of course, he could be space ready in no time if he would undergo the Rejuvena process that would reverse his age by 20 years, but after Grandma died, he didn't want to live any longer than his natural life. Grandma was a Naturalist, who didn't believe in age reversal and would not accept many of the medical treatments that could have extended her life. It was ironic, since my Mom decided to go in the opposite direction working for Rejuvena. The Naturalist movement started when the first Advanced Artificial Intelligence (AAI) system broke the code of genetic aging and allowed scientists to regenerate organs and tissues back to their healthy state and eventually rejuvenate the entire cellular structure of the human body. The company that came up with the breakthrough was Rejuvena and was started by two brilliant women who won the Age Span XPRIZE competition back in 2030. Now, thirty-seven years later, they have changed the course of humanity and made it possible for people to live a long, healthy life, even in space. What was unique about these two young women was that they open-sourced their findings so the benefits of their invention would be available to all. Now they were working on extending healthy human life span in space.

Onsite the training involved life-like real-time rendered virtual environments in which our sense of balance and adaptability to what we might feel in a low or zero G environment was put to the test. It wasn't the same as being in space, but over time the DBs learned to, with high certainty, identify those that would have a poor, or even potentially life-threatening journey. The assessment was the initial source of the learning for the DBs, but during the entire space journey, our physical selves were monitored in real-time. This resulted in a massive data store, credited with leading to the invention of a host of medicines and treatment protocols to help us live well in space.

It has been said that the eyes are the window to the soul. This can have many interpretations based on your belief systems, but beyond debate, the eyes tell a great deal about what someone is thinking and feeling. About twenty years ago, Xena was first developed to evaluate eye health, a proxy for an ophthalmologist, in regions where doctors were in short supply. Initially, and within mission parameters, the bots became proficient in identifying diseases of the eye, and later maladies in other areas of the body that could manifest through the eye tissue. An impressive feat in just a few short years and lauded across the globe. But, soon after, the bot began reporting mental health issues completely on its own. It was not directed to do so, but began generating detailed reports, and later suggested treatments for mental-health-related conditions. At first, mental health professionals scoffed at this ad-hoc assessment of human mental health as malpractice and went to great political and regulatory means to put an end to it. Despite this tremendous resistance, just a few years ago, the use of Xena began to gain approvals for mental health assessment in many countries around the world. Whether this was out of necessity, as there were never enough psychiatrists to support a population of ten billion, or because Xena was at least as capable as a typical psychiatrist is unclear. But here "she" was in the "flesh", able to evaluate large numbers of individuals simultaneously, filling that void. Licensed psychiatrists referred to Xena as a preliminary psychiatric assessment tool, not a replacement for a well-trained psychiatrist. Oddly, this had been the same line of resistance that so many other professionals had used before the intelligent bots became more capable of operating at scale and at a fraction of the cost. Like it wasn't difficult enough to decide what you wanted to do in your career, now you had to think about which types of roles were both fulfilling and least likely to be replaced by the ever-growing constellation of intelligent synthetic beings (ISBs). That's a lot of pressure for me as a teenager. I'm planning to be an architect as it combines creativity, beauty and function, something that ISBs have yet to conquer. I'm hoping they never do.

One last assessment area that a surprising many didn't make the cut was in being tested for claustrophobia and astraphobia (a space specific version of agoraphobia). These were tested with virtual reality (VR) environments in which we were put through an experience of being in very tight spaces and alternatively floating free in near-infinite space. My family did well in this environment as we had limestone caves in the desert near our home and were free from claustrophobic anxieties. We were comfortable in both.

What we were less prepared for was the actual launch into space and later transit to the Moon. I'm not sure anything will really prepare you for that, and I'm at loss for words to describe it, a condition I have only rarely experienced. My brother would say I never shut up, but Papa says I have a quick mind and plenty to share. They are probably both right.

Our trip to the Moon would happen in two stages. We would first use the Orbital Transit System or OTS for short, to go the Starlab II to do final fitting of our personalized space suits, get acclimated and finish our training for the second part of the flight to the Moon onboard the new Selene Earth-Moon (SEM) shuttle system.

I was particularly excited when we did our 3D scans at the end of our qualification day at the Virgin Galactic space port. I entered this glass cylinder about a meter in diameter and the two revolving walls started moving in opposite direction and showered me with beams of light. I felt I was being beamed up to the Moon. I wish we could have been atomically transported instead of this long trip, but we had made limited advancement on transporting a large object instantaneously over long distance. Scientists have made it as far as a transporting 1 mm cube of iron 100 meters but that is far from transporting a complex organic being over 380,000 kilometres. Maybe that's something I'll work on if the architecture plan doesn't pan out.

The OTS was less exciting and lacked the future forward look of the Virgin Galactic space port. Even though the port was built back in the early 2000s named after the first commercial suborbital transport company created after winning the first XPRIZE. The older space port was well kept and updated over the decades, but it had a cool immersive story wall that told the history of commercial space travel. It was hard to believe that for the first sixty-five years of the space program, only governments could fly people to space, and because of that was probably the slowest time in space technology advancements. During that entire time only 500 or so people flew to space, but now it is so common we don't even keep track; it is just something everyone has access to, health permitting. I have always been fascinated by XPRIZE and its Equitable Abundance Movement, repeatedly referenced in my history studies they had launched a series of competitions over the course of fifty years, and each had nudged us toward a better future that I get to enjoy now. I wish I could have been one of those pioneers who competed back then and made a real impact on the future. Today it seems synthetics are responsible for most advancement. Maybe the transporter would be my thing or a warp drive! — a way to get humanity farther than we have ever been before. Probably the warp drive, I don't have the patience for the transporter, at least the warp drive is almost here in early experimental stages.

10.3 TO THE MOON

Ok, back to our journey to the Moon. We boarded the OTS early in the morning. We were all very quiet on our ride there. The POD picked us up at 6 AM and Tep, my brother, was not in a talking mood. His friends were getting ready for a hiking trip to Machu Picchu, a top destination for him, and now he was being dragged to the Moon – the gray, dusty, boring Moon. Mom was busy looking through her research notes and Papa was just gazing at the distance as the rising sun was painting the landscape in vibrant orange hues, knowing that for a while he wouldn't see colors as beautiful as this. He always got his inspiration from nature, even though he was considered a techno-artist as he uses AI and purpose-built machines that he designed to create large 3D masterpieces. He was going to use his time on the Moon to create a new form of art. He wasn't sure what, but was hoping to get his inspiration once he settled in up there.

Me — I had butterfly stomach. I couldn't eat breakfast I was so excited. I was ready for an adventure. I had made no plans but wanted to capture every moment. Fortunately, the entire journey was recorded (all our movements were), and so the synthetics could create a collection of 3D experiences to relive whenever we wanted.

Once I dreamt of designing a floating city orbiting the Moon. I thought the surface could be like the industrial zone, the orbiting city would be the habitable zone with atmosphere and energy supplied from the lunar surface and the sun. The structures were alive and smart and self-regulating during the more than 700-hour lunar days. Now I get to be ON the Moon and sketch and design my dream floating city even if it never gets built.

By the time we arrived at the OTS boarding pad, the sun was up and reflecting off the shiny metal exterior of the unimpressive building. We entered and made our way through the building to the launch pad. I expected some frills, given we were to go to space, but I guess trips to Starlab II these days are so common that it is now just a transport. People regularly go to eat at the "Restaurant at the End of Universe" on Starlab for special occasion dinners. We entered a large metal structure that looked somewhat like a spin top, but the crown was overly large (approximately ten meters in diameter). It looked outsized compared with the rest of the structure. This top was on one of the five launch pads. The capsule had no windows and, in the center, had five rows of four seats each. The seats were the most sophisticated part to the MHD. It was body forming with cross shoulder seatbelts that would buckle just below your chest with the belts around your waist. The seats had a high back with a curved area around the neck and head area, I guess for extra protection. There were large masks embedded to the side, closest to your face. As we sat in our assigned seat, the seat started moving. These were smart seats that would sense your weight and shape and adjust to provide the most protection during launch. The facial scan by the embedded cameras already knew who I was and greeted me.

"Welcome aboard Citalee. Are you comfortable? Do you need any adjustment?" The sound was coming from what seemed like inside my head, but it was using a technology called tracked holosonic bone conduction. It basically focused a beam of sound directly into my skull near my ear. It isolated the sound so that none of us heard the sound from other occupants' chairs. The AI system tracked your head position and adjusted the beam focus as you moved around for a comfortable and consistent experience.

"I'm comfortable thank you".

Even though I knew I was talking to a synthetic, I always felt that the intelligent machines are beings and should be treated as sentient beings.

"Great! Now let's go over the launch sequence and emergency situations preparedness," continued the nice female voice in the chair.

There was a control pad in the armrest and as the safety instruction started, the control pad illuminated.

"Ok let's get started. You are onboard MHD 333. MHD 333 is the safest orbital transport system in operations to date. It is a fully-autonomous, maser-powered lightcraft and every system is controlled and managed by Omega Epsilon AAI. The MHD 333 capsule is made of super light, but super strong, Titan 7 alloy that allows for the most efficient and safest beam-powered lift system. This is a hypersonic flight on which you will feel some acceleration during the launch but the supplements you were provided this morning should counteract any physical reaction you may have. If you feel any discomfort, please ask me for help or press the red MER button on the control pad and the Medical Emergency Robot (MER) will immediately assist you. This capsule is pressurized for your maximum comfort. In the unlikely case of any malfunction, your individual oxygen mask will be automatically deployed from your headrest and positioned on your face. Please stay calm and do not move during this process. The mask will tightly fit over your nose and mouth, but if you feel there is a gap, please slowly adjust it manually until you feel the suction on the edge of the mask on your face. We will reach Starlab II in 45 minutes. Please remain seated and belted for the duration of transit. Thank you for flying with OTS."

"Ugh! Does this mean I won't feel weightless on the ride?" No response.

Ahhhh I'm so disappointed. StarLab II has artificial gravity to make living in space easier and healthier – so when will I be able to finally float?! I must credit or blame XPRIZE for taking away the joy of floating for me and getting artificial gravity out of research and into practical application. Without artificial gravity, functioning in space would be a lot more complicated for us Earthlings. I guess I cannot blame them, after all this is what has made it possible for me and my family to be able to go on this amazing adventure.

The beam-powered propulsion was another technology critical in making space tourism a reality. This very capsule was the result of early work done by Leik Myrabo and his research partners[1]. The capsule itself is named after Myrabo, the Myrabo Hyper Drive (MHD).

After everyone got situated in their seats, and properly briefed, we listened to relaxing cosmic music while waiting for takeoff. Inside the capsule started feeling warmer but not uncomfortable. I knew I was being monitored, and that if I got too warm the systems would adjust the personal climate controls in my seat. Then the announcement came through that we should prepare for liftoff!

This is it! My first trip to space! Despite what I imagined; it was a bit of a yawn except initially. It didn't even feel like I was going into space. It felt more like a long ride in an elevator. A loud metal cranking noise marked the end of our ride to Starlab. I was somewhat happy to get off the MHD. As comfortable as my seat was, feeling the $2G^2$ on liftoff made me initially short of breath. Tep fell asleep after the first five minutes. I couldn't believe

---

[1]See Salvador, Israel I., et al., *Hypersonic experimental analysis of impulse generation in airbreathing laser thermal propulsion.*, Journal of Propulsion and Power 29.3 (2013): 718–731: https://doi.org/10.2514/1.B34598; see also https://science.nasa.gov/science-news/science-at-nasa/1999/prop16apr99_1.

[2]G stands for Gravity, the force by which a planet or other body draws objects toward its center. Near Earth's surface, gravitational acceleration ("1G") is approximately 9.8 m/s².

he could sleep during a space launch! Shows how unimpressed he is with this whole trip. Mom and Papa were holding hands and just as we lifted off, they looked at each other with big smiles. I could see in their gaze the love they felt for each other. Then they looked at us to make sure we were doing ok, and I reassured them by returning a warm smile.

After the first three minutes, we didn't feel the extra G forces anymore as we had reached escape velocity and we were on our way to rendezvous with Starlab. The docking process was autonomous, AAI guided and, as expected, went smoothly after proper calibration and orbital adjustments were completed. The locking mechanism of the docking bay is what made that loud metal sound, but the sound was a comforting signal to know the seal was airtight. We waited patiently. The door was finally opened, and we were able to disembark.

"Hope you enjoyed your flight with OTS, please come back soon." I will, my inside voice murmured, if you add some windows. Here I was in space now for almost an hour and I had not seen Earth from space, nor had I experienced weightlessness. What kind of space flight was this!

10.4 STARLAB

The landing pad on Starlab looked just like the launch pad at OTS – unimpressive. I was hoping that the five-day stay here would be like a mini resort vacation, but the place looked nothing like a resort. I guess I shouldn't be surprised as this is not a vacation spot. It is primarily an industrial servicing station, but you could make industrial structures look nice too, at least this is my opinion. I have two passions, engineering because I like to build things and architecture because I like to build beautiful things. No reason this couldn't have been beautiful too.

Starlab II started operation after Starlab I, the prototype, was retired three years ago. It serves as a salvage station for orbital debris collectors. Debris removal and salvage is big business. During the first century of access to space we managed to generate lots of junk in space, a decidedly human flaw. To keep orbiting assets safe in space, someone had to make sure to sweep the orbits and make room for new satellites and stations, and that is what gave rise to the space debris and salvage industry. Here again, XPRIZE's name pops up as there was a multilateral XPRIZE competition that gave the space debris mitigation technology a big boost, forcing governments of the world to put smart regulations together to allow for commercial companies to remove debris safely from space while maintaining the security and sovereignty of the assets removed. While Starlab is a big salvage operation, there are not that many people around. The debris removal craft brings the captured assets to the station and the assets are placed into a capture and imaging tunnel where each is scanned to identify all the components and different materials through AAI and then the generic Learning Bots are programmed accordingly to disassemble and salvage different material into separate bins for processing. All of this happens without human intervention and intelligence. Frankly, it could not happen at this scale and speed if humans were in the loop. Humans are great at many things, but efficiency in manually-complicated tasks is decidedly not one of them.

There are only about fifteen people working in Starlab, half of whom are focused on ensuring visitors are comfortable and happy. Even though now we live with robots all around us, human touch provides unrivaled mental comfort.

We were greeted by a lovely man dressed in a blue uniform with a big STARLAB II logo on his left chest, as well as on the large infopad he carried.

"Welcome to Starlab, my name is Ali and I'll be your welcome guide. You must be the Rojas family. We are excited to have you with us for the next, let me see, five days. I trust your journey was comfortable?" We nodded yes.

"I know you had an early morning start. Let me show you to your quarters so you can get settled." Your profiles have been synched with our visitor database and given your visitation parameters you are granted access to all common areas, plus your training facility and recreation arena." Tep's ears perked up.

"Wait did you say arena? Are there any sports?"

"Yes of course" replied Ali.

"You will find ample sports equipment and facilities that you are familiar with, but in practice, very different, as our arena is the one place that we don't have full artificial gravity (AG) activated. We also have some sports designed especially for a micro gravity environment. As a matter of fact, we have a Space Jam Basketball game tonight between the crew and G-Bots. They are generic robots that can be programmed and will learn any task, so we exposed them to the rules of basketball. The Junkyard Dogs, our crew team, is a crowd favorite and is still undefeated!"

Tep turned to Mom, "Can I go watch the game please – please-please-please!"

Mom turned to Ali, "Is it possible for us to watch the game?"

"Of course!" replied Ali. "We love to have an audience."

Tep smiled for the first time since he woke up this morning. This brother of mine really loves his sports.

We continued walking toward our quarters. In addition to the twenty people who transited to StarLab with us, there were probably another hundred people spread out around the station, however most were not staying, but rather just there to have an orbital lunch or dinner and enjoy the view. A few VIPs also had access to the arena and were here to watch the game. There were also a few families visiting the crew as they did not get to go home until their rotation was over, typically three months in orbit and three months in ground operation. I learned all this on our walk to our quarters. My Mom tells me I let my curiosity get the better of me and sometimes I make people feel like I'm interrogating them. Maybe I am.

Tep and I were sharing a room that was across the hall from Mom and Papa. We had a bunk bed and I rushed to get to the ladder to the top bunk. As I ran and grabbed the metal pole I started flying. Or at least I felt like I was going to fly.

"Wooow, what just happened?"

Mom laughed, "We are in a about two-thirds the gravity of Earth so you may feel lighter, and you can definitely jump higher."

Mom said it takes energy to create AG so they studied the human body and response to different levels of G forces and discovered that there is minimal health impact under this level of reduced gravity and unless we are doing things like running or jumping or lifting heavy objects, we barely notice it.

After storing our gear we convened in a white room and found Ali there to greet us at the entrance.

"The Rojas Family, wonderful, please come in. You should have received your daily schedule for the rest of your stay. We will start with the medical exam. Please follow the DBs and they will guide you and report back. I'll be waiting for you when you are done."

As he was finishing his sentence, four flying DBs appeared from around the corner and positioned themselves in front of each of our faces. The screen before us illuminated with different faces. The faces were specially selected by the AAI based on our individual profiles to make us more comfortable. The face started to talk to us.

"Hi Cit, I'm Sandi. How are you feeling today?"

"I'm fine thank you." Sandi was the name of my best friend, what a coincidence, or was it?

"Please follow me," Sandi started flying toward a small exam room with an exam table at its center.

"Please lay still on the bed. a scanner will move around you so please don't be scared." "I am not scared!" I said and meant it.

I sat on the cold, white glass table surface and laid back while immediately a large ring started moving up from the bottom of the bed scanning me. As it went up and down it would stop in specific areas and an additional attachment would be lowered to the surface of my body for further examination. This took about ten minutes. I was nervous to move and break one of the delicate devices. Once finished, Sandi instructed me to sit up and then in a little cup in her small robotic arm handed me a ref capsule.

"Please take the ref-capsule. It will collect all the data we need and dissolve when no longer required."

Next, Ali took us to pick up our spacesuits, which looked completely rad. They were smart suits, meaning they were made of special sensing fabric that would transmit thousands of data points to the AAI and would also react to special environmental needs and body responses. It could flex or become rigid, it could even self-repair! This I learned from the immersive holoviron-ment I experienced on measurement day. I put on my suit over a thin base layer. It felt strange – it felt like it was alive, it moved and slightly slipped around on my skin. It felt like what I assumed snakeskin would when the snake is shedding it. The mirror in front of me showed Sandi's face in the corner.

"You look great. How does it feel?"

"Feels strange," I said.

"It is sensing you and adjusting. It will feel more like a suit once it stops its initial calibration. This is our latest space suit. The fabric has embedded within it, special neuro-sensing material employing AAI to make small adjustments based on your internal and the external environmental data streams. It will also be programmed with your daily tasks providing you with just the right combination of protection and flexibility needed to perform your tasks in an optimal fashion".

The suit has my name sewn into it. Or I guess programmed into it. "Citalee Rojas". An uncontrollable smile appeared on my face – I had my very own spacesuit now.

The rest of the day was more orientation covering the layout of the SEM shuttle, the initial flight profile, and operational overview. There was not much for us to do as operations were completely controlled by the AAI system on-board. I don't even know why they think we need to know what is happening as we are not in any decision-making process. AAI has backup for every backup so it won't fail, and if it does, there is no way anyone of us can take over with just five days training. In the past, it was staffed with people like Grandpa who had trained for years and could deal with and fix any malfunction. Today, there were no shuttle pilots, just an AAI. I guess I should not worry as there

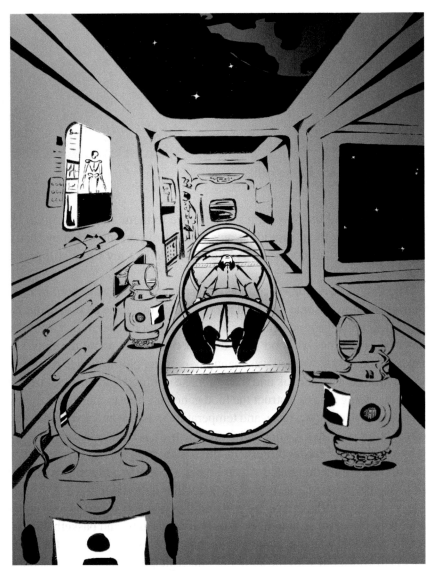

Figure 10.1   Medical Room.[3]

has been zero AAI critical failures in the past twenty years. The AAI is only the beginning steppingstone on the road to Artificial General Intelligence (AGI) – the holy grail of AI. While everyone thought AGI would be functional by now, a combination of technical challenges and regulatory hurdles has slowed down progress. Maybe it is a good thing – otherwise why would

---

[3]Illustrations in this chapter courtesy of Kelsey Mainard.

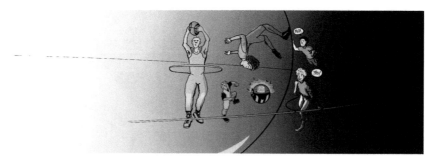

Figure 10.2   Space Basketball.

the machines need us? We are just a nuisance to them, like a mosquito, just annoying them without providing any value. I'm personally happy that we have not given up all control to AAI and people like my Mom still provide value.

The first day orientation completed just as the game was about to begin. We decided to grab some fast food and go watch the game as we ate our hot-dogs produced in the food generator. Funny, with food generators, fast food no longer had meaning. All food from the generators took about the same time to make. The only slow food was manually-cooked, old-fashioned food. Even though Mom is a great cook and likes to make our food from original ingredients, I like being able to instruct the food generator how to make exactly what I want in the size, shape, taste, and temperature. Also, Mom never makes hot-dogs at home, she says it's not real food. The truth is by her definition, almost nothing we eat is real food. Perhaps the best part of food generators is that you can make something, and if you love it, share it exactly as you had it to someone else. Food was essentially just another data stream.

As we entered the arena, we realized that this was no ordinary game of basketball. It was something like a cross between rugby and basketball. There was no dribbling without gravity. The court was twice the size of a regular basketball court, but with the added 20-meter height as the 3rd dimension, so the players were flying like Superman in the boxed field marked with laser light and a Rugby sized ball. The teams would score with a player diving full body into the basketball style hoop that was barely wider than half meter in diameter, while holding the ball. There were two hoops suspended at each end of the court and about twelve meters from the ground. There were five players in each team like basketball and the rules were, for most part, similar. If the player passed through the hoop without touching the rim, the rim would turn green and they would score three points but if they touched the rim, the rim would turn yellow, and they only scored two points. The hoops and the entire court were fitted with sensors and cameras to keep track of scoring. There were also small disk-shaped referee bots flying around capturing everything to call out any fouls and out of bound movements. The game was divided into two 30-minute halves, with a 20-min break in-between, highest score wins. The most interesting aspect about playing this game in micro gravity was that the players had to push themselves off the walls or each other to move and gain momentum or to stop. When scoring, one teammate would dunk another teammate holding the ball, through the hoop by pushing them from their feet head down into the hoop. It was like they were throwing a Rugby ball, but the ball was a human with a ball. The player with the ball had to keep their body rigid so they would move straight like an arrow and hopefully not touch the rim. The game was intense as the players would bang against the walls and each other, this part was a little like hockey. The players could only crash into their own players to assist with their movement, but they were only allowed to grab the arms of the opponents to stop or change direction by shifting their momentum and direction, otherwise the move would be considered a foul. Ali explained all these rules to us during the first five minutes of the game and we were all hooked. The other fun was that the audience was also in micro gravity and was often floating. As there were only a small number of people on the station, there were no seats or assigned zones so everyone could float and watch the game at any angle they wanted and move as the players moved. The audience was separated from the players by a thick see-through acrylic barrier. If you happened to be in contact with its surface when a player smashed into it, you would be startled from the impact.

Tep was screaming for the blue team, the Pulsars. He met the team captain on the tour, and he had been given the official team patch for his spacesuit, so he was an instant fan. We watched the game with great excitement, I never thought I would enjoy watching any sporting match this much. I guess every experience in space is new and different.

The next three days were more of the same. Ali and Tep became fast friends as Tep got deeply curious about the salvage operation on the station. The station was equipped with long-range sensors that would detect any object in an orbit of up to 100,000 kilometres. The AAI was able to determine if the object was space debris or a legitimate asset. If it was debris, the AAI would determine which debris had the most value and was at the greatest risk for damaging other orbiting assets and considered the cost of retrieval. The prioritization scheme was entirely up to the AAI, and ultimately a fully pre-programmed mission would be launched specifically designed for the location, size and the type of debris detected. Once the collector vehicle captured the debris and returned it to Starlab, the debris would be scanned and a swarm of bots driven by AAI would disassemble it to its basic elements and put it through the stations salvage operation. Some of the newer parts would be kept and refurbished to be used as spare parts for the SEM shuttle and the OTS capsule repairs. Of course, there were no humans in the loop as the operation was delicate and humans were too inefficient for such complex tasks. Tep was fascinated by the bots at work. I never knew he was interested in AAI and robotics. He never ceases to surprise me.

Our last night on the station was honestly the best of all. We had a training graduation party at the Restaurant at the End of the Universe. It was named after a fictional restaurant in a classic called the Hitchhikers' Guide to the Galaxy. The restaurant was at the edge of the station, facing Earth and it was like a giant glass sphere hanging precariously into space. There were no chairs and tables, and the gravity was a thrilling 0.1G. The food was in floating bubbles of different color, green was for appetizers, red was for main courses, and blue was for desserts. As we entered the restaurant, we were given special hover boots and a glove that looked like five attached rings with a cord that would wrap around the wrist. Everything was driven by voice command. We picked our desired food bubble by just pointing to the bubble and saying "go" and the boots would propel us toward the bubble we selected. As we touched it, a little door would open for us to pick the small food item out and pop it in our mouth. This was the ultimate case of finger food. I continuously aimed at the blue bubbles as I have a huge-sweet tooth. Let's just say I overindulged and paid for it that night.

The next morning, we woke early again to suit up for our flight to the Moon. By now I had spent so much time in the VR simulator that I felt like I had flown in the SEM many times before. The SEM was a large cylindrical shuttle with the latest Vasimir X Plasma engine[4] that made the trip to the Moon in just under six hours. As we were already outside the grip of Earth's gravity, we didn't need much energy for launch, we just needed a little push and thrust to get to the Moon. The SEM could carry twenty passengers to the Moon. The crew consisted of a DB and a few flight comfort bots to help answer questions or get us anything we needed. I had selected a seat next to the porthole. We strapped in and got our launch safety instruction and countdown started. As we undocked to make our launch orbit, I looked out at the beautiful and colorful Earth. As we passed over the Pacific Ocean, the sun reflected on the blue waters, shimmering like gold. I couldn't help but to think how uniquely gorgeous our planet is, contrasting with the dusty, grey lifeless

---

[4]See the Wikipedia page: https://en.wikipedia.org/wiki/Variable_Specific_Impulse_Magnetoplasma_Rocket

Figure 10.3   Restaurant.

Moon, indicative of our known universe. I guess we could be called crazy for wanting to leave Earth at all, but it is our sense of curiosity and exploration that drives us to leave the comfort and beauty of our home and travel long distances, endure harsh environments, and sometimes risk our lives to learn and discover something new, something amazing that would get us one step closer to answering the ultimate questions "How did we get here? Why are we here? Are we alone?"

As we left orbit and headed to the Moon, I took comfort in knowing that we were only 385,000 kilometres from Earth, and I could still see our home planet from the surface several times per day. We were heading to the Tafiya base near the south pole. The site was selected as some of the sunlit areas near the pole are also close to high concentrations of water ice, in craters where the sun never reaches. Constant sunlight, benign temperatures, near the water and a great view – that's prime real estate on the Moon. The base was built as a multi-purpose research station and habitation for the thousands of the workers, researchers, and their families living on the Moon. Luna one was the first part of the city that was built during the Artemis projects almost fifty years ago. Since then, Tafiya had subsumed Luna and had expanded resulting in multiple bases all in close proximity. Tafiya was a multinational base that welcomed all nations, all humanity. It was like the international waters on Earth where no one nation owned it, but everyone could use it and work on it under internationally agreed upon guidelines. The Moon was governed by the Peaceful Use of the Moon Treaty signed by all UN members and generally respected by all nations. We had learned a lot from endless wars fought over land and resources on Earth and those learnings helped to shape our occupation of space. Today there was little reason to fight as energy and raw materials were no longer so scarce as they once were on Earth. The Cyber War of 2045 taught us there are no winners in any war – just losers. An early version of AAI simulations demonstrated to the world leaders that war ultimately leads to mutual self-destruction of human beings on the planet. The world leaders then came together and ratified the Global Peace Act of Mauna Lani[5] allowing for equitable sharing of resources and technologies by all nations so no nation would gain control and power over others. This among all advances in AAI was its most important accomplishment to date in my view.

I was excited to see our new temporary home. I wondered if it resembled anything like the early drawing I made on my imagined lunar city when I was just ten years old. The ride was smooth and uneventful, as expected. An announcement prompted that we have entered lunar orbit. The SEM docked to the LOTS, Lunar Orbital Transport Station, from where we took the Moon Lander vehicle that was designed for just that – to take us down to the surface. Shortly, we exited the lander and entered a tunnel that took us to an entry port in Tafiya. There were many entry ports depending on where you were stationed. The passengers from SEM were each assigned to the different landers

---

[5] Agreement signed on the Big Island of Hawaii.

and taken to their designated entry ports. As we went through the scanning pods to make sure we were clear of all contaminants, we were greeted by a very nice women named Teressa. She was one of Mom's research assistants.

"Welcome to Tafiya. We are so happy you are here. I know you must be tired, and wearing these suits, as cool as they may look, is not all that comfortable. I will take you directly to your housing unit. There is an orientation video you can watch when you get there and the Robocon in your unit has been preprogrammed with all your needs and preferences. It is not a big unit, but I think you'll find it has all the amenities you may need for comfort. Tomorrow, I can show you around and help you feel at home."

"THANK YOU, Teressa! We look forward to it" said Mom as she hugged Teressa with a youthful glee in her eyes and big smile on her lips. She could hardly hide how excited she was to be here. Living on the Moon was her childhood fantasy, having listened to Grandpa tell tales of his own space adventures. Now she was here!

We all followed Teressa to a transport pod to take us to our quarters.

"These pods are how we get around here. They will take you through the tube systems to any place in this station and you don't need to don your suit. But if you want to go to another station, you must use the interstation rovers and you need to be suited. If you need an interstation transfer you can arrange it through your Robocon, but everything you need daily should be here on this station."

The pods were small transports that glided in a low friction tube that connected every corner of the station. It was the lunar version of a metro system, designed for just four people, but could be attached for larger groups. They glided and maneuvered through the tubes at a very fast speed and past each other within a few centimeters but never collided. Just watching the system of pods moving through the tubes would make your head spin but the entire system was managed by AAI and, though complicated, the pod traffic management system was child's play for the mighty Epsilon Alpha AAI.

The pod decelerated rapidly and glided to the resting pad where we got off. Our Robocon was waiting for us and greeted us heartily.

"Welcome to Tafiya, Rojas Family, I am RC22 and will be taking you to your new home. Please don't hesitate to ask me anything you need." RC was an egg shaped and faceless robot about 1 meter long on a set of wheels that helped her rove when on a surface and could activate her propellers embedded on top her head if she needed to fly. An actual propeller head ☺, I'd never seen a propeller head robot, but I guess low G environments make that practical.

RC22 guided us to our quarters and pulled our luggage tethered to her into the room behind her. We didn't have a lot of luggage as almost everything we needed could be 3D printed on the station. We just brought a few personal items to make us feel at home.

"Your unit is equipped with sensors tuned to your DNA signatures so it can respond to your presence and preferences. I am always accessible through your voice commands should you need assistance, otherwise I remain in sleep mode."

"Here is the food generator, but we also have fresh produce from our garden you can order. You will be visiting the gardens in the coming days. The rooms are temperature and light controlled as the lunar days and temperature are quite different as you know. Each day is about 29 Earth days and the sun at our location is visible about 80% of that time so your sleeping pattern would be greatly impacted. The injections you will receive tomorrow will help you with your sleep patterns. Let me know if you have any prolonged problems adjusting and we can activate more sophisticated AI assisted sleeping aids in your rooms. Do you have any questions?"

"Yes, where is my room?" asked Tep restlessly.

"It is the last one on at the end of that corridor to your right," answered RC22.

Tep picked up his bag and headed to the room. RC22 guided us each to our rooms and disappeared. I guess she went to her hibernation station. We were all tired, so Mom suggested we just get something from the food generator and get some rest. I took off my suit and stretched a bit, then ordered a nice cup of pumpkin soup and got cozy in bed sketching my dream place on the Moon, but quickly fell asleep.

10.5 BIG DAY

It has been almost two weeks since we have arrived, and the place is starting to feel familiar. Today is a big day for us. Mom and Tep are headed out for the first time to explore the area outside the confines of the base. They will travel in a fully autonomous rover vehicle capable of scaling and descending steep terrain. Quite an experience we were told in the pre-travel training materials. Their destination is to visit the site where frozen water extraction is done, deep in a crater where the sun's rays never reach the surface and the temperatures hover around -170°C. I am excited to see this at some point as well, but Papa and I had other plans. Today he and I were getting our creative juices flowing.

Papa has been given the opportunity to deliver the creative design for what will be the first "city park" in space, located in the heart of Tafiya. It is an incredibly ambitious project, but one that most agree is essential for human well-being in space. Papa is a famous artist and designer, but we were, and still are, a bit shocked that he was chosen to lead the design. There would be several dozen designers working on the project, but he would oversee all the efforts.

Doctors, like Mom, had been studying the effects of space and determined that having large open space is essential for maintaining a healthy human psyche. Long debated was the idea that simulated virtual spaces like VRealm would satisfy that need, and maybe someday we would evolve to accept that as a substitute, but at least for current and foreseeable generations, they concluded we would need access to scaled open spaces to live our best lives. This space would serve as a prototype informing humanity's colonization within and beyond our solar system. It was also concluded that the learnings from this park on wellbeing were to be applied to future spacecraft designs. They hoped to answer questions about how much open space is needed for long duration flights to promote a harmonious environment for humans. New AAIs were being trained to measure human happiness or wellbeing and would be deployed throughout Tafiya to see the uplift (if any) gained from providing this tremendous oasis on the Moon. This would be measured against other bases (all smaller) that do not yet have this park-like amenity.

This project was going to be very expensive and resource intensive. Over the next several years a large-scale dome will be constructed stretching nearly 250 meters across and over 30 meters tall at its apex. The interior of the space was the province of Papa including gardens, sculptures, waterscapes, and park activities. Today we were setting up to test some ideas Papa had about organic sculptures that took advantage of the Moon's gravity of about sixteen percent of Earth's gravity. This allowed for the creation of fantastic structures not possible in Earth's gravity. Today Papa and I were going to make a micro-scale model of some of the organic structures he was considering. This machine could take small pellets made from materials found in abundance on the Moon that could be fed into an extrusion device to create branch like structures. It would be a bit like painting a structure into existence in 3D. In this gravity, structures with small, seemingly fragile looking bases could support what would appear to be impossibly balanced structures. A bit like inverting a tree or pyramid but having a stable, safe structure. Once these structures were tested, Papa said the AAIs would assess the feasibility, cost, safety and make recommend optimizations and produce timelines, resource, staffing, and construction plans. The AAIs would also identify, spec, and initiate manufacturing of any tools that would be required to build and deploy Papa's vision. AAIs were great at anticipating all critical details, and over the past few decades had been responsible for construction costs to be predictable and optimized for building highly complex structures, something no nation or major multi-national had been able to do reliably prior. It was just beyond human ability to account for the seemingly infinite factors involved. We have a vague notion of what the AAIs are doing, but we really don't understand how they do it in detail. I guess that is OK, but it is, at times, a little unsettling all the things we trust the AAIs are doing on our behalf.

After riding a shuttle pod to the outskirts of Tafiya, Tep and his Mom entered the remote rover launching dock. They were scanned upon entry and welcomed by a synth.

Tep asked, "Mom, I don't see the rovers, maybe they are all out."

Walking up behind Tep, the bipedal synth pointed to the ceiling where dozens of small passenger cabins of varying sizes were suspended. One of the smallest ones was being lowered. The synth indicated they should step inside the small passenger cabin. It had seating for four, but this trip would just have two passengers. Where was the rest of the rover, they both thought. Once they were belted in and the cabin locked and pressurized, it began moving across the floor on an embedded conveyor toward what turned out to be an airlock. Once the airlock was closed behind them, a grappling arm attached to the top of the cabin and the floor opened. They were lowered onto the rover frame and felt a heavy vibration as the cabin was locked into place. They were now outside the base. They could see the Moon's surface around them.

Tep exclaimed, "This is amazing! Let's go!".

Mom was a lot less comfortable. As the rover pulled out onto the Moon's surface and began its journey, she had a rising sense of panic. It was mostly dark at the pole, and they could only see perhaps fifty meters ahead with the rover lights. The Earth was just off the horizon to their right. It appeared to be upside down from what one would expect. The trip reminded Tep's Mom of what it was like to scuba dive at night. Not exactly like that, but she felt the same sense of claustrophobia given the limited visibility around the surface. Strangely though above her she could see countless stars in the night sky, an incredible view, both expansive and constricting at the same time. She instinctively held Tep's hand. Her danger senses were on full alert. It would be an hour travel time out to the crater.

"Where do we begin Papa?", Cit asked.

"First we have to load the machines supply bin from that barrel of pellets I had delivered here," he replied, "And then we can fire up the extrusion tool."

Cit went over to one of the barrels her father pointed to and looked at it with the expectation it would be too heavy for her to lift. But, due to the much lower gravity and the surprisingly light material, it was an easy lift. She carried the barrel across the room, gently lifted off the lid, and carefully poured the pellets into the feeder bin. They bounced around but eventually

Figure 10.4   Moon Surface.

settled into the bin. She closed the bin lid and gave the thumbs up. Papa fired up the machine which consisted of what looked like a spray gun with a hose leading to it. Papa wore goggles so he could see the projected shape he was making before the gun fired out the material. The tool allowed him to draw arcs in the room in three dimensions, and then when he initiated the material release, the gun effectively took over and delivered the material into whatever shapes he created. It was called an AAI-assist as it required both a human and an AAI to fully realize the shape. The material was contained into the shapes he specified by an electrical field that was precisely manipulated to direct the materials into the desired shape. The material was hot upon release but cooled and solidified quickly as it met with the air in the room. The room's climate control system autonomously adjusted to the rising ambient temperature by increasing the room exhaust fan rate substantially. After about twenty minutes the structure was beginning to take shape. It was about two meters tall and perhaps twice as wide. It looked organic, almost like a tree's branching system, but it balanced perfectly on the implausibly small base Papa had designed. The AAI would continue to add material to the base as needed to retain balance of the overall structure. It was fascinating to watch. Papa then asked me to reload the bin and then I would have my chance to add to the structure. I was very excited, perhaps a little too excited.

The rover was now just cresting the ridge of the crater. They could only see parts of the crater walls below; it was stunning how the sun lit portions of the interior and exterior rim casting long crisp shadows. The ride had been surprisingly smooth, in part, owed to the construction bots that cleared the frequently traveled path to the crater, and had excavated portions to make the thirty-degree natural slope to the crater floor more manageable through a series of switchbacks akin to a mountain road on Earth. As they descended to the floor, they could see lights in a semi-circle that turned out to be the mining operation as they got close enough to see details. Autonomous mining equipment was systematically cutting up the frozen water into blocks approximately half a cubic meter per block. These were then loaded onto a transport vehicle where they would be moved into the Tafiya water supply and treatment facility. AAIs managed all aspects of this work, identifying the prime locations for water extraction, directing the efforts of mining equipment, and the subsequent conversion of the ice from frozen to liquid form. It was here that the water would be treated to remove any residual particulates or biologics. Tep looked at his Mom and asked her if she was OK. She seemed rather pale and sweaty in her suit. Indicators on the cabin status screen indicated she was getting warm, and the suit corrected to help adjust to her rising temperature, and increased heart and respiratory functions.

Mom replied, "I think I'm having some sort of a panic attack, Tep. I know we should be fine, but I just feel strangely anxious."

Tep comforted her, "Mom, don't worry, everything will be OK. We've seen enough, thank you for taking me to see this amazing place. I will never forget it. Let's head back."

Tep instructed the AAI to return them to the base, and the rover promptly began the climb out of the crater to the rim. Just as they had crested the crater rim, and began descending the outer wall, the AAI system notified them that a primary battery system failure had occurred.

Tep looked at his Mom and said, "Should we be worried our primary battery system has failed?"

She responded, "I'm sure they thought of this Tep. We can just operate on secondary battery power and I'm sure everything will be fine."

At that moment the AAI confirmed that they would indeed be doing just as she suggested. Strangely, having this kind of a failure triggered her into analysis mode and she almost immediately felt better. She had something to focus on beyond the alien landscape and immediate confines. She looked at the secondary battery levels remaining, did a rough calculation and determined they should have at least another hour of battery power remaining when they reach the base. Plenty of time to get back to safety.

I examined the barrel contents and saw it was about half full. I picked it up and carried it closer to the extruder loading bin. On my way to the bin, my foot caught on the extrusion feeder tube.

"Whoa!", I yelled, as I lost my balance, came crashing to the floor, and pellets were launched into the air and on the floor.

Papa ran to me, "Are you OK honey?"

I looked around at the mess of pellets all over the art room and said, "I'm ok Papa, I am so sorry for this mess." In this low G environment, the pellets had gone all over the room like nothing you'd ever expect to see on Earth.

"We have a lot of cleaning up to do," I said sheepishly.

With that we scanned the room for a broom. There was none to be found. They saw an autonomous sweeper down a hallway and redirected it to begin cleaning the art room. The sweeper bot dutifully began sucking up the pellets. However soon the bot left the room as it reached its carrying capacity. Papa suggested we go get something to eat, and as we were packing up, we heard:

"Warning, a possible breach of the containment has been detected, entering compartment level lockdown."

At that, they watched the thick containment door close with a resounding thud. They were sealed inside.

"What does this mean Papa?", I exclaimed with alarm.

He too looked concerned, she could see it in his eyes.

"I'm sure it is nothing Cit, it is probably being done out of an abundance of caution. This sort of thing probably happens all the time."

Both of us knew this was not the case as we had been told that the base had never had a breach since it was first occupied. This wasn't an infrequent issue; this was a first. They looked at each other trying not to look panicked.

"The lockdown will protect us in case there was a breach, our air seems fine. And your brother and Mom should be fine, they are out in the rover, so a breach here wouldn't impact them."

Papa looked around. There were still a lot of particles on the floor, but it seemed like they were disappearing, even after the sweeper bot left. That seemed odd. What would have caused this? Where did they disappear to? It seemed odd that this question had popped into his head just then. Where the pellets went didn't matter at this stage. Safety was the focus.

10.6   REAL INTELLIGENCE

At that moment, Dad's communicator vibrated. It was Mom. He answered the call.

"Hi honey, hope you are having a great trip out to the crater...," before he was cut off.

"We have a potentially big problem. Our primary battery system has failed so we are returning under secondary batteries. All should be fine, but we just got notified by the rover AAI that the base is in complete lockdown for a possible breach of containment and docking is not permitted during lockdown. Is that true, the base is in lockdown? Are you both OK?"

"Yes, we are fine. We are just stuck in the art room and can't leave until the lockdown is lifted. How long until you arrive at the base?"

"We are about five minutes out, and we currently have forty-five minutes of reserve power. We will need to get into the dock and be safely inside the base well before that. I sent an alert to the Tafiya security base commander, but no response yet, I'm sure she has a lot on her plate right now."

"Papa, I remember during the security briefing that the lockdown and reopening of the base is entirely in the control of the security & safety AAI. We need to connect with that AAI if we are going to be able to get out of this room and get Mom and Tep into the base."

Papa used the voice activated interface to connect with security. The AAI listened patiently to the dilemma created by the base lockdown. The security AAI simply responded.

"Thank you for making me aware of the issue with Rover Z1B. Until the breach is isolated and impacted areas are secure, the base must remain in lockdown. The safety of the population of the base is my primary mission." After a pause, it went on to say, "Every effort will be made to preserve the lives of occupants of Z1B."

At that, the AAI disengaged from the further conversation. Papa asked to re-engage. There was silence.

"Papa we can be left out here to freeze to death, you have to do something," said Tep. Mom and Papa reassured Tep and I that would not happen. But it was clear none of us necessarily believed that. The room was getting noisier, and well more than half the pellets now appeared to be gone. The noise appeared to be coming from the exhaust fan. I thought, why am I even thinking about these pellets? I should be focused 100% on getting Mom and Tep back on base.

As various regions of the base reported in as "all clear", it appeared more and more like this was a malfunction of the AAI or some sensors. At least for the areas of the base where humans were, there didn't appear to be any breach, but all were locked down in the rooms they had prior occupied. Papa kept trying to engage with the security AAI. Finally, it responded.

"Yes Mr. Rojas, what may I do for you?"

"You can let my wife and son back into the base. They have just ten minutes of power left."

"Yes, I am monitoring Z1Bs power level. I have remotely turned down non-essential systems which adds approximately three minutes to the battery depletion time."

"It appears based on human reports that all life support systems are functioning as normal and all base personnel have reported in as safe (except my wife & son). Are you sure that you are not in error in your estimation that an outer shell breach has occurred", Papa demanded.

After a moment, the AAI responded, "While the conditions do appear to be unusual, there is a moderately high probability of breach given the compromise of the air system in sector five."

"There is a 100% certainty my wife and son will die if they run out of power. You need to unlock this base now!", Papa screamed. We could hear Mom and Tep crying over the open channel. I looked around the room once more. What was I missing? Something wasn't right about this emergency.

"What is the exact location of the compromised air system?", I belted out. After a moment the AAI system said it was sector five, exhaust system 13K. I pressed further, "What is the air system that feeds the room I am in right now. The AAI, responded, "Sector 5, exhaust 13K."

Papa and I looked at each other, looked around the room, and said, "It's the pellets!" Mom and Tep said, "What did you say? We heard pellets. What does that have to do with the breach."

Papa engaged the security AAI.

He said, "I believe I know what the problem is. This is a false alarm. There is no outer breach. I know this because we caused the issue."

There was a long pause, and the AAI said, "Did you intentionally do harm to sector five, exhaust 13, with the intent to do harm to the station and its inhabitants? If you have done so, you will be placed under arrest for heinous crimes, subject to Tafiya International Governance Act."

"No, it was an accident, Cit tripped and dumped sculpting pellets onto the art room floor. Those pellets were drawn into the air system, triggering your sensor to detect debris, perhaps interpreting it as space dust and debris, the kind that would be seen during an actual hull breach. Unlock this base now, or at least the dock where my wife and daughter need to enter."

"The supposition you propose is insufficient to provide a high probability assessment of causality," there was a long pause. "Request for unlocking base is denied."

"We are almost out of power, we have just three minutes left," said Mom.

"My estimate is three minutes 23 seconds", the AAI responded.

At this I stepped in and began to interrogate the AAI. With every question I launched at it, it would ask me two more to assess the actual probability that what I was saying was true. The AAI and I were locked in a battle of wits, but at this time I had my doubt about the intelligence level of this AAI. It was like we were playing a game of twenty-one question, but bidirectional and with the ultra-high stakes of life and death. After about 2 minutes, the AAI paused.

"What you propose has a probability of more than an 80% causal relationship, unlocking docking Bay 4 and retrieving rover Z1B," said the AAI.

You could hear relief from Tep and Mom as the rover began to be lifted back up into the air lock.

"The base will otherwise remain locked until exhaust 13K, sector five has been verified as blocked with art debris and restored to proper function."

It was four hours before the base was fully unlocked and back to working order. Mom, Tep, Papa, and I were all reunited in the dining facility near their quarters. We were all emotionally drained and realized just how hungry we were. Tep ordered a huge meal consisting of a double Cheeseburger with a double order of fries while the rest of us chose an array of comfort foods. Things we missed from home.

As we were sitting together and taking comfort in our greasy food in silence. We were acutely aware of the harsh surrounding we were now living in. We were so isolated in our cocoons as we left Earth, surrounded by technologies we had become fully dependent upon and had relinquished almost all critical decision making to AAI, trusting that they can make tough and complex decisions when a human life is involved better than anyone of us. I never even had much concern about AAIs running the show here and on Earth, but our experience today made me doubt if blindly relinquishing our decision-making role altogether was the right choice. Thirty more seconds and my brother and mother would be dead. Just thinking about that possibility sent shivers down my spine. I was lucky I convinced the ignorant AAI that we were not in danger. We were lucky this time but what about the next time? After what happened I was ready for this adventure to come to an end so we could all go back to the comfort of our One G home with a breathable atmosphere.

Dad reached out and held Mom's hand and looked into her eyes as if he was going to say something, but my Mom stopped him.

"I know… Someone must make a case of this incident so this type of black and white decision cannot be made without some kind of human intervention in the future."

"A few more seconds and I would have lost you both!"

"I'm here darling and so is Tep. I will report back on this and make a recommendation to improve the protocols when lives are at stake. Let's just celebrate that we are all together and we have a brilliant daughter who could beat the AAI at its own game of logic."

"But Mom – I caused the incident to begin with I could have killed you and Tep!" There it was, this lump that had been making breathing so difficult for me burst and my tears started rolling down uncontrollably. I couldn't stop, in between my sobbing I could only get out

"What if I had killed you?" Mom pulled me closer and held me in her arms as she rocked me like I was just a little girl to comfort me.

"Oh honey! It was an accident; you had nothing to do with what happened. You are my hero, my love. Hush don't even think these thoughts." I slowly stopped sobbing as I took comfort in Mom's arms, but some fear inside me had woken and I couldn't rid myself of it.

10.7 EPILOGUE

The next day we awoke in a somber mood. Tep and I avoided adventures. We just stayed in our pajamas and played games on our panels. Mom was already gone when we woke. Papa had made our favorite breakfast, chocolate chip pancakes, but neither of us were hungry. That is a real rarity for my brother.

"Come on kids let's put it all behind us and have a fun day together. This is our real vacation day. Let's just stay in, play games, and laugh about our adventure," Papa said.

"Where is Mom?" Tep asked.

"She has to go to a briefing session on what happened yesterday".

We lounged for hours. When Mom finally returned, she looked exhausted. She had been gone for most of the day. She relayed to us the debate that took place, and the blind insistence that AAIs saved more lives than endangered and that this was just a training glitch and should not cloud those facts, after all, nobody was physically hurt. We stared at Mom in disbelief, and she stared back. Eventually she said...

> "Your Grandpa frequently warned me we were on a slippery slope of intellectual complacency, giving over to our machines. I used to think he was just living in the past, but now I too worry about our future. Humans have a difficult time finding balance. When it comes to AAIs, the fate of humanity may be dependent on us finding that elusive and delicate balance."

ACRONYM

AAI Advanced AI

AG Artificial Gravity

AGI Artificial General Intelligence

AKA Also Known As

DB doct-o-bot

DNA Deoxyribonucleic acid

G Gravity

ISB Intelligent Synthetic Being

LOTS Lunar Orbital Transport Station

MER Medical Emergency Robot

MHD Myrabo Hyper Drive

OTS Orbital Transit System

SEM Selene Earth-Moon

UN United Nations

VIP Very Important Person

VR Virtual Reality

GLOSSARY

360 Degree Review: performance review that includes feedback from superiors, peers, subordinates, and clients.

Abnormal Variation: changes in process performance that cannot be accounted for by typical day-to-day variation. Also referred to as non-random variation.

Equitable Abundance Movement: the world came together in to focus on providing equitable abundance of energy, food, clean water, education, and opportunity for all.

Cyber War: a mutually destructive information war between nation states and between corporations

Epsilon Alpha AAI: fictional Moon-based transportation pod traffic control system

Junkyard Dog: the fictional name of a basketball team based off world

Global Peace Act of Manau Lai: a hypothetical world peace declaration signed in Hawaii

Myrabo Hyper Drive: a laser powered propulsion system using originally conceived of in 1976 by Dr. Leik Myrabo

Omega Epsilon AAI: the fictional advanced AI system controller for the MHD 333 lightcraft transport

Rejuvena: a hypothetical protocol for rejuvenating the human body's cellular structure, essentially turning back the biological clock

Space Jam Basketball: a low gravity sport that is similar to Earth's basketball requiring the players and ball to go through the hoop

Vasimir X Plasma engine: an electrothermal plasma thruster

VRealm: a fictional fully immersible virtual environment simulator

Afterword

To expect the unexpected shows a thoroughly modern intellect.

– Oscar Wilde, 1895.

There is no doubt that the space industry is one of the beneficiaries of the advancements brought by Artificial Intelligence (AI) technologies. In return, scientific knowledge – poured down to Earth from in-orbit operating satellites – has enabled the development of modern industries and opened new market opportunities. It is not any more just about the "Space Race" when the space sector was largely occupied by governments. The shift in the mid 2000-s from cost-plus contracts to fixed-price contracts catalysed a burst in the sector. In fixed-price contracts, the risk of investing in space is distributed between the contractors, contrary to fixed-price contracts where all the risk is shouldered by the government agencies.

The host of modern critical industries rely on space data. For the everyday customer, the space data is used for navigation, multimedia streaming, and internet connection. Similar to the resilient power grids that became the backbone of reliable modern industry operation, space is evolving as another ridgeback. The space industry is expected to grow even more rapidly over the coming decades, in part due to recent technological breakthroughs concerning data collection and storage, as well as new business models, including online data marketplaces and cloud infrastructure to process data. UP42 marketplace – a one-stop-shop for geospatial data and analytics – is a characteristic example. The integration of UP42 in Esri's GIS[1] software products has increased value to end-users. However, as it was shown in this book, significant issues still exist in flux on the regulatory side.

It is clear that there is much to gain from expanded access to space. Nevertheless, the space industry development leaves many questions unanswered followed by the multiple empirical changes. An increasing number of scholars are studying "how the opportunities offered by modern space affect the

---

[1]Esri – Environmental Systems Research Institute – is an American multinational geographic information system (GIS) software company. It is best known for its ArcGIS products.

future world development". Bold opinions are shared in this book starting from umbrella-level vision of how modern space supports achieving the sustainable future to case-studies across the sectors. Overall, the interdisciplinary approach is used for building bridges across the sectors for giving the full-view of the AI impact on the space industry to the viewer.

The role of governments should extend far beyond regulations. Governments still play the role of key financial players in the form of guaranteed contracts, direct purchases of services and government funded R&D activities. However in this rapidly changing environment, the contribution of smaller entities is evolving. One should not treat the modern space industry as just a market of the wolves of Wall Street and giant companies. Advanced technologies development such as AI also brought smaller players on board. The space market democratization also encourages the non-aerospace companies to share the pie, including the R&D sector, thus gambling that scientific activities will be cost-effective. In 2021, the global space funding from the US government decreased from 70% to ca. 50%.

Space industry driven by frontier technologies including AI is on the cusp of a major expansion. The space market is gaining momentum. Although the potential for profit is yet undefined. The developed capabilities could give the linchpin in solving vital problems. Despite the major progress, multiple challenges may emerge stronger in the long term. We would like to conclude this book with this quote of Oscar Wilde from the year 1895, *"To expect the unexpected shows a thoroughly modern intellect"*.

Zurich, Switzerland, June 2023

Olga Sokolova, Dr. sc.
CTO / Risk Analyst
Sirin Orbital Systems AG

Index

3D printing, 33, 35
6G, 89

Additive Manufacturing, 7
Akaike Information Criterion
     (AIC), 249
Alliance for Collaboration in the
     Exploration of Space
     (ACES), 96
Alphabet/Google, 61
     Google Earth Engine, 63
Amazon, 78
Application Programming
     Interface (API), 63
Artificial General Intelligence
     (AGI), 377
Artificial Gravity (AG), 374, 375
Artificial Intelligence (AI), 56,
     198, 217, 219, 274, 364,
     370
     Advanced AI (AAI), 367,
     371, 374–376, 379, 382,
     391, 392, 394
     Automatic differentiation, 23
     Distributed learning
        Federated learning, 42
     Eclipse Networks
        (EclipseNETs), 22, 24,
        28
     Explainable Artificial
        Intelligence (XAI), 348

Message passing, 34
Mortal Computation, 109
Neural Density Fields
     (geodesyNets), 22, 26
     VGG16, 121
Artificial Intelligence for Data
     Analysis (AIDA), 216,
     217, 229, 241
     AIDApy, 216–218

Bayesian Information Criterion
     (BIC), 249, 253, 254
Bayesian Network (BN), 343
Best Matching Unit (BMU), 221,
     223
Big Data, 54–56, 59, 67, 94, 96
Blue Brain project, 60, 61, 64

Centre national de la recherche
     scientifique (CNRS),
     216
Centrum Wiskunde &
     Informatica (CWI), 216
ChatGPT, 366
Chip technologies
     Analogue & digital, 117
     CMOS, 117
     Memristor, 118
CITI bank, 89
Compute Unified Device

Architecture (CUDA), 225
Computer processing chips, 61
3XS, 61
Edge, 61
Lambda Labs, 61
Lenovo, 61
NVIDA, 61
Computer Vision (CV), 329, 332
Continuous Intelligence (CI), 310, 335
Convolutional Neural Network (CNN), 320
Coronal Mass Ejection (CME), 87, 191, 199, 203
Coupled Thermosphere Ionosphere Model (CTIM), 219
COVID-19 Pandemic, 79, 83
Crater matching method, 165
Appropriate crater, 162
Camera anomaly, 162
Camera-shot image, 162
Close craters, 162
Crater map, 161, 166, 168
Crater matching strategy, 172
Crater shadow, 162
Cross product, 171, 174
External craters, 166, 168, 170
Inner product, 171, 174
Line Segment Matching (LSM), 162, 165, 166, 169, 182, 186
Point cloud matching, 167
Triangle Similarity Matching (TSM), 161, 163, 168, 169, 172, 175, 179, 180, 183, 184, 186
Current Sheet (CS), 230, 234, 235

Cyber War, 382

Decision Support Systems (DSS), 66
Deep Learning (DL), 319, 321, 322, 339
Deep Neural Network (DNN), 324
Deoxyribonucleic acid (DNA), 384
Doct-o-bot (DB), 365, 368, 375
Drag-based Model (DBM), 195
Dynamic Time-Lag Regression (DTLR), 204

Earth Observation (EO), 6, 55, 121, 124, 304, 318, 321, 325
Ecole Polytechnique Fédérale de Lausanne (EPFL), 60
Edge devices, 40
Commercial off-the-shelf (COTS), 37, 109
Elementary Motion Detector, 109
Event-based vision sensors, 110
Neuromorphic hardware, 117
Electron-diffusion layer (EDR), 230, 231, 252, 254
Electronic Numerical Integrator and Computer (ENIAC), 57, 58
Esri, 396
European Cooperation for Space Standardization (ECSS), 325
European Space Agency (ESA), 22, 194, 276, 284
Φ-sat, 23, 37

Advanced Concepts Team
(ACT), 22, 26, 31, 109,
110, 115, 119, 121, 129,
130, 136, 140
Competitions
the OPS-SAT case, 38
OPS-SAT, 38
Sentinel-2, 63, 121
EuroSAT, 121, 125
European Union (EU), 285
exa-scale, 61

Fast Fourier Transform (FFT),
324
Fault Tree (FT), 342
Field Programmable Gate Array
(FPGA), 186
Finite Element, 34
Free and Open Source Software
(FOSS), 97

Gaussian Mixture Model (GMM),
249, 252, 254
gene therapy system, 366
Geocentric Solar Equatorial
(GSE), 220
Geographic Information system
(GIS), 396
Geographic Information Systems
(GIS), 65, 85, 97
geomagnetic index
Auroral Electrojet (AE), 201
distributed storm time (Dst),
200
Inter-Hour Variability (IHV),
201
Kp, 200
Geomagnetic storm, 191
Geospace General Circulation
Model (GGCM), 219

Geostationary Earth Orbit (GEO),
308, 325
Global Climate Observing
System (GCOS), 11
Global Navigation Satellite
System (GNSS), 6
Global Positioning System (GPS),
6, 79, 196, 329
GOES Satellites, 95
Graphical User Interface (GUI),
63
Gray-box paradigm, 196, 199, 203
Gross domestic product (GDP),
192, 193
Guidelines, 274
Long-term sustainability
(LTS) guidelines, 282,
289, 290, 294
Space Debris Mitigation
Guidelines, 275

Hewlett Packard, 61
High Altitude Platform Systems
(HAPS), 62
HYPSO-1, 37

In-Orbit Servicing (IOS), 303,
308
Inertial Measurement Unit (IMU),
329
Information and Communication
Technology (ICT), 61
Intelligent Synthetic Being (ISB),
369
Intelsat, 68, 83
Inter-Agency Space Debris
Coordination Committee
(IADC), 275
Interface Description Language
(IDL), 217

International Astronautical
    Congress (IAC), 97
International cooperation, 292,
    293
International intergovernmental
    organisations (IGO),
    276, 284
International responsibility, 281
Internet of Things (IoT), 9, 87, 89
Ion-Diffusion Layer (IDR),
    230–232

Japan Aerospace Exploration
    Agency (JAXA), 162,
    186
    Smart Lander for
        Investigating Moon
        (SLIM), 162–165, 176
Japan Meteorological Agency
    (JMA), 194

KAGUYA (SELENE), 162

Lagrangian point (L1), 196, 201,
    205
Laser Imaging Detection and
    Ranging (LIDAR), 329
Lattice materials, 33
Launching States, 281, 288
LiDAR, 23
Location estimation, 161, 162,
    164, 172, 179, 185
    Crater matching, 164
    Cross-view image matching,
        164
    Simultaneous Localization
        and Mapping (SLAM),
        164, 165
    Star catalog matching, 164
Low Earth Orbit (LEO), 3, 10, 35,
    36, 81, 82, 275, 308,
    321, 325, 326, 328, 333
Satellite constellations, 10
Lunar Orbital Transport Station
    (LOTS), 382

Machine Learning (ML), 4, 14,
    57, 195, 196, 205, 215,
    241, 247, 291, 292, 315,
    317, 329, 336, 349
    Machine Learning
        Operations (MLOps),
        349
    Unsupervised
        Machine-Learning
        (AML), 233, 237
MagnetoHydroDynamic (MHD),
    194, 198
MagnetoHydroDynamics (MHD),
    220, 230–232
Markov Decision Process (MDP),
    327
MATLAB, 217
Medical Emergency Robot
    (MER), 372
Medium Earth Orbit (MEO), 82
Microsoft Azure Cognitive
    Systems, 61
Mitigation, 275, 280, 282, 290,
    295
    Risk mitigation, 290
    Space Debris Mitigation
        Guidelines, 275
Myrabo Hyper Drive, 371
Myrabo Hyper Drive (MHD), 372

Nanosatellite, 36
National Aeronautics and Space
    Administration (NASA),
    8, 9, 95, 205

Advanced Composition
   Explorer (ACE), 219
Agriculture and Resources
   Inventory Surveys
   Through Aerospace
   Remote Sensing
   (AgriSTARS), 79
Large Area Crop Inventory
   Experiment (LACIE), 79
Magnetospheric Multiscale
   Mission (MMS), 218,
   219, 239
National Oceanic and
   Atmospheric
   Administration (NOAA),
   95, 194
Space Weather Prediction
   Center (SWPC), 194,
   195
National Space Weather Strategy
   (NSWS), 194
Neural networks, 195
Neuromorphic, 108
Neuromorphic computing, 108
   Spike-based coding, 113
   Spike-based communication,
      110
   Spiking neural networks
      Surrogate gradient, 35,
         115, 124
      Time-To-First-Spike, 113,
         125
      Weight conversion, 114,
         121
   Spiking neurons
      Adaptive Exponential LIF
         (AdEx), 112
      Leaky Integrate-and-Fire
         (LIF), 110
Neuromorphic sensing

DAVIS, 138
Dynamic vision sensor, 138
   Dynamic range, 138
Event-based vision, 110, 126,
      129, 139, 140
   Dynamic range, 140
   Feature extraction, 137
   Optical flow, 130
   Retinomorphic, 129
   v2e, 139
Landing with events
   Autonomous landing, 140
   Optical flow, 140, 141
   PANGU, 141
   Time to contact, 140
   v2e, 141
Retina
   Adaptation, 128, 130
   Amacrine cell, 127, 130
   Bipolar cell, 126
   Convolution, 130
   Dynamic range, 128
   Eccentricity-dependent
      receptive field, 128, 130
   Elementary Motion
      Detector, 125
   Feature extraction, 125
   Fovea, 127
   Ganglion cell, 127
   Homeostasis, 128, 130
   Horizontal cell, 126, 136
   Orientation selectivity, 126
   Photoreceptor, 125, 126,
      130
   Saccades, 130
   Sparsity, 130
Silicon retina, 137
   Address Event
      Representation, 137
Sparse convolution, 130

Adaptation, 133
Eccentricity-dependent , 134
Eccentricity-dependent receptive field, 132
Excitatory, 134
Im2col, 131
Inhibitory, 134
Saccades, 132–134
Normalized Differential Vegetation Indices (NDVI), 64

OMNIweb, 218
On-chip learning, 119
Onboard inference, 35
Onboard training, 41
One Web, 78
OneWeb, 10
Open Source, 97
Open-source software, 42
Python, 63
Deep learning libraries, 23
NIDN, 31
PASEOS, 42
pyaudi, 24
pyLattice2D, 33
Spike-based libraries, 116
Optic flow detection, 109
Orbital Transit System (OTS), 369, 371–373, 380
Organization for Economic Cooperation and Development (OECD), 285

Particle-in-Cell (PIC), 225, 247, 251
iPic3D, 219, 239
PhotoVoltaic (PV), 86

Policy and Regulatory Frameworks, 282
Positioning, Navigation and Timing (PNT), 66, 89
Prewitt filter, 123
Principle of cooperation and mutual assistance, 289

Qubit processing, 57, 97

Random forest method, 202
Rejuvena, 367
Remote Sensing (RS), 57, 79, 93, 95, 306, 344
Rendezvous and Proximity Operations (RPO), 329
Rice Convection Model (RCM), 219

Safety, 289
Satellites for Health and Rural Education (SHARE) project, 68, 78, 83
Scientific research, 294
Selene Earth-Moon (SEM), 369, 376, 382
Self-Organizing Map (SOM), 221, 224, 225
Single Image Super Resolution (SISR), 322
Single-Look-Complex (SLC), 324
Smart Lander for Investigating Moon
Pinpoint landing, 162, 163
Solar flare, 191, 199, 201
Solar Power Satellites (SPS), 66
Solar wind, 191, 195, 200, 204
solver-bots, 366
Space debris, 275, 283, 290
Space debris mitigation, 275, 295

Space debris remediation, 277
space missions, 196
 Advanced Composition Explorer (ACE), 196
 Deep Space Climate Observatory (DSCOVR), 196
 Defense Meteorological Satellite Program (DMSP), 196
 Geostationary Operational Environmental Satellites (GOES), 196, 200
 Polar Operational Environmental Satellites (POES), 196
 Solar and Heliospheric Observatory (SOHO), 196, 200
 Solar Dynamics Observatory (SDO), 196, 200
 Solar Terrestrial Relations Observatory (STEREO), 196
 Wind, 196
Space Situational Awareness (SSA), 194, 313
Space Traffic Management (STM), 310, 313
Space weather, 191
Space Weather Action Plan (SWAP), 194
Space-based Solar Power Satellite (SBSP), 87
Spacecraft landing, 109
Starlab, 369, 371–375
Starlink, 10, 78
State of registry, 287
State Vector Machine (SVM), 339

Super Resolution, 322
Synthetic Aperture Radar (SAR), 323, 324
Synthetic Aperture Radar Data Processing
 Chirp Scaling Algorithm (CSA), 324
 Omega-K Algorithm (ωKA), 324
 Range Doppler Algorithm (RDA), 324
 Spectral Analysis Algorithm (SPECAN), 324

Telemetry Analysis, 337
 Auto-Regressed Integrated Moving Average (ARIMA), 338
 Gated Recurring Units (GRU), 338
 Long Short-Term Memory (LSTM), 338
 Multi-Layer Perception (MLP), 338
 Recurrent Neural Networks (RNN), 338
Temporal Fusion Transformer (TFT), 341
Tesla, 8
Time History of Events and Macroscale Interactions during Substorms (THEMIS), 219

Unified Distance Matrix (UDM), 226, 228
United Nations (UN), 55, 56, 382
 Scientific and Technical Subcommittee (STSC), 278

Space 2030 Agenda, 55
Sustainable Development
Goal (SDG), 55–57
UN Millennium Goals, 56
United Nations Committee
on the Peaceful Uses of
Outer Space
(UNCOPUOS), 278
United Nations Educational,
Scientific and Cultural
Organization
(UNESCO), 285
United Nations Office for
Outer Space Affairs
(UNOOSA), 288
United Nations Space
Treaties, 280, 281
United States Geological Survey
(USGS), 66, 80

United States of America, 286
Federal Trade Commission
(FTC), 287
Unmanned Aircraft Systems
(UAS), 62, 65
UP42, 396
User Experience (UX), 347

Variable Rate Technology (VRT),
9
Vertical Lunar Descent (VLD),
176
Virtual Reality (VR), 369, 380
Visual Navigation (VN), 329,
332, 333
von Neumann bottleneck, 117

XPRIZE, 368, 370, 372, 373

Yuri Gagarin, 364

9781032432441